Biophysical Techniques in Drug Discovery

Drug Discovery Series

Editor-in-Chief
David Thurston, *King's College, UK*

Series Editors:
David Fox, *Vulpine Science and Learning, UK*
Ana Martinez, *Centro de Investigaciones Biologicas-CSIC, Spain*
David Rotella, *Montclair State University, USA*
Sarah Skerratt, *Vertex Pharmaceuticals, UK*

Editorial Advisor:
Hong Shen, *Roche Innovation Center Shanghai, China*

Titles in the Series:
1: Metabolism, Pharmacokinetics and Toxicity of Functional Groups
2: Emerging Drugs and Targets for Alzheimer's Disease; Volume 1
3: Emerging Drugs and Targets for Alzheimer's Disease; Volume 2
4: Accounts in Drug Discovery
5: New Frontiers in Chemical Biology
6: Animal Models for Neurodegenerative Disease
7: Neurodegeneration
8: G Protein-Coupled Receptors
9: Pharmaceutical Process Development
10: Extracellular and Intracellular Signaling
11: New Synthetic Technologies in Medicinal Chemistry
12: New Horizons in Predictive Toxicology
13: Drug Design Strategies: Quantitative Approaches
14: Neglected Diseases and Drug Discovery
15: Biomedical Imaging
16: Pharmaceutical Salts and Cocrystals
17: Polyamine Drug Discovery
18: Proteinases as Drug Targets
19: Kinase Drug Discovery
20: Drug Design Strategies: Computational Techniques and Applications
21: Designing Multi-Target Drugs
22: Nanostructured Biomaterials for Overcoming Biological Barriers
23: Physico-Chemical and Computational Approaches to Drug Discovery
24: Biomarkers for Traumatic Brain Injury
25: Drug Discovery from Natural Products
26: Anti-Inflammatory Drug Discovery
27: New Therapeutic Strategies for Type 2 Diabetes: Small Molecules
28: Drug Discovery for Psychiatric Disorders
29: Organic Chemistry of Drug Degradation
30: Computational Approaches to Nuclear Receptors
31: Traditional Chinese Medicine

32: Successful Strategies for the Discovery of Antiviral Drugs
33: Comprehensive Biomarker Discovery and Validation for Clinical Application
34: Emerging Drugs and Targets for Parkinson's Disease
35: Pain Therapeutics; Current and Future Treatment Paradigms
36: Biotherapeutics: Recent Developments using Chemical and Molecular Biology
37: Inhibitors of Molecular Chaperones as Therapeutic Agents
38: Orphan Drugs and Rare Diseases
39: Ion Channel Drug Discovery
40: Macrocycles in Drug Discovery
41: Human-based Systems for Translational Research
42: Venoms to Drugs: Venom as a Source for the Development of Human Therapeutics
43: Carbohydrates in Drug Design and Discovery
44: Drug Discovery for Schizophrenia
45: Cardiovascular and Metabolic Disease: Scientific Discoveries and New Therapies
46: Green Chemistry Strategies for Drug Discovery
47: Fragment-Based Drug Discovery
48: Epigenetics for Drug Discovery
49: New Horizons in Predictive Drug Metabolism and Pharmacokinetics
50: Privileged Scaffolds in Medicinal Chemistry: Design, Synthesis, Evaluation
51: Nanomedicines: Design, Delivery and Detection
52: Synthetic Methods in Drug Discovery: Volume 1
53: Synthetic Methods in Drug Discovery: Volume 2
54: Drug Transporters: Role and Importance in ADME and Drug Development
55: Drug Transporters: Recent Advances and Emerging Technologies
56: Allosterism in Drug Discovery
57: Anti-aging Drugs: From Basic Research to Clinical Practice
58: Antibiotic Drug Discovery: New Targets and Molecular Entities
59: Peptide-based Drug Discovery: Challenges and New Therapeutics
60: Drug Discovery for Leishmaniasis
61: Biophysical Techniques in Drug Discovery

How to obtain future titles on publication:
A standing order plan is available for this series. A standing order will bring delivery of each new volume immediately on publication.

For further information please contact:
Book Sales Department, Royal Society of Chemistry, Thomas Graham House, Science Park, Milton Road, Cambridge, CB4 0WF, UK
Telephone: +44 (0)1223 420066, Fax: +44 (0)1223 420247
Email: booksales@rsc.org
Visit our website at www.rsc.org/books

Biophysical Techniques in Drug Discovery

Edited by

Angeles Canales
Universidad Complutense de Madrid, Spain
Email: ma.canales@quim.ucm.es

ROYAL SOCIETY OF CHEMISTRY

THE QUEEN'S AWARDS FOR ENTERPRISE: INTERNATIONAL TRADE 2013

Drug Discovery Series No. 61

Print ISBN: 978-1-78262-733-3
PDF eISBN: 978-1-78801-001-6
EPUB eISBN: 978-1-78801-286-7
ISSN: 2041-3203

A catalogue record for this book is available from the British Library

The Royal Society of Chemistry is a charity, registered in England and Wales, Number 207890, and a company incorporated in England by Royal Charter (Registered No. RC000524), registered office: Burlington House, Piccadilly, London W1J 0BA, UK, Telephone: +44 (0) 207 4378 6556.

For further information see our web site at www.rsc.org

Printed in the United Kingdom by CPI Group (UK) Ltd, Croydon, CR0 4YY, UK

Preface

This book is the result of the combined efforts of scientists from academia and industry to provide an overview of the biophysical methods applied in drug discovery today, both the established techniques and the recent developments, with applications in this field.

Biophysical methods are required for different reasons at different stages of the drug discovery process, such as the screening of new ligands for a certain receptor, the validation of hits found by using biochemical and cellular assays and the optimization of drug properties in the hit to candidate phase. In addition, these methods are used to characterize the mechanism of action of a drug. This knowledge is of paramount importance in order to efficiently focus efforts and reach further phases of the drug development process.

Techniques and questions addressed span from the first protein crystal structures to the delivery of orally available drugs. Today, the scope of drug modalities has expanded to include peptides, proteins and antibodies. Pharmaceutical chemical matter collections and chemical technology platforms engage many different drug discovery strategies. Each of these approaches, fragment based, targeted or phenotypic drug discovery, requires its own specialized screening modalities.

We hope that this book provides an overall sense of the encompassing influence and necessity of biophysics in drug discovery. It is an exciting time as we are able to connect atomic level understanding through the solid and liquid phase analysis of proteins and attempt to identify the most relevant targets for diseases. The level of complexity of the opportunities: protein complexes, post translational modifications, and network analysis has only begun to be fully appreciated. Biophysics has become the connection between cross functional partners and disciplines en route to a drug. Please enjoy the individual perspectives of the scientists who have contributed to this book.

MariJean Eggen and Angeles Canales

Contents

CHAPTER 1

Impact and Evolution of Biophysics in Medicinal Chemistry

M. EGGEN*[a] AND J. SCHINDLER*[b]

[a] Discovery Chemistry Research & Technology, Lead Generation and External Innovation, Eli Lilly and Company, Lilly Corporate Center, Indianapolis, IN 46285, USA; [b] Discovery Chemistry Research & Technology, Quantitative Biology, Eli Lilly and Company, Lilly Corporate Center, Indianapolis, IN 46285, USA

1.1 Introduction

Fifty five years ago, Max Perutz and John Kendrew received the 1962 Nobel Prize in Chemistry. Their pioneering work delivered the first 6 Å structure of the protein Myoglobin, and represented the foundation for modern structural biology.[1–3] This first atomic-level picture heralded the complexity of the protein universe and provoked questions about how to predict the driving forces that allow proteins to rapidly fold into biologically relevant conformations. This was the beginning of biophysics. In 1969, the first enzyme structure of the extracellular nuclease of *Staphylococcus Aureus* as an enzyme-inhibitor complex was solved by F. Albert Cotton and coworkers.[4]

Thus began significant motivation to devise and evolve computational methods to predict protein folding. "Force fields" in modern computer simulation draw their information primarily from the ~40 000 protein families represented in the Protein Data Bank (PDB), where structures obtained by X-ray crystallography, nuclear magnetic resonance (NMR) spectroscopy and, increasingly, cryo-electron microscopy (cryo-EM) are deposited. Today, X-ray crystallography has been extended further to characterise inherently disordered proteins and protein aggregation.[5–8]

Analysis of structural information has been applied to study protein equilibria and dynamics. The driving forces of (1) hydrogen bonding, (2) van der Waals interactions, (3) backbone conformational preferences, (4) electrostatics, and (5) hydrophobic interactions leading to the precise folding of proteins produce the minimum energy conformation, as can be seen crystallographically.9 Beyond the minimum energy conformation determined through x-ray crystallography, additional lower energy minima representing the dynamic movement of proteins can be studied by other biophysical techniques such as protein NMR and proton deuterium exchange mass spectrometry (HDX-MS).

Small molecules must be identified and optimized to productively interact with the protein's disease-relevant conformation or surface. This requires several pieces of information: (1) disease biology, (2) an understanding of the protein form or forms within the cell, and (3) an understanding of the protein surface and the potential interactions and MOA that can occur with the small molecule. Many of the same forces that impact protein conformation influence the low energy conformations and physical properties of drugs and can be assessed through small molecule crystallography, analytical methods and computational minimizations. Ultimately, the small molecule design team, the chemist, computational scientist, and structural biologist are challenged to identify opportunities for optimal molecular recognition from the protein. Optimization requires an understanding of interaction geometries and approximate contributions identified through crystal structures and demonstrated through affinity data.[10] However, molecular interactions behave in a highly "non-additive fashion" and are context-dependent. Solvation of the protein and small molecule, long-range interactions and conformational changes of the protein (see Figure 3, ref. 9) all influence binding energetics.

Design of the small molecule must focus on specific intermolecular reactions based on available structural information. The most important and well defined are hydrogen bond interactions directly with the protein or through structural water.[11–13] Weaker hydrogen bonds are also sometimes available when an aromatic ring can act as H-bond acceptor.[14] Although H-bonds are among the strongest, these interactions should not be the sole focus of the designer. Orthogonal multipolar interactions of C–F to C=O such as the Bürgi–Dunitz angle,[15] halogen bonds that leverage sigma-hole anisotropy, and cation–Pi interactions are also opportunities in the design of a molecule.[16]

Ultimately, the amount of hydrophobic surface buried through Van der Waals interactions upon ligand binding appears to best correlate with binding affinity.[17] This concept holds for a diverse set of protein–ligand complexes, including protein–protein interactions.[18] All other types of interactions are highly context-dependant and their utility must be tested within a given target. The thermodynamics and kinetics of the target interactions can be evaluated to understand progress and will be discussed in a later section.

Intramolecular interactions within a protein as well as stabilizing intermolecular interactions with an inhibitor were key to delivering the first biophysical technique, X-ray crystallography. The power of interpreting a low-energy atomic-level picture of a molecule bound to an enzyme created a paradigm shift in drug discovery. However, the dynamic nature of proteins requires an understanding of both their minimum energy states and also the status of their movement in other conformations. Further, these states must be measured and interpreted amidst the complex and dynamic processes in the cell. Techniques such as molecular imaging can provide direct measurement of the location and progression of processes. Connecting these techniques to non-invasive measurements of patient disease pathology and response in the evaluation of neurodegeneration,[19] cardiovascular disease[20] and oncology[21,22] can leverage molecular imaging through PET-CT and MRI scans.

Biophysics is utilized throughout drug discovery strategies from target identification and selection, construct selection, screening, ligand optimization, drug development and ultimately imaging within our patients. The following sections will provide context for the techniques presented in this book.

1.2 Evolution of Biophysics in Medicinal Chemistry

1.2.1 Phenotypic Drug Discovery

The earliest drug identification and approval were driven by small molecule phenotypic drug discovery (PDD) approaches.[23–25] This strategy involved identifying and optimizing efficacy in animal physiological and behavioural models for human diseases, and *ex vivo* tissue-based or phenotypic cellular assays. While many drugs used clinically today were discovered agnostic of their target protein, work in this fashion increased the necessary investment in the systematic iteration and synthesis of lead molecules focusing on potency and phenotype. Without specific target and mechanism of action information, agents often displayed poly-pharmacology, which could be good for disease modification or bad for toxicology. Coupled with extended times in early stage discovery, the advent of protein crystallography, and the potential for off-target liabilities that wouldn't be identified until entry into the clinic, drug discovery attempted to minimize these challenges through targeted therapies, that is, target directed drug discovery (TDD) (Figure 1.1).[26]

1.2.2 Targeted and Fragment-based Drug Discovery

With the advent of a deeper understanding of the proteins involved in a disease pathway through cell biology tools such as siRNA, proteomics,

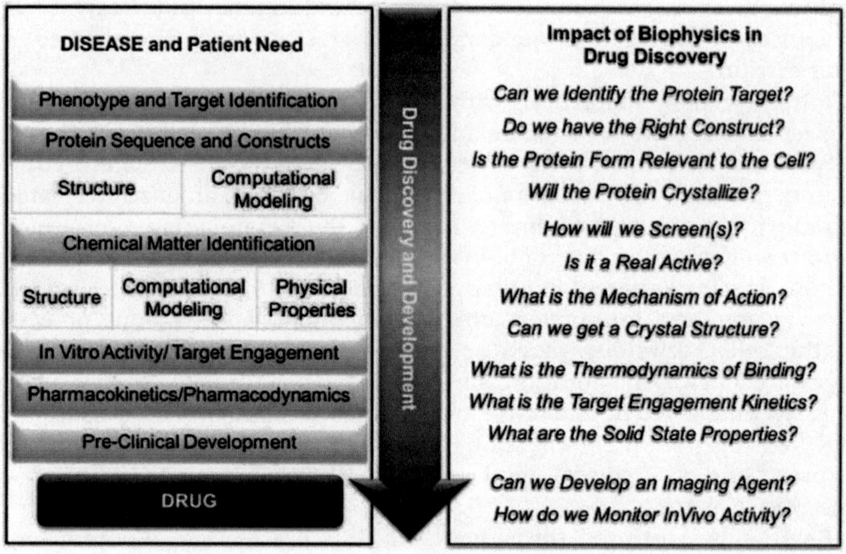

Figure 1.1 Drug discovery pathway to a medicine with examples of key questions that biophysics can address experimentally and influence in molecule identification, progression and design.

knock-outs, and human genetics (mutations found in patient populations), drug discovery efforts have focused more on specific proteins believed to be involved in a disease. This knowledge has led to a paradigm shift in drug discovery that went from dosing a limited set of compounds in animal models to screening 250 000 to 1 million compounds in biochemical assays in a specific protein. Typically, the biochemical assays were performed in formats that looked at competition binding or enzyme activity against recombinant proteins.

Drug discovery targets, in response, moved into more complex protein targets such has epigenetics, protein–protein interactions (PPI), membrane-bound proteins and transporters, and protein aggregation. This resulted in gaps or greater challenges in screening approaches, and protein expression, production and purification. The new target space ultimately affected structural biology approaches. As an example, discrete protein target EZH2 (Section 1.5.2), actually a component of a complex of proteins, requires this complex for activity and stability. In addition, protein complexes could in theory be inhibited by binding in different proteins and read-out as competitive, non-competitive, uncompetitive, mixed or irreversible in nature. In the case of protein–protein interactions, one may simply be interacting with a protein and disrupting a productive, functional interaction with a second protein.

This paradigm shift in targets also resulted in re-thinking how compound libraries were computationally selected for screening. For example, PPI targets required larger molecules that were not present in most traditional libraries based on Lipinki's rules,[24] and when starting points were found in traditional screening, there was very little room to maintain potency while fixing the physical properties. To overcome gaps in compound collections and property challenges, some drug discovery groups implemented a new strategy and proposed to screen compound "fragments" with MW < 300 Da. Fragment-based drug discovery (FBDD) takes advantage of Jencks' description of Gibb's free energy and attributes of the binding energy between enzyme and substrate.[25] By beginning with fragments, a greater chemical space could be sampled with smaller sets of molecules while optimizing biological activity and physical properties in parallel.[26] This introduced a strategic shift in how to approach drug discovery and an opportunity to identify chemical matter without a large compound collection.

The challenge with FBDD is that the initial compound actives have highly efficient per heavy atom (non-hydrogen) target binding but low measured binding constants. Given their low affinity, high concentration biochemical assays (enzyme or competition binding) were used, but did not behave or were still not sensitive enough to pick up fragments in all targets or target classes. Sensitive biophysical methods had to be identified. Teams also needed orthogonal methods to confirm the fragment as a legitimate active, identify a potential MOA, and to provide clues about vectors in order to decorate the molecule for efficient, key interactions within the larger protein active site.

1.2.3 Phenotypic Drug Discovery 2.0

A renewed interest in PDD strategies was influenced by a focus on the complexity of patients and their disease states in oncology, autoimmune disease and neurodegeneration. Coupled with technological developments at the interfaces of disciplines scientists have been able to deliver complex cell culture systems and gene-edited disease relevant mutations for phenotypic screening assays.[27–29] Advances in mass spectrometry methods such as affinity selection (AS-MS), cellular thermal shift assays (CETSA) and cDNA expression microarray technologies have allowed probing of complex pathways. Improvements in cell-based and model-organism based automated screens,[30–32] along with the development of new biophysical methodologies have enabled improved screening, faster identification of targets and profiling of the MOA in complex biological samples at the genomic, proteomic and phenotypic levels.[33]

1.2.4 Evolving Compound Collections and Chemical Technologies

Highly elaborate compounds that became part of compound collections were developed during work on well-characterized targets and focused target class research. Re-entry into PDD to identify an alternative molecular MOA and the most relevant but broadened diversity of targets made it clear that there were gaps in the available chemical matter. Scientists have developed different chemical matter strategies and technologies to identify starting points, identify their cellular targets, and select opportunities for enablement of structural information. The lack of chemical starting points for new targets has led to the initiation and expansion of Open Innovation models for accessing novel chemical matter and enabling potential collaborations.[34] The breadth of molecular weight, property space, and molecular topology has expanded to cover FBDD, traditional compound collections and renewed interest in natural products and peptides.[35] For example, given the low affinity of initial fragments, measured biochemical activity typically ranges from low mM to high μM. Chemical technologies such as phage display,[36] DNA-encoded libraries[37,38] and covalent technologies have further expanded the available tools. These developments, as a whole, elevate the requirements for biophysical methods to cover broad sensitivity ranges and be compatible with diverse physical properties. The new target space coupled with FBDD and the regenerated interest in natural products and peptides have led to an explosion of biophysical approaches to support chemical discovery, and these can be grouped into four areas: (1) biophysical screening, (2) validation of screening hits, (3) chemical optimization, and (4) structural biology. Most significantly, biophysical methods have become a key component in drug discovery from selectively interacting with a protein to demonstrating target engagement in the cell (Figure 1.2).

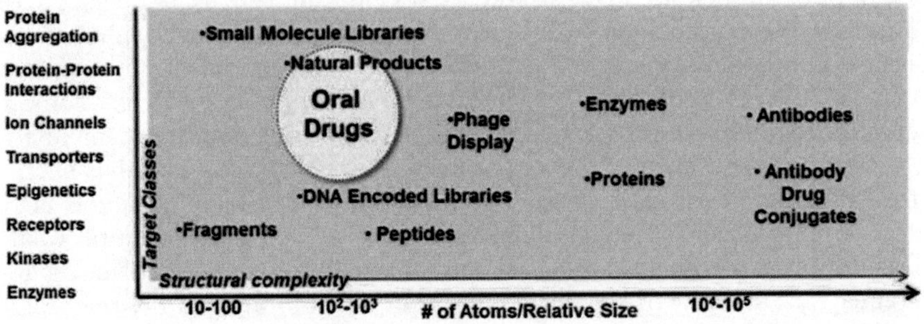

Figure 1.2 Diversity of molecular targets and the chemical matter that is leveraged in modern medicinal chemistry. The complexity encountered by drug discovery teams is both in target chemical libraries and chemical technologies. Chemical matter spans from fragment-based libraries of <300 MW to information obtained from antibodies. The diversity in chemical technologies has also greatly expanded. The target class and biology related to disease influences the tools and compound libraries that are used and ultimately the optimization and delivery of the drug.

1.3 Biophysical Screening Approaches

1.3.1 Protein- and Ligand-based NMR

Two biophysical screening approaches that enabled FBDD to be established are the "protein-observed" and the "ligand-observed" nuclear magnetic resonance (NMR) methods.[39] Both NMR approaches are carried out against a small library (15 K) of low molecular weight compounds or fragments. While these two techniques measure the binding of compounds differently, they both bridge the gap in necessary sensitivity by detecting weak ligand binding. Protein-observed NMR and ^{19}F NMR can pick up binding constants in the range of 1 mM to 10 mM, respectively, and are among the most sensitive biophysical techniques.

In protein-observed NMR, relatively small proteins are ^{15}N and ^{13}C double labelled, however other novel approaches observing and labelling tryptophan or other individual residues that may be prevalent in binding sites are also being developed.[40,41] Protein size becomes the main limitation over 40 kDa and is dependent on protein behaviour and increasing complexity. In protein NMR, the structure is first solved through a suite of experiments including ^{15}N-HSQC and 3D NMR that determine hydrogen bonds, dihedral angles and distance constraints through the nuclear Overhauser effect (NOE), to assign specific residues and connectivity. Once individual residues are assigned, binding of the ligand to the protein is observed through ^{15}N or ^{1}H-amide chemical shift changes observed where ligands bind or indirectly influence movement of the protein. Protein-observed NMR has two additional features. Firstly, the region of the protein that the ligand binds to can be mapped out using a double-labelled (^{15}N, ^{13}C) protein and following the NOEs of the protein in the presence of the compound. When these data are linked to a crystal structure, one can begin to optimize interactions with the minimum energy conformation of the protein while understanding something about protein movement. Linking the fragment hits to structure is a key requirement for FBDD. Given the low affinity of the initial fragments, very little to no cell or biochemical activity may be measurable, therefore linking the hits to structure is key to driving rapid SAR design to improve target affinity.

The second key feature of this screening approach is the ability to differentiate between compounds binding to the same or overlapping region of the protein. Analysis of the individual compound patterns of shift in the protein structure and comparison with ligands of known MOA enables computational modelling and design. If two fragments are binding to adjacent sites of the protein, NOE experiments can demonstrate their proximity and enable compound merging and linker optimization. This approach has delivered rapid improvements in binding affinity, although the precision of vectors required can sometimes lead to synthetic and design challenges. This approach has been used to screen difficult targets such as BACE, Hsp90, MCl-1 and BCl.[42–44]

Ligand observed-^{19}F NMR provides an alternative. In ^{19}F NMR, a protein is added to mixtures of fluorinated fragments and monitored for changes in the intensity of previously assigned individual ^{19}F signals. Binding leads to increases in relaxation rate and reductions in signal intensity. In this case, there is limited structural information, but some may be determined or inferred through additional ligand to protein related NMR experiments and competition experiments with ligands of known MOA. However, caution should be exercised since this does not exclude the compound from being allosteric competitive. Thus, one could argue that it is critical to have a robust structural component to support the ^{19}F NMR screening approach as it may be difficult to optimize compounds in the absence of the structure.[45]

1.3.2 Thermal Denaturation Techniques

The next approach takes advantage of work done almost 30 years ago that demonstrated that productive ligand binding to a protein can stabilize that protein from thermal denaturation. Differential scanning calorimetry (DSC) has been used to measure the binding constants of substrates, metals, cofactors, as well as protein–protein interactions.[46] All thermal denaturation techniques use the principle that a ligand-induced conformational change or stabilization of a protein is coupled by ligand binding and an enzyme or receptor melting reaction. If the ligand binds to the native form of the protein, an increase in the melting point or thermal shift is observed. Typical thermal shift studies are carried out with DSC equipment that monitors the change in heat capacity as proteins undergo a temperature-induced melting transition. To use this nearly universal phenomenon to screen compound libraries, drug discovery needed to develop a plate-based instrument that was coupled with a temperature control system and fluorescence detection.[47] In this approach, protein unfolding or denaturation upon heating exposes hydrophobic regions of the protein. Low quantum yield, fluorescent reporter dyes preferentially bind to the lipophilic regions and upon binding, increase in fluorescence. This method offers a simple approach for most screening laboratories.

One example of implementing a thermal denaturation approach is pseudo-kinase domains. Pseudo-kinases have recently been recognized as players in cancer, autoimmune diseases and diabetes.[48] Significantly, one can also envision different screening approaches for multi-substrate enzymes. For example, for a two-substrate mechanism, one could screen in the absence and presence of one substrate and find compounds that are dependent upon the substrate binding. This screening approach is most interesting if one knows that the enzyme mechanism follows an ordered substrate binding (Figure 1.3). Substrate A must bind prior to substrate B in order to form the ternary complex. Catalysis then takes place followed by the ordered release of the products (Q and P). One can see that an inhibitor (I) can bind to many different forms of the enzyme (E, EA, EQ and EP). With this knowledge, one can design screening strategies to find compounds that bind

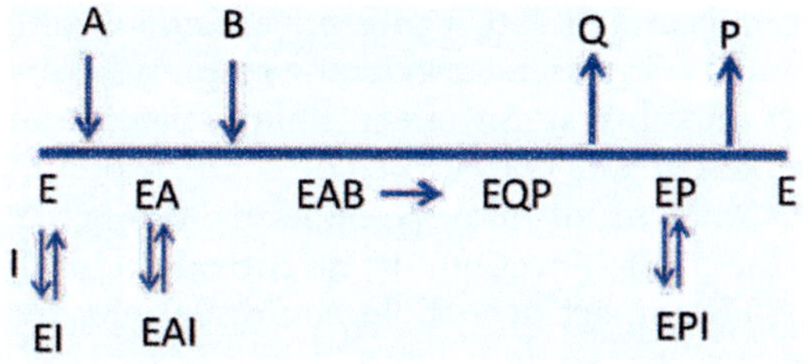

Figure 1.3 Scheme describing ordered mechanism for substrate binding. Substrate A binds to free enzyme (E) first followed by the binding of substrate B. This is followed by catalysis and the ordered release of products Q and P.

to these different forms of the enzyme. An example of this approach applying TDF is for the enzyme Dihydrofolate reductase (DHFR), which binds NADP+ prior to binding folate.[47] While TDF offers a straightforward approach to realising this strategy, other biophysical approaches may also be adapted to screen in this manner.

Higher transition melting temperatures (T_m) identified through TDF, demonstrate stabilization and increased probability of crystallographic enablement of the ligand–protein complex. This approach is extremely useful when the ligand or substrate of a receptor or enzyme is not known and is agnostic of mechanism. The compound library for screening in a thermal shift assay can be fragment-based and can greatly diversify the collection of relevant compounds that can be evaluated and ultimately optimized by the medicinal chemist.

Thermal shift data have also been used to characterize protein constructs for crystallographic enablement studies. It became crucial to understand what buffer conditions are optimal for protein stability. TDF is a good tool to use in these studies. Not all screening approaches are compatible with buffers used in crystallography. In addition, compound affinity may be dependent upon the buffer. To bridge the possible disconnects in going from one biophysical approach to a structure, one must test compounds in the buffers used in crystallography. TDF is a great approach for this.

1.3.3 Surface Plasmon Resonance

Another method that can provide fragment hits is surface plasmon resonance (SPR). SPR is a microfluidics biosensor-based approach. In addition to fragment screening, SPR is a popular approach for providing assessment of peptides coming from phage display studies. Recently, a new approach known as Peptidream has attracted a great deal of attention in drug discovery for PPI targets. Peptidream is a technique for incorporating unnatural amino acids into a peptide, providing a unique chemical space for peptides when compared to traditional peptide phage display approaches.[49] SPR has been used to assess the binding affinity of peptides coming from this peptide screening approach.

In SPR, a protein is attached to a gold surface and a solution of molecules is flowed over the protein surface producing response units (RU) and kinetic information. One must invest upfront to identify the best linker and location for access to the biologically relevant region of the protein. It is preferable to have known substrates or inhibitor compounds to test the behaviour of the immobilized protein. In addition, it is critical to understand that the immobilization may impact accessible conformations that allow binding. This can be the case for some proteins that are dimers or undergo large conformational changes upon binding ligands. To truly understand the protein target behaviour, groups have used SPR and isothermal titration calorimetry (ITC) in parallel. ITC is label free and measures the heat produced when a ligand binds to a protein. ITC can also provide stoichiometry and binding constants, which can be compared to the values determined with SPR. ITC and SPR should tend to agree in most cases. If one begins to see a disagreement then this might raise a flag in terms of using SPR.

SPR is a technique that requires expertise, experience and diligence in running screening studies.[50] This being said, SPR is a powerful biophysical technique and screening approach that can provide a binding constant (K_d), stoichiometry and binding kinetics. One key aspect of this data package comes from the ability to understand the stoichiometry. A compound that has a stoichiometry of $\gg 1$ can be flagged as an aggregator and dropped from further follow up studies. Compounds that do not display simple binding kinetics or are irreversible can be detected. Together, compounds can be funnelled into structural enablement, functional assays and chemical design.

In addition to being a data-rich screening approach, SPR screens can be carried out in parallel against 2 or more proteins, depending on the instrument. This feature allows one to have a good understanding of selectivity profiles, which may aid in determining which scaffold to take forward. The richness and value added from SPR screening has no current rival among the other biophysical screening techniques.

1.3.4 Affinity Selection and Size Exclusion Chromatography Mass Spectrometry

Some proteins may not be amenable to immobilization techniques used in SPR. A number of groups have worked towards a true label-free approach to screening. One approach that has begun to grow is affinity-based MS techniques. The use of MS techniques started out by looking at the binding of a ligand and protein in the gas phase, but has moved to solution-based methods. In the solution-based method, compound and protein are allowed to equilibrate and then the unbound and bound compound are separated either through ultrafiltration, as in affinity selection (AS), or by size exclusion chromatography (SEC). The amount of bound compound is measured using ESI-MS. This approach avoids manipulation of the protein and does not require a labelled ligand for competition binding, making this approach a true label-free technique. For example, AS-MS was used to screen the G-protein coupled receptor (GPCR) M2R. This approach led to the discovery of orthosteric antagonists and potential allosteric ligands.[51] Thus, affinity mass spectrometry approaches are well suited to identifying ligands for orphan receptors or enzymes that have no activity such has pseudokinases. In addition to screening, AS-MS can also provide binding constants as well as MOA data.

One caveat of AS-MS is that this approach will not find weak binders. Typically, the cut off is around 10 µM, so AS-MS is not a biophysical method that can be generally applied for FBDD. However, it does enable one to screen traditional chemical libraries and natural products efficiently with increasingly larger sets of compound mixtures. While the rate of actives is generally low, they have good binding affinity and may be less prone to nuisance mechanisms, allowing a team to focus. In the past, the issue around natural products was identifying the hit in a mixture or potentially an extract, as most screens of natural products were carried out in cell or biochemical assays. With the advent of AS-MS screening, identification of ligands should be faster. This now allows us to tap into the chemical space of natural products and chemical biology, which can be useful for PPI and other more complicated drug targets that are being identified.

Each of these approaches has their pros and cons and places in drug discovery, and they have been discussed in a number of reviews on drug discovery. In most cases, the comparison is made based on the number of compounds that can be screened and the "sensitivity" of the assay. However, it is clear that each approach should be judged on what targets (receptors, enzymes, PPI, *etc.*) and chemical libraries can be screened in an effective manner. The screening approach should also be able to effectively follow up on hits. Smaller drug discover groups may pick one screening approach that is amenable to a specific target class and library, while larger groups may have the capacity and instrumentation for multiple and often parallel screening strategies, depending upon the target. However, multiple screening strategies running in parallel must also match a drug discovery team's size and capacity to triage, analyse and implement chemical plans.

One key theme that each of these techniques share is their ability to align with crystallography. Therefore, the protein constructs in these screening approaches should be the same as those used in structural work, and they need to be purified and examined by MS for any post-translational modification such as phosphorylation, protease degradation, aggregation, and glycosylation. This will eliminate disconnects when co-crystal structures are obtained with screening hits. In addition, buffer conditions can alter the binding affinity of compounds. Crystallography uses buffer screening to find the best conditions for crystallography, which may affect compound binding. It is reasonable to suggest that the buffer conditions in one technique may not be reasonable for another, and one needs to stay aware of protein constructs and the screening conditions. Another key factor is understanding the biology of the system and if biophysical screening approaches will result in compounds that translate into the cell context. Often, the synthesis of proteins for biophysical or biochemical studies that mimic the cell can be difficult to reproduce. Truncated constructs are used, which may give rise to no lead matter or to compounds that do not translate to the cell. In addition, one may miss an opportunity to find a unique mechanism of action (MoA) by not using the correct protein. Finally, biophysical screening *vs.* enzymes in the absence of understanding the kinetics of the enzyme from substrate binding to product release is wrong. Enzyme catalysis can lead to different forms of the enzyme, which can also lead to different places for compounds to bind. While biophysical screening is a direct binding method and might weed out some false positives seen in biochemical assays, it may also miss compounds that bind to enzyme forms. Once again, all of this speaks to understanding the protein, the protein in the context of the cell and what type of MOA is needed to be effective *vs.* a disease.

1.4 Mechanism of Action and Determination of Kinetics and Thermodynamics

1.4.1 Mechanism of Action-evading Nuisance Mechanisms

False positives from biochemical screens can be high due to one of two reasons, compound interference or promiscuous mechanisms. Firstly, the compound may interfere with the assay format because of fluorescence. However, enzyme reactions can now be followed using LC–MS methods such as "Rapid Fire".[52] Significantly, this development allowed a number of groups to characterize, screen, and carry out MOA studies for epigenetic and metabolic targets for oncology. While this LC–MS approach mitigates assay artifacts, it will not identify non-drugable mechanisms.

Over the past 10 years, a number of laboratories began to publish work reporting that an alarming number of screening hits worked by non-drugable mechanisms, and the terms "promiscuous inhibitors" or "pains compounds" were born.[53,54] The initial discovery was found using fundamental, mechanistic enzymology, and was dependent on the buffer conditions or enzyme concentration. In addition, this class of compounds displayed uncompetitive or non-competitive inhibition *versus* all of the substrates. An uncompetitive inhibitor is especially difficult to explain for an enzyme that uses only one substrate. The series derived from these inhibitors did not display structure–activity relationships with a range of potencies, and as the name implied, were not selective. Overall this led to a great deal of wasted chemical effort along with failed co-crystallization attempts with the protein target. Many groups began to avoid non- or uncompetitive screening actives. However, in the absence of an experiment with the appropriate technique, one could be throwing away true starting points with unique mechanisms that could be leveraged for improved selectivity or chemical space properties.

A potential issue with a competitive mechanism is that the efficiency and potency can be diminished by mass-action competition with a prevalent substrate. To overcome this challenge, many groups implement the parallel approaches of enzymology and biophysical studies into the follow up. Now, most screening groups use biophysical approaches, such as SPR, ITC and AS-MS, in parallel with mechanistic enzymology. With the addition of SPR for follow up, non-competitive inhibitors can be tested for super stoichiometry and be eliminated based on these results. Additionally, direct binding constants from ITC or SPR can be compared with the enzyme binding constant (K_i value), and a deeper understanding of what drives affinity can be uncovered. For example, in the case of uncompetitive inhibitors for a two-substrate mechanism, such as Dihydrofolate Reductase, one might run SPR, ITC or AS-MS in the presence or absence of one of the substrates. Binding of the compound only in the presence of the substrate would confirm the uncompetitive mechanism of the compound. The same fundamental enzymology also confirms the behaviour of constructs to be used in experimental approaches for structural biology.

The argument for favouring biophysical approaches comes from false positives in biochemical screening. However, a biophysical screen should never replace an enzyme screen when applicable. An example of this is the case of Inosine-5′-monophosphate dehydrogenase2 (IMPDH2). IMPDH2 is an enzyme that catalyzes the reaction of NADH and IMP to make NAD and XMP (Figure 1.3). A natural product screen for this target was carried out in a cell-based assay and mycophenolic acid (MPA) was found to bind to IMPDH2. This compound is uncompetitive *versus* both NADH and IMP. Crystallography carried out in the presence of IMP and XMP showed that the compound binds to the NAD(H) binding site. Attempts to crystalize MPA in the absence of substrates was unsuccessful. Kinetic analysis indicated that MPA inhibits IMPDH2 by trapping the covalent products of the hydride transfer reaction.[55] Thus, to find this compound from a screen, one would have to be following enzyme turnover, and a biophysical screen alone would miss this MOA. Once again, this example points to the fact that enzyme catalysis can lead to different forms of the enzyme and different ways for compounds to bind along the dynamic process.

1.4.2 Residence Time, Target Engagement and Kinetics

Medicinal chemists optimize the K_d value of their compounds while maintaining reasonable physical properties for exposure and solid-state performance. A potent drug combined with good pharmacokinetics (PK) typically requires lower doses and overall exposure, maximizing performance and potentially minimizing side effects. However, a molecule's K_d value is the ratio of two components (eqn (1.1)), the dissociation rate (k_{off}) and the association rate (k_{on}). The k_{on} is expressed in $M^{-1}\,s^{-1}$ and the k_{off} is expressed in s^{-1}.

$$K_d = \frac{k_{off}}{k_{on}} \tag{1.1}$$

From the dissection of K_d, one can see that two compounds could have the same K_d value but may have very different binding kinetic profiles. The identification of performance disconnects, the advent of more integrated and accessible biophysical data, and the desire for selectivity have driven medicinal chemistry to evaluating the use of detailed target kinetics in drug design.

For the medicinal chemist, structure–kinetic relationships (SKR) unlike SAR, are challenging to incorporate in design, and the measurement of these rates across analogues within a scaffold would be required. To obtain enough data for optimization across multiple scaffolds, one would want a label-free method. A number of groups have heavily relied on SPR, which provides real time k_{on} and k_{off} rates using relatively little protein. Several examples now exist of rigidification, inclusion of halogens and shielding of water and H-bonds, which demonstrate interactions that can influence residence time. Also, the behaviour of targets that involve significant protein movement can influence the kinetics as well.

An example of hidden information within the K_d value that can influence drug design is p38 kinase compound "BIRB796".[56] This compound was reported to have an IC_{50} value of ~200 nM. Upon implementation of a 4 hour pre-incubation of the compound with p38 kinase, the IC_{50} shifted 200 fold to 1 nM. This time-dependent shift in potency upon pre-incubation is driven by the k_{on} value of BIRB796, which was measured to be 8.5×10^4 M^{-1} s^{-1}. The k_{on} value is significantly slower (100-fold) than what would be expected for diffusion-limited binding, which sits in the range of 10^6 to 10^8 M^{-1} s^{-1}. The slow binding event for this compound is a sign that the protein undergoes conformational change in order to accommodate the compound. This is demonstrated in the co-crystal structure of p38 and BIRB796, where a phenylalanine of p38 kinase moves to a DFG out position to create a pocket for the compound to bind. This type of binding is often called an induced fit binding mechanism. What is significant about this kinetic mechanism is that in most cases the dissociation of the compound can be quite slow, and the pocket may not be identified by an apo–protein crystal structure. The dissociation rate for BIRB796 is 8.3×10^{-6} s^{-1}, which translates into a $t_{1/2}$~24 hours ($t_{1/2}=0.692/k_{off}$). Structural studies might provide some evidence, such as a conformational change upon compound binding, but cannot predict the binding kinetics. In other cases, a structural approach may not be in place for a given target.

With the enablement of SPR, teams can routinely determine the binding kinetics of compounds allowing them to unlock SKR and differentiate compounds. Eventually, SPR will be limited by very slow off rates, which behave in a pseudo-irreversible manner on the experimental timescale and require a method to regenerate the protein surface. With longer $t_{1/2}$ values, other methods such as enzyme kinetics or radio–ligand binding might be needed to determine the kinetics. It becomes critical to cross-check from enzyme to cell, with cellular washout experiments. If one does not observe the translation to the cell, the question of the relevance of the biochemical data must be addressed. In addition, the regulation rate of protein synthesis in the cell must also be understood.

The kinetic profile of drug candidates has been suggested to be critical to obtain a more durable response. This rationale comes from the fact that the K_d value determined *in vitro* is done in a closed system in which the drug and target are present at constant concentrations. However, *in vivo*, the drug is in an open system in which drug concentration is not constant. It will depend both on the rate of drug elimination, the effect compartment, and the initial level of receptor occupancy. Thus, one may argue that what drives a durable response of a drug is the half-life of the drug–target complex or its "residence time".[57] As the drug is cleared, the fraction of target inhibited by the drug can still be high enough to block the activity to a significant extent. This can last until the drug dissociates from the target or the target undergoes turnover. Thus, some drug discovery teams have optimized k_{off} values during lead optimization.

Understanding the location and behaviour of the individual drug target as well as requirements for safety and protein half-life can influence strategies to deliver safe and selective drugs. Another strategy could be to drive selectivity *versus* target isoforms. The key strategy for COX-2 inhibitor programs was to be highly selective *versus* COX1. Celecoxib has a K_d of 2.3 nM for COX2 and a $t_{1/2} = 50$ min but shows little or no binding to COX-1.[58] Crystallographic studies indicated that the selectivity and slow dissociation kinetics of this drug were due to key interactions in a "side pocket" of COX2 that was not present in COX-1. This data resulted in optimizing compounds to take advantage of this structural difference in these two iso-enzymes. Other selectivity examples include kinases. A number of kinase inhibitors with very long $t_{1/2}$ values have been discovered, once again with strategies of selectivity and overcoming the large concentration of ATP in the cell.

Target biology and desired pharmodynamic effects can also require careful consideration of association rates. Thrombin inhibitors are a key example in this case. Kinetic studies demonstrated that the more successful compounds had fast association rates (10^7 $M^{-1}s^{-1}$), indicating that optimization is not always driven by the dissociation rate.

Understanding the rate of new protein production is important. If the rate of new protein synthesis is very rapid, a drug with slow dissociation would have no effect. In the case of the kinase ITK, its $t_{1/2}$ for protein synthesis in CD4 + T cells is 1 hour, indicating that the protein is rapidly turned over in these cells. However, when an irreversible inhibitor was used, the synthesis rate of the protein changed to a $t_{1/2}$ of 22 hours. In this case, the inhibitor–ITK adduct blocked the degradation of the protein and turned off the signalling to resynthesize ITK.[59] While this is an example of an irreversible inhibition, it demonstrates the importance of understanding the target biology in order to ultimately select and deliver kinetics parameters that lead to a durable effect in the clinic.

The relationship between drug–target $t_{1/2}$ value and the PK of the drug is the next factor to consider. If the drug–target residence time is much longer than the terminal half-life of the drug, the residence time can be a factor in having a durable pharmacology. If the residence time of the drug is close to or less than the terminal half-life of the drug, then the effect of residence time is less of a factor in driving the pharmacology of the drug. The pharmacology will be driven by a simple mass action balance effect.

Overall, it is difficult to predict what kinetic parameters are needed. A number of groups have simply done experiments in animal models using compounds that have different residence times. In these cases, compounds were carefully optimized to a very similar potency and PK performance compounds that had longer residence times showed a more durable response.[57,60] Overall, the key learning from the data is to begin to examine the binding kinetics and thermodynamics early on in the lead optimization of the compounds.

1.4.3 Thermodynamics in Optimization

The long-lasting clinical action of drugs depends on the overall pharmacokinetic profile, k_{off} rates and receptor occupancy that represent recognition and selectivity for a target, and rebinding as the drug comes off a target and binds to another nearby. In this case, investigating both mechanisms under relevant conditions requires the use of tissues or intact cell systems that physiologically mimic the relevant conditions. We can also break down the equilibrium binding constants into thermodynamic signatures using ITC. The binding constant of a compound to a protein is driven by the change in free energy or ΔG. This is given by the relationship of $K_a = e-(\Delta G/RT)$. From this, one can see that a high negative value of ΔG will result in a lower binding constant. However, the change in Gibb's free energy, ΔG, is a combination of the change in enthalpy (ΔH) and change in entropy (ΔS), as described in eqn (1.2).

$$\Delta G = \Delta H - T\Delta S \tag{1.2}$$

The ΔH value is related primarily to the formation of hydrogen bonds, and van der Waals interactions, while ΔS is related to desolvation and conformational change or flexibility. As previously shown with binding kinetics, one can break down the K_d of a compound into its thermodynamic parameters. This is possible given the invention of isothermal calorimetry (ITC) and micro-calorimetry instruments that are highly sensitive and can be partially automated to provide the ability to test multiple compounds.

Medicinal chemists want to link structural features to thermodynamic parameters to drive the binding affinity of lead compounds. Once again, two different scaffolds can have the same K_d value but have different ΔH and ΔS values. The work by Freire and co-workers taught us how ΔH and ΔS might help a chemistry team drive forward with the "best" compound and how design changes influence K_d values.[61] In this work, Freire determined the thermodynamic signature of approved drugs for HIV-1. Each next generation drug had improved affinity mostly driven by an improved enthalpy or ΔH value. For example, comparing the thermodynamic profile of Indinavir to Tipranavir showed that the binding of Indinavir is actually endothermic ($\Delta H = +1.8$ kcal mol^{-1}) and unfavourable. In contrast, the binding of Tipranavir is exothermic ($\Delta H = -0.7$ kcal mol^{-1}), indicating favourable binding. This relatively small difference in the ΔH values between the two compounds actually drives the affinity of Tipranvir to be 70-fold more potent than Indinavir. The binding of Indinavir is actually driven by the ΔS value, which would suggest that the binding component for this compound is mostly driven by hydrophobic effects. Unfortunately, it is much easier to drive affinity through ΔS *vs.* ΔH. Fragment-based drug discovery strategies utilize calculations such as Ligand Efficiency (LE) and Lipophilicity Ligand Efficiency (LLE) to identify the most efficient optimization of binding and lipophilicity per heavy atom, to design true optimization of both activity and compound characteristics.

This example teaches that we should understand the thermodynamics that make up the binding constants of lead molecules as lead optimization begins. Teams have begun to examine the profiles of compounds very early in a project. This approach can be very effective if a team has multiple scaffolds and has structural enablement. With the thermodynamic approach, one can begin to determine what scaffolds might have liabilities in balancing ΔH and ΔS, and the potential opportunities for a given target. This approach has also been used to examine fragment hits as starting points.

1.5 Structural Biology of Complex Targets—Overcoming Challenges with Biophysics

1.5.1 Introduction

The complexity of biologically significant drug targets and their pathways is evolving very rapidly. Integral membrane proteins provide challenges of stability and handling as individual proteins. Often, the extracellular portion of a receptor is used as an initial model for assays for drug design. Extensive effort is being put into stabilizing and developing techniques to deliver models with similar profiles with known ligands, which is relying heavily on dynamic modelling. Another example is large protein complexes, which are required to display relevant biological activity, selectivity and influence on kinetics. Achieving high-resolution structures while maintaining relevant active conformations becomes more and more complex. Teams are leveraging the interplay of protein engineering, enzymology and additional solid state and dynamic techniques such as NMR, proton deuterium exchange mass spectrometry (HDX MS), small angle X-ray scattering (SAX), and cryo-EM. Both challenges require significant investment in protein engineering, biophysics and maintaining connectivity between protein activity and cellular context.

1.5.2 Protein Complexes

The polycomb repressive complex 2 (PRC2) is an epigenetic target that mediates gene silencing through methylation of histone H3 lysine 27 (H3K27). Although it was understood that the catalytic properties of the complex reside in the C-terminal SET domain of EZH2, EZH2 itself was neither stable, nor catalytically active.[62] Cyclic peptides targeting the interface of EED and EZH2 caused degradation of the complex in a cellular context and biochemical systems required at least three components for catalytic activity. Specific information on how the complex was spatially associated for function was not available although individual components from different sources were being reported.[63]

Solution of the three required components of the complex took significant upfront investment demonstrating the interplay of protein engineering, enzymology and several structural approaches and dynamic assessments. HDX-MS has been leveraged to understand conformational changes, ligand binding, protein dynamics and the selection of constructs. Identification of dynamic regions was used to systematically remove flexible regions. Small angle X-ray scattering (SAXS), a technique that assesses the compactness of a particle, was further relied upon during construct selection. Ultimately, a catalytically competent and crystalline protein complex was identified.[64]

A second approach utilized information from individual components crystallized in complex with SUZ12 peptide and the histone peptide substrate. The team further leveraged information about the kinetics of binding and methylation of mutants of the PRC2 complex. The resolution of cryogenic-electron microscopy (cryo-EM)[65] maps have improved significantly, and in 2014, structures at near-atomic resolution were being published for viruses, ribosomes, mitochondria, ion channels, and enzyme complexes. In the case of PRC2, cryo-EM provided the first picture of the pentameric complex at sufficient resolution.[66] Interaction mapping through chemical cross-linking and mass spectrometry provided a highly informative picture of the spatial relationships of EZH2, EED, SUZ12, RbBP4 and AEBP2.

1.6 Conclusion and Introduction to Book Chapters

Welcome to the world of "Biophysics in Drug Discovery". It has been recognized by the scientific community that biophysical techniques influence all areas of drug discovery. Techniques have evolved and questions addressed from the first protein crystal structures to the delivery of orally available drugs. Today, the scope of drug modalities has expanded to include peptides, proteins and antibodies. Pharmaceutical chemical matter collections and chemical technology platforms engage many different drug discovery strategies. Each of these approaches, fragment-based, targeted or phenotypic drug discovery, require their own specialized screening modalities, confirmation of actives and determination of target or mechanism of action, optimization of physical properties and an understanding of the

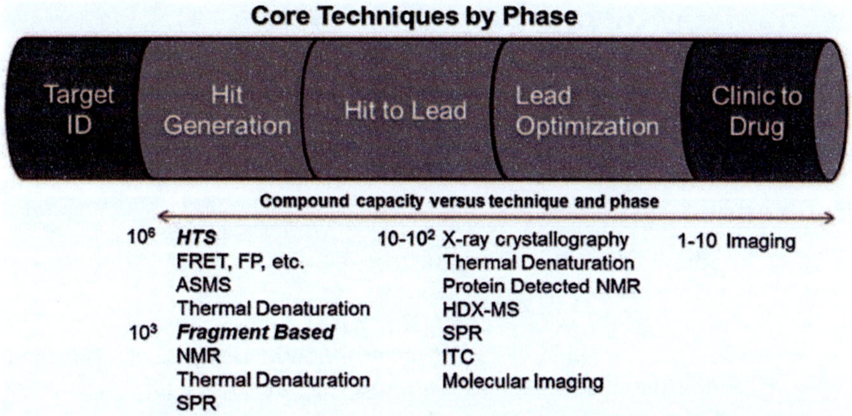

Figure 1.4 Biophysical technique integration throughout drug discovery. Biophysical techniques are impactful throughout the target and drug discovery process. Shown here are core techniques and approximate capacities. In general, the capacity of a given technique is inversely proportional to the richness of information obtained. Techniques of greater capacity are employed early on in drug discovery screening campaigns with the exception of X-ray crystallography and related technologies used to identify and select constructs. X-ray crystallography has been effectively used in FBDD screens. This technique must also be invested in early in order to deliver crystallography capabilities for maximum impact. Imaging technologies must integrated relatively early on in lead optimization in order to have greatest impact in the preclinical and clinical settings.

thermodynamics of specific protein interactions. Target engagement kinetics and thermodynamics are informing the optimization and design of ligands. Chemical probes and biophysical techniques are connecting the protein of interest to its location and pathways in the cell. Many more biophysical techniques are being extended into non-invasive imaging, confirming target occupancy, duration of occupancy, patient selection, patient tailoring, treatment response and target engagement (Figure 1.4).

We hope that we have provided an overall sense of the encompassing influence and necessity of biophysics in drug discovery. It is an exciting time as we are able to connect atomic level understanding through to the solid and liquid phase analysis of proteins and attempt to identify the most relevant targets for diseases. Understanding the level of complexity of the opportunities: protein complexes, post translational modifications and network analysis, has only begun to be fully appreciated. Biophysics has become the connection between cross-functional partners and disciplines en route to a drug. Please enjoy the individual perspectives of the scientists who have written the rest of the book as experts in designing, leveraging and analysing different biophysical experiments and data, connecting individual proteins to cellular and *in vivo* contexts, and driving drug discovery.

Acknowledgements

The authors would like to acknowledge support from Eli Lilly Discovery Chemistry Research & Technology, and Dr Cynthia Jesudason and Dr Guillermo Cortez for their helpful discussions and critical questions.

References

1. J. C. Kendrew, *et al.*, *Nature*, 1958, **181**, 662.
2. J. C. Kendrew, *et al.*, *Nature*, 1960, **185**, 422.
3. M. F. Perutz, *et al.*, *Nature*, 1960, **185**, 416.
4. A. Arnone, C. J. Bier, F. A. Cotton, V. W. Day, E. E. Hazen Jr., D. C. Richardson, J. S. Richardson and A. Yonath, *J. Biol. Chem.*, 1971, **246**, 2302.
5. A. Aguzzi and T. O'Connor, *Nat. Rev. Drug Discovery*, 2010, **9**(3), 237.
6. M. Stefani and C. M. Dobson, *J. Mol. Med.*, 2003, **81**(11), 678.
7. F. G. De Felice, M. N. Vieira, M. N. Meirelles, L. A. Morozova-Roche, C. M. Dobson and S. T. Ferreira, *FASEB J.*, 2004, **18**(10), 1099.
8. R. E. Tanzi and L. Bertram, *Cell*, 2005, **120**(4), 545.
9. K. A. Dill and J. L. MacCallum, *Science*, 2012, **338**, 1042.
10. C. Bissantz, B. Kuhn and M. Stahl, *J. Med. Chem*, 2010, **53**, 5061.
11. R. Taylor, O. Kennard and W. Versichel, *J. Am. Chem. Soc.*, 1983, **105**, 5761.
12. J. P. M. Lommerse, S. L. Price and R. Taylor, *J. Comput. Chem.*, 1997, **18**, 757.
13. I. Nobeli, S. L. Price, J. P. M. Lommerse and R. Taylor, *J. Comput. Chem.*, 1997, **18**, 2060.
14. S. K. Panigrahi and G. R. Desiraju, *Proteins*, 2007, **67**, 128.
15. K. E. Riley and K. M. Merz Jr., *J. Phys. Chem. A*, 2007, **111**, 1688.
16. R. Paulini, K. Müller and F. Diederich, *Angew. Chem., Int. Ed.*, 2005, **44**, 1788.
17. H.-J. Boehm and G. Klebe, *Angew. Chem., Int. Ed.*, 1996, **35**, 2588.
18. B. Vallone, A. E. Miele, P. Vecchini, E. Chiancone and M. Brunori, *Proc. Natl. Acad. Sci. U. S. A.*, 1998, **95**, 6103.
19. R. H. Mach, *J. Nucl. Med.*, 2014, **55**(8), 1221.
20. K. C. Allman, *J. Nucl. Cardiol.*, 2013, **20**(4), 618.
21. K. A. Wood, P. J. Hoskin and M. I. Saunders, *Clin. Oncol.*, 2007, **19**(4), 237.
22. G. Treglia, R. Sadeghi, A. Del Sole and L. Giovanella, *Clin. Transl. Oncol.*, 2014, **16**, 770.
23. D. C. Rees, M. Congreve, C. W. Murray and R. Carr, *Nat. Rev. Drug Discovery*, 2004, **3**, 660.
24. C. A. Lipinski, F. Lombardo, B. W. Dominy and P. J. Feeney, *Adv. Drug Delivery Rev.*, 1997, **23**, 3.
25. W. P. Jencks, *Proc. Natl. Acad. Sci.*, 1981, **78**(7), 4046.
26. D. A. Erlanson, S. W. Fesik, R. E. Hubbard, W. Jahnke and H. Jhoti, *Nat. Rev. Drug Discovery*, 2016, **15**, 605.

27. F. Sams-Dodd, *Drug Discovery Today*, 2005, **10**(2), 139.
28. J. A. Lee and E. L. Berg, *J. Biomol. Screening*, 2013, **18**(10), 1143.
29. J. W. Scannell, A. Blanckley, H. Boldon and B. Warrington, *Nat. Rev. Drug Discovery*, 2012, **11**(3), 191.
30. M. Bickle, *Anal. Bioanal. Chem.*, 2010, **398**(1), 219.
31. T. Y. Chang, C. Pardo-Martin, A. Allalou, C. Wahlby and M. F. Yanik, *Lab Chip*, 2012, **12**(4), 711.
32. B. Isherwood, P. Timpson, E. J. Mcghee, K. I. Anderson, M. Canel, A. Serrels, V. G. Brunton and N. O. Carragher, *Pharmaceutics*, 2011, **3**(2), 141.
33. U. Rix and G. Superti-Furga, *Nat. Chem. Biol.*, 2009, **5**(9), 616–624.
34. M. Alvim-Gaston, T. Grese, A. Mahoui, A. D. Palkowitz, M. Pinerio-Nunez and I. Watson, *Curr. Top. Med. Chem.*, 2014, **14**, 294.
35. B. L. DeCorte, *J. Med. Chem.*, 2016, **59**(20), 9295. Underexplored Opportunities for Natural Products in Drug Discovery.
36. G. P. Smith and V. A. Petrenko, *Chem. Rev.*, 1997, **97**, 391. Phage Display.
37. M. A. Clark, R. A. Acharya, C. Carico-Muendel, S. L. Belyanskaya, D. R. Benjamin and N. R. Carlos, *et al.*, *Nat. Chem. Biol.*, 2009, **5**(9), 647. Design, Synthesis and Selection of DNA-encoded small molecule libraries.
38. R. M. Franzini and C. Randolph, *J. Med. Chem.*, 2016, **59**, 6629. Chemical Space of DNA-Encoded Libaries.
39. S. B. Shuker, P. J. Hajduk, R. P. Meadows and S. W. Fesik, *Science*, 1996, **274**(5292), 1531. Discovery high-affinity ligands for proteins: SAR by NMR.
40. K. E. Arntson and W. C. K. Pomerantz, *J. Med. Chem.*, 2016, **59**, 5158. Protein observed fluorine NMR, A biorthogonal Approach to small molecule discovery.
41. A. K. Urick, J. Espinosa, J. P. Calle, H. Hu and W. C. K. Pomerantz, *Chem. Biol.*, 2016, **11**(11), 154. Comparative analysis of ligand observed 1H NMR and protein observed 19F NMR for fragment screening.
42. A. M. Petros, J. R. Huth, T. Oost, C. M. Park, H. Ding, X. Wang, H. Zhang, P. Nimmer, R. Mendoza, C. Sun, J. Mack, K. Walter, S. Dorwin, E. Gramling, U. Ladror, S. H. Rosenberg, S. W. Elmore, S. W. Fesik and P. J. Hadjuk, *Bioorg. Med. Chem. Lett.*, 2010, **20**(22), 6587. Discovery of a potent and selective Bcl-2 inhibitor using SAR by NMR.
43. A. Friberg, D. Vigil, B. Zhao, R. N. Daniels, J. P. Burke, P. M. Garcia-Barrantes, D. Camper, B. A. Chauder, T. Lee, E. T. Olejniczak and S. W. Fesik, *J. Med. Chem.*, 2013, **56**(1), 15.
44. Y. S. Wang, C. Strickland, J. H. Voigt, M. E. Kennedy, B. M. Beyer, M. M. Senior, E. M. Smith, T. L. Nechuta, V. S. Madison, M. Czarniecki, B. A. McKittrick, A. W. Stamford, E. M. Parker, J. C. Hunter, W. J. Greenlee and D. F. Wyss, *J. Med. Chem.*, 2010, **53**(3), 942.
45. R. S. Norton, E. W. W. Leung, I. R. Chandrashekaran and C. A. MacRaild, *Molecules*, 2016, **21**, 860.
46. J. F. Brandts and L. N. Lin, *Biochemistry*, 1990, **29**(29), 6927.

47. M. W. Pantoliano, E. C. Petrella, J. D. Kwasnoski, V. S. Lobanov, J. Myslik, E. Graf, T. Carver, E. Asel, B. A. Spring, P. Lane and F. R. Salemme, *J. Biomol. Screening*, 2001, **6**(6), 429.
48. I. S. Lucet, J. M. Hildebrand, P. E. Czabotar, J.-G. Zhang, N. A. Nicola, J. Silke, J. J. Babon and J. M. Murphy, *Bio-Protoc.*, 2014, **4**(11), e1135.
49. P. C. Reid, Y. Goto, T. Katoh and H. Suga, *Methods Mol. Biol*, 2012, **805**, 335.
50. A. M. Giannetti, *Methods Enzymol*, **493**, 169.
51. D. A. Annis, E. Nickbarg, X. Yang, M. R. Ziebell and C. E. Whitehurst, *Curr. Opin. Chem. Biol.*, 2007, **11**(5), 518.
52. C. C. Ozbal, W. A. LaMarr, J. R. Linton, D. F. Green, A. Katz, T. B. Morrison and C. J. Brenan, *Assay Drug Dev. Technol.*, 2004, **2**, 373.
53. J. L. Dahlin, J. W. M. Nissink, J. M. Strasser, S. Francis, L. A. Higgins, H. Zhou, Z. Zhang and M. A. Walters, *J. Med. Chem.*, 2015, **58**(5), 2091.
54. J. Baell and M. A. Walters, *Nature*, 2014, **513**, 481.
55. M. A. Fleming, S. P. Chamber, P. R. Connelly, E. Nimmesgem, T. Fox, F. J. Bruzzese, S. T. Hoe, J. R. Fulghum, D. J. Livingston, C. M. Stuver, M. D. Sintchak, K. P. Wilson and J. A. Thomson, *Biochemistry*, 1996, 35(22), 6990.
56. C. Pargellis, L. Tong, L. Churchill, P. F. Cirillo, T. Gilmore, A. G. Graham, P. M. Grob, E. R. Hickey, N. Moss, S. Pav and J. Regan, *Nat. Struct. Biol.*, 2002, **9**(4), 268.
57. R. Copeland, *Nat. Rev. Drug Discovery*, 2016, **15**(2), 87.
58. W. F. Hood, J. K. Gierse, P. C. Isakson, J. R. Kiefer, R. G. Kurumbail, K. Seibert and J. B. Monahan, *Mol. Pharmacol.*, 2003, **63**(4), 870.
59. C. W. Zapf, B. S. Gerstenberger, L. Xing, D. C. Limburg, D. R. Anderson, N. Caspers, S. Hans, A. Aulabaugh, R. Kurumbail, S. Shakya, X. Li, V. Spauling, R. M. Czerwinski, N. Seth and Q. G. Medley, *J. Med. Chem.*, 2012, **55**(22), 10047.
60. G. Dahl and T. Akerud, *Drug Discovery Today*, 2013, **18**(15–16), 697.
61. E. Freire, *Drug Discovery Today*, 2008, **13**(19–20), 869.
62. Z. Han, X. Wing, M. Hu, Y. Zhang, P. Liu and J. Chai, *Structure*, 2007, **15**, 1306. Structural Basis of EZH2 Recognition of EED.
63. W. Kim, G. Bird, T. Neff, G. Guo, M. A. Kerenyi, L. D. Walensky and S. H. Orking, *Nat. Chem. Biol.*, 2013, **9**, 643.
64. A. Brooun, K. S. Gajiwala, Y.-L. Deng, W. Liu, B. Bolanos, P. Bingham, Y.-A. He, W. Diehl, N. Grable, P.-P. Kung, S. Sutton, K. A. Maegley, X. Yu and A. E. Stewart, *Nat. Commun.*, 2016, 11384.
65. E. Callaway, *Nature*, 2015, **525**, 172.
66. N. Justin, Y. Zhang, C. Tarricone, S. R. Martin, S. Chen, E. Underwood, V. D. Marco, L. F. Haire, P. A. Walker, D. Reinberg, J. R. Wilson and S. J. Gamblin, *Nat. Commun.*, 2016, 11316.

CHAPTER 2

Ligand-detected NMR Methods in Drug Discovery

MARÍA ÁNGELES CANALES*[a] AND JUAN FÉLIX ESPINOSA*[b]

[a] Department of Química Orgánica I, Universidad Complutense de Madrid, Avd. Complutense s/n, 28040 Madrid, Spain; [b] Centro de Investigación Lilly, Avda. de la Industria 30, 28108, Alcobendas, Madrid, Spain
*Email: ma.canales@quim.ucm.es; jfespinosa@lilly.com

2.1 Introduction

Although the initial role of NMR spectroscopy in drug discovery was the structural characterization of compounds isolated from natural sources or obtained by synthetic methods, this technique has evolved from a purely analytical tool into a powerful biophysical technique for detecting ligand binding. NMR spectroscopy is particularly well suited to examining intermolecular interactions and its application to drug discovery is simply a special case, wherein one of the molecules is a target of therapeutic interest and the other is typically a small compound with the potential to become a drug after several cycles of optimization.[1–4] NMR spectroscopy is probably the most versatile biophysical technique available to the pharmaceutical industry. It can be used for structure elucidation, screening of compound libraries, hit validation, identification of ligand binding modes and structural studies of protein–ligand complexes that are useful for hit optimization.[5–28]

NMR methods for the detection of ligand binding are based on the perturbation of one or more NMR parameters upon complex formation and can be either "protein-detected" or "ligand-detected", depending on the resonances being observed. Protein-based methods have been widely used and are very reliable for ligand identification.[29–32] Such experiments provide detailed structural information to guide compound design but are highly demanding since they require the expression of significant amounts of ^{15}N-, and/or ^{13}C-, and/or ^{2}H-labeled proteins, together with the acquisition and assignment of 2D- and 3D-heteronuclear correlation spectra. Protein-based NMR methods are thus hampered by low-throughput and an upper size limit of the proteins that can be studied. In contrast, ligand-detected NMR experiments, which do not require labeled proteins and typically involve 1D-NMR experiments, have much higher throughput and can be used with larger proteins. Ligand-detected methods, however, provide much less structural information about the ligand–protein complex.[33–36]

For a favourable binding event, the rate constant for dissociation of the ligand from the receptor (k_{off}) is the limiting factor that defines the kinetics.[37] If k_{off} is sufficiently fast relative to the NMR time scale, an average signal is observed. On the other hand, high-affinity ligands with slow k_{off} show two separated signals, corresponding to the free and the bound states. Ligand-detected experiments are typically performed under "fast-exchange" conditions, using a large excess of the small molecule over the protein. Under these conditions, averaged signals are observed. For practical purposes, these signals appear at the chemical shifts of the free ligand since this parameter depends on the molar ratio. A key feature of ligand-observed NMR methods is that other NMR parameters (relaxation, NOEs), unlike chemical shifts, are not linearly averaged but are weighted towards the bound state. The ligand binds to the receptor and acquires protein-like properties and is then dissociated back into solution, leaving the binding site available for other ligands. Within the NMR time scale, this process is repeated many times, ultimately resulting in the transfer of "information" from the bound to the free state. As a consequence, although the bound fraction at any particular time can be quite low, many molecules bind to the receptor over the course of the NMR experiment, resulting in non-stoichiometric responses for parameters such as relaxation or NOEs that are informative about the bound state of the ligand.

The goal of this chapter is to explain the principles of the most frequently used ligand-observed NMR experiments. These experiments are used in the drug discovery process by many pharmaceutical companies for screening, especially when weak affinity ligands need to be identified, for hit validation and for the study of protein–ligand interactions in solution. Examples of the use of ligand-observed NMR methods in specific medicinal chemistry projects will be discussed in the last part of the chapter.

2.2 NMR Methods

2.2.1 Relaxation

Small molecules are characterized by long transverse relaxation times (T_2), which lead to narrow linewidths. The relaxation times of large macromolecules are much shorter and their corresponding signals are consequently much broader. A small molecule ligand that binds to a receptor acquires the behaviour of the macromolecule and also experiences fast relaxation. Under fast-exchange conditions, many ligands occupy the binding site on the protein, increasing the relaxation rate of the ligand. The ligand signals thus become broader and the effect is proportional to the molecular weight of the protein.[38] Additional broadening arises from chemical exchange between the free and bound states, as occurs in any equilibrium between two species. Although binding can be directly inferred from signal broadening in a ^{1}H spectrum, it is more convenient to use a T_2-filtered experiment (^{1}H-CPMG pulse sequence) to convert signal broadening into signal reduction, which allows the easy detection of changes in the relaxation times of a ligand. $T_1\rho$-filtered experiments, in which a spin-lock block instead of spin-echo train is used, can be used as an alternative to the T_2-method. T_1-filtered experiments, on the other hand, are not especially useful since the longitudinal relaxation times of proteins are not dissimilar to those of small molecules.

Several approaches that involve modification of either the protein or the ligand have been used to enhance ligand relaxation and increase the sensitivity of relaxation-based NMR experiments:

(a) In target immobilized NMR screening (TINS), the receptor is immobilized on a solid support, such as a sepharose-based polymer, which is compatible with acquisition of high-resolution NMR spectra, leading to an apparent increase in the molecular size of the receptor.[39] Signal broadening relative to a control spectrum for the polymer without the coupled protein indicates interaction with the protein. This technique allows a single protein sample to be used to carry out a complete ligand screen, thus minimizing protein consumption and maximizing speed and throughput.[40]

(b) In the SLAPSTIC (spin labels attached to protein sidechains as a tool to identify interacting compounds) method,[41,42] a paramagnetic label is added to the protein in a position adjacent to the ligand binding site to act as an extra source of relaxation. The label should be located close enough to the binding site to affect ligand relaxation, but far enough away to avoid any interference with the ligand binding. Nitroxide radicals bearing an unpaired electron are typically used as spin labels and can be linked to a nucleophilic residue in the protein.

(c) A fluorine atom can be introduced into the ligand to exploit the efficient relaxation of fluorine nuclei, which is enhanced by contributions from chemical shift anisotropy (unsymmetrical distribution of the electrons in C–F bonds that stimulates relaxation) and from the exchange term (large differences in chemical shifts between the free and bound ligand). As well as its favourable relaxation properties, ^{19}F occurs at 100% natural abundance and its NMR sensitivity is not too dissimilar to that of ^{1}H (0.83), obviating the need for water suppression or relaxation filters to remove protein signals. A significant proportion of compound collections held by pharmaceutical companies contain fluorine atoms and libraries of fluorinated molecules are also now available commercially.[43–45]

2.2.2 Transferred-NOESY experiments

The sign and intensity of another easily observable NMR phenomenon, the nuclear Overhauser effect (NOE) between two proton resonances, show a strong dependence on molecular size. The intramolecular NOEs of small molecules are positive and weak, whereas those involving proteins are negative and intense. If the k_{off} of a ligand is sufficiently fast with respect to the mixing time of the NOE experiment, the strong and negative intramolecular NOEs between ligand protons that develop during the residence time of the ligand in the binding site are transferred and detected in the signals of the free ligand.[46–48] The resulting NOEs are usually dominated by the bound state and, therefore, the binding of a small molecule to a macromolecule will produce a change in the sign of the NOE (Figure 2.1). Compounds that bind to a protein show negative transferred NOEs (same sign as the diagonal peak in a 2D NOESY experiment) in the presence of the protein. These are easily distinguished from the weak positive NOEs of compounds that do not interact with the target protein. NOESY experiments usually include a short T_2-filter to suppress protein background signals from the spectrum to facilitate interpretation.

2.2.3 Saturation Transfer Difference

Saturation in NMR spectroscopy equalizes populations between spin states and has been traditionally used to achieve solvent suppression or, in NOE difference experiments in which a resonance is "irradiated", to produce a

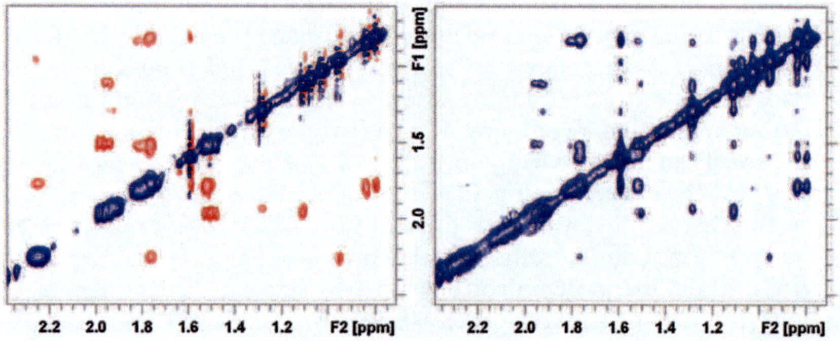

Figure 2.1 Left: 2D NOESY spectrum for a small molecule in D_2O solution. NOE cross peaks are of opposite sign relative to diagonal signals (positive NOEs). Right: tr-2D NOESY spectrum for the same compound in the presence of tubulin. NOE cross peaks are of the same sign as the diagonal signals, indicating that the compound is bound to tubulin.

change in the intensity of other resonances through dipole–dipole interactions. The use of saturation transfer to detect ligand binding was first reported in 1999 and, since then, the saturation transfer difference (STD) method has become the most widely used ligand-based NMR experiment in drug discovery.[49–54] In essence, the STD method exploits the NOE, specifically, the intermolecular NOE between an irradiated receptor and a bound ligand.

In the STD experiment, saturation of specific resonances is achieved by selective irradiation of a region of the spectrum (typically between 0 and −1 ppm) that contains shielded protein resonances but is free of ligand resonances. The saturation is rapidly and efficiently spread from this subset of resonances to the entire protein through intramolecular dipole–dipole interactions and transfers to any bound ligand through intermolecular dipole–dipole interactions. In a second experiment, an "off-resonance" region, free of any signals, is irradiated with the same pulse sequence to generate a reference spectrum, which is subtracted from the spectrum obtained in the first experiment. Although saturation of the ligand occurs only in the bound state, if k_{off} is faster than the T_1 relaxation time, this saturation is accumulated in the free state because of multiple association and dissociation steps occurring during the course of the experiment. As a result, the STD experiment yields a difference spectrum that contains signals from compounds that bind to the protein, while signals of non-binding compounds are cancelled out by subtraction (Figure 2.2). The influence of the dissociation constant, K_d, the off-rate constant, k_{off}, and saturation times in the STD effect has been extensively discussed in several reviews[12,36,55] and will not be covered here.

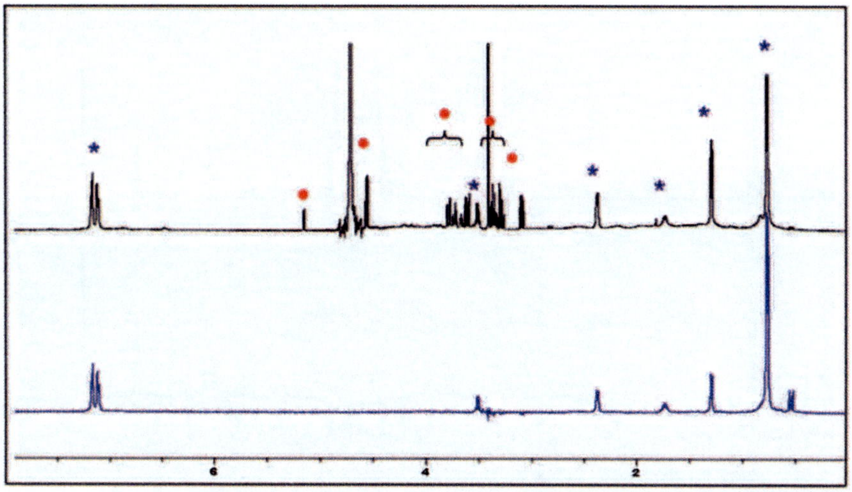

Figure 2.2 Top: ^{1}H-NMR spectrum of a mixture of ibuprofen (blue asterisks) and glucose (red circles) in the presence of bovine serum albumin (BSA). Bottom: Saturation Transfer Difference (STD) spectrum after irradiation at −1 ppm using a train of gaussian pulses. Only the ibuprofen signals are seen, indicating that ibuprofen binds to BSA whereas glucose does not.

2.2.4 Water-LOGSY

The Water-LOGSY experiment is an alternative NOE-based experiment, in which bulk water is selectively inverted and the magnetization of the hydrated water is transferred to all solutes.[56,57] The perturbed water transfers magnetization to the protein *via* an intermolecular NOE from the solvating water molecules or *via* exchange with labile proton (NH, OH or SH) resonances and this is then transferred to any bound ligand. Bound ligands also receive magnetization directly from bound water molecules within the active site of the protein. As a result of the relationship between NOE and molecular size, as outlined above, macromolecules and bound ligands exhibit strong negative NOEs, whereas small compounds that do not bind to the protein only receive magnetization from solvating water in the free state and give rise to weak positive NOEs. This outcome differs from that of the STD experiment, where only signals of bound compounds are observed (Figure 2.3).

2.2.5 Pseudo Contact Shift measurements

The standard NMR parameters that contain structural information, NOEs and J couplings, provide only local information. In molecules with extended shapes, this hampers conformational analysis and binding studies. In these systems, the use of lanthanide-chelating conjugates has emerged as a novel approach to obtain additional parameters with structural information, such

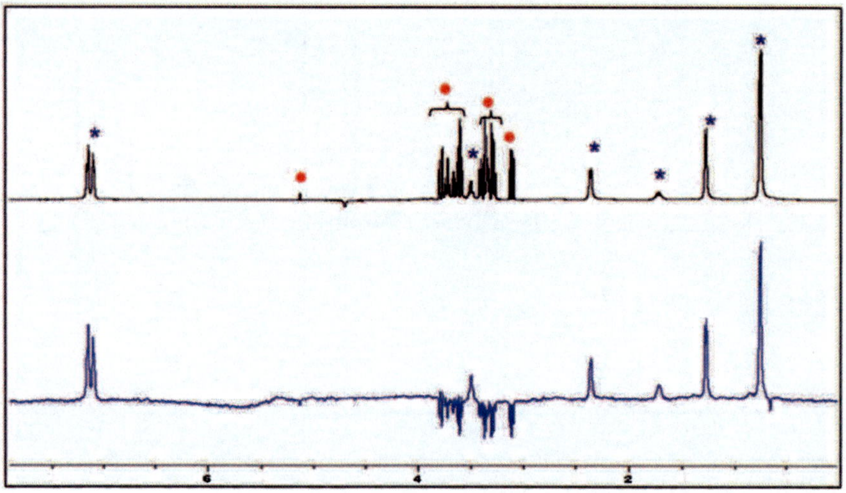

Figure 2.3 Top: ^{1}H-NMR spectrum with water suppression of a mixture of ibuprofen (blue asterisks) and glucose (red circles) in the presence of bovine serum albumin (BSA). Bottom: Water-LOGSY spectrum in which signals of ibuprofen (BSA ligand) display the opposite sign to signals of glucose (not BSA ligand).

as pseudocontact shifts (PCSs) and paramagnetic relaxation enhancements (PREs). The presence of a lanthanide nucleus in a molecule gives rise to large chemical shift changes in the NMR signals of nuclei in the vicinity of the lanthanide. These PCSs are caused by the dipolar interaction between the unpaired electron of the lanthanide and nearby nuclei. The effect is distance dependent, as can be seen in eqn (1.1).

$$\Delta\delta_{\mathrm{PCS}} = \frac{1}{12\pi r^3}\left[\Delta\chi_{\mathrm{ax}}(3\cos^2\theta - 1) + 3/2\Delta\chi_{\mathrm{rh}}\sin^2\theta\cos^2\varphi)\right] \qquad (1.1)$$

where $\Delta\delta_{\mathrm{PCS}}$ is the difference in chemical shifts measured between diamagnetic and paramagnetic samples, *r* is the distance between the metal ion and the nuclear spin, and θ and ϕ are the polar coordinates describing the position of the nuclear spin with respect to the principal axes of the magnetic susceptibility tensor ($\Delta\chi$). The effect is proportional to the anisotropy of the magnetic susceptibility tensor and independent of the isotropic component of the χ tensor. It is therefore sufficient to describe the PCS as a function of the axial and rhombic components of the $\Delta\chi$ tensor, $\Delta\chi_{\mathrm{ax}}$ and $\Delta\chi_{\mathrm{rh}}$. The PCS decreases less rapidly with distance than the NOE. It is, therefore, possible to use these parameters to quantitatively obtain medium- and long-range distance constraints. The lanthanide also induces PREs in the surrounding nuclei. The PRE decreases with r^{-6}, as does the NOE.

2.3 Competitive NMR Experiments

The ligand-based NMR methods described above work only when many ligands cycle on and off the protein during the mixing time or during the saturation or relaxation period. For compounds with high affinities (<1 μM), the number of dissociations on the NMR timescale is usually insufficient for effects to be transferred to the free state and ligand-based methods cannot be used. Another potential problem with these methods is that compounds that bind to the protein in a non-specific manner appear as positives in the NMR assays. Both drawbacks can be overcome using competitive NMR experiments in which the displacement of a reporter ligand, often referred to as a "spy" molecule, is monitored.[58–60] In this case, the requirement for rapid exchange on the NMR timescale applies only to the *spy* molecule. If the change in the NMR parameter observed for the spy molecule is reversed on addition of the test compound, this reflects displacement of the spy molecule from the binding site by the test compound. A requisite for this type of experiment is that a resonance of the spy molecule should appear in a region of the spectrum that is free of signals from test compounds, solvents or buffers. Ligands with a ^{13}C or ^{19}F-label, especially fluorinated compounds, which are readily accessible and have significant advantages in relaxation-based experiments, are attractive options to avoid overlap.[61] In a related competition experiment, a high affinity compound known to bind in the target site is added to a sample containing an NMR hit. This valuable complementary approach can be used to ascertain whether the NMR hit under analysis is displaced by a more potent ligand and to distinguish fragments that bind to a specific target site on the protein from non-specific binders that are of no interest.

Competition-based experiments thus extend the range of affinities over which ligand-based methods can be used and also allow non-specific binders to be excluded. These methods, nevertheless, have two major disadvantages when used for screening: (a) ligands cannot be directly identified from mixtures and a deconvolution step is needed and (b) secondary binding sites cannot be explored. Another limitation of competition-based methods is that a suitable spy molecule may be difficult to identify for some targets, especially those involved in protein–protein interactions. It is also worth noting that allosteric effects cannot be discounted when displacement of a spy molecule is observed, since binding of a test compound at a remote site could decrease the affinity of the spy molecule at the primary binding site. Although competition most likely indicates that the two ligands bind at the same site, this assumption cannot be guaranteed. In cases where a high-affinity ligand is not available and competition experiments are not possible, STD effects may help to distinguish between specific and non-specific binding. Differences among the STD effects of ligand protons suggest a single binding mode, with a group of protons close to the protein and another group of protons exposed to the solvent. In contrast, similar STD effects would be expected for all proton resonances of non-specific compounds that interact with the protein in multiple binding modes.[62]

2.4 Binding Quantification

The formation of a protein–ligand complex causes a change in the chemical shifts of both ligand and protein resonances. The changes in protein chemical shifts have been fully exploited in heteronuclear correlation NMR spectroscopy of labeled proteins in a process known as structure–activity relationships by NMR (SAR by NMR).[30] Changes in ligand chemical shifts upon complex formation are also indicative of binding, although this effect is usually too small to be detected because of the high ligand:protein ratios used in ligand-based NMR experiments. Detection of changes in the chemical shifts of ligand resonances is only possible if a moderate excess of ligand over protein or equimolar or sub-equimolar ligand:protein ratios are used, which significantly increases protein consumption relative to NOE- or relaxation-based approaches. The major advantage of observing changes in the chemical shifts of a ligand is that dissociation constants can be measured using titration NMR experiments.[63] Relaxation-based experiments, in which line broadening in a ^{1}H spectrum or signal intensity in a ^{1}H-CPMG experiment are monitored as a function of protein concentration, are usually better options, and require smaller amounts of protein.[64,65] STD experiments can be also used for measuring dissociation constants of protein–ligand complexes if the STD effect is plotted against the ligand concentration, although this methodology is more time-consuming.[66] Other approaches using competition experiments have been designed to assess the relative affinities of ligands rather than to measure accurate binding constants since, in many cases, a rank order of affinities is sufficient to establish SAR.[67,68]

2.5 NMR in Drug Discovery

2.5.1 Screening

NMR is not a high-throughput technique and the number of compounds that can be screened in a conventional enzymatic or binding assay is far higher than the number that can be achieved using ligand-based NMR methods. There are, however, circumstances in which an NMR screen becomes the preferred option, especially when very simple molecules, often described as "fragments", are desired as an initial step in the route to more elaborated molecules. Fragments offer some advantages since they allow exploration of a greater amount of chemical space and can be more easily accommodated into a protein binding site than more complex compounds.[69–72] Fragments typically have weak affinities (0.1–10 mM) and may not be detected by conventional enzymatic assays or other biophysical techniques. Unlike bioassays, NMR is able to detect weak ligand–protein interactions and can be used to identify initial hits with binding constants in the mM range.[23,73,74] On the downside, NMR methods require higher protein concentrations and longer acquisition times than other screening technologies and mixtures of compounds are usually screened to increase throughput. Dedicated libraries for NMR screening typically contain commercially and synthetically available compounds of low molecular mass (<300 Da) and high aqueous solubility that are designed to have maximum diversity.[75–77] The fragments are usually drug-like scaffolds derived from frameworks found in known drugs.[78] Each test mixture typically contains 5–10 non-reacting compounds, each with an isolated NMR signal for easy interpretation of spectral data. When carrying out screening using ^{19}F-labeled compounds, each test mixture can contain more than 10 compounds because of the superior spectral dispersion of ^{19}F-NMR (Figure 2.4) and examples using mixtures of 36 compounds have been reported.[79] Libraries screened using NMR typically contain only a few hundred or a few thousand compounds. The NMR hits are elaborated in follow-up strategies designed to increase binding affinity and improve molecular properties.

An additional advantage of NMR screening is that it has the potential to identify fragments that bind in a protein site that is proximal to the primary binding site, thus enabling a fragment-linking strategy.[80] In principle, fragments that bind in adjacent sites could be identified from an initial NMR screen. In practice, however, a second screen in the presence of the first ligand is preferred because the two binding sites may not be totally independent and binding of the first fragment may induce a conformational

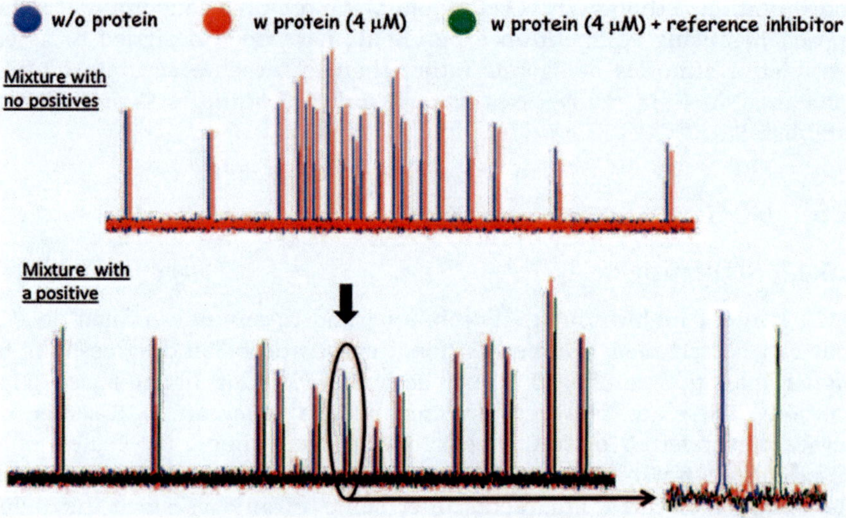

Figure 2.4 ^{19}F-CPMG spectra recorded with proton decoupling for mixtures of fluorinated compounds without protein (blue), with protein (red) and with protein and a known high-affinity ligand for the protein. The spectra are slightly offset for clarity. Top spectra: No signal reduction for any of the compounds is observed upon addition of the protein, indicating that none of the compounds in this mixture bind to the protein. Bottom spectra: A significant reduction in signal intensity is observed for one of the compounds upon addition of the protein indicating that this compound binds to the target protein. The signal intensity is recovered when a known high-affinity ligand is added, indicating specific binding of the hit compound to the target protein.

change in the protein. The binding affinity of the second fragment may be increased when the first fragment is bound (cooperative binding) or the secondary site may only be formed when the primary site is occupied. The simplest approach for a second-site screen is to repeat the initial screen using a saturating concentration of a ligand that binds in the primary site. Although a fragment found in the initial screen could be chosen as the primary site ligand, it is better to optimize the initial hit to increase affinity. Increases in molecular size should, however, be avoided to prevent occlusion of the second pocket. Other more sophisticated, but very effective, second-site approaches require the preparation of spin-labeled analogues of the primary ligand. When the labeled ligand occupies the primary binding site, a second fragment binding to a peripheral site experiences enhanced relaxation.[81] As an example, adenine analogues bearing TEMPO substituents have been used to detect allosteric ligands for protein kinases.[82,83] Replacement of the ATP-bound magnesium ion by a paramagnetic manganese ion has also been used for the identification of ligands that bind to a second site in kinases.[84] In a different situation, it may be desirable to replace a moiety of a ligand to achieve an increase in potency or selectivity.

A compound lacking this moiety can used as the primary ligand for an NMR-based screen to identify fragments that bind to the same subsite and could be used as a replacement for the undesirable substructure.

2.5.2 Hit validation

High throughput screening assays are prone to false positives because of assay interference or aggregation[85] and an NMR step can be included in the screening process to confirm that hits from the biochemical assay do indeed bind to the target. Ligand-based NMR methods are usually combined with other orthogonal biophysical techniques, and compounds that show no binding to the protein by any technique are discarded.[86] As a follow-up to fragment-based NMR screens, analogues of the hits (from commercial sources or corporate collections) can be used in further NMR experiments to develop preliminary SAR around each hit. In some cases, alternative orthogonal biophysical techniques that provide accurate K_d measurements, such as surface plasmon resonance or isothermal titration calorimetry, may be preferred for follow-up of NMR hits. These techniques are, however, unsuitable for use with weak-affinity fragments that can only be detected by ligand-based NMR methods.

2.5.3 Structure-guided Hit Optimization

Once it has been established that a particular compound binds to a protein of interest, the next step is to gain knowledge about the interaction of the ligand with the biological receptor. Although ligand-based methods cannot compete with protein-based methods in terms of the structural characterization of the protein–ligand complex, they can still provide information about the binding mode and bound conformation of the ligand and/or the location of the protein binding site. This information can be used to build a computational model of the ligand in complex with a protein of known structure that can then guide the design of more potent compounds.

The binding mode of a ligand can be inferred from the extent of saturation on each proton signal in a STD experiment, which can be quantified using the saturation amplification factor defined as follows:[87]

$$\mathrm{STD_a} = [(I_o - I_{sat})/I_o]L$$

where I_o and I_{sat} are the intensity of the ligand signal in the "off-resonance" and "on-resonance" experiments, respectively, and L is the concentration of the ligand in excess. The protons of ligands in direct contact with the protein receive the highest degree of saturation and show the strongest signals in the STD spectrum, providing the group epitope mapping that reports on the orientation of the ligand relative to the receptor. The group epitope mapping can be improved if the differences among relaxation rates of protons are taken into account and the amplification factor of each proton is divided by its nonselective longitudinal relaxation rate (T_1).[88] Another approach, referred to as DIRECTION (difference of inversion recovery rate with and without target irradiation), uses the difference in T_1 values of ligand protons with and without irradiation of the protein. The differences in T_1 values are related to the proximity of the protons to the receptor and reveal which part of the ligand is in contact with the protein surface.[89] A complementary method, referred to as SALMON (solvent accessibility, ligand binding, and mapping of ligand orientation by NMR spectroscopy), is based on water-LOGSY and determines the orientation of a protein-bound ligand by mapping the accessibility of the ligand to solvent.[90]

For molecules with an extended shape and several repeating units that cannot be individually distinguished by conventional experiments, it is possible to derive the binding epitope using lanthanide-chelating conjugates, as demonstrated for a complex biantennary N-glycan.[91] Characterization of this molecule is challenging since the molecular pseudosymmetry leads to isochronous chemical shifts in both branches. Covalent attachment of a lanthanide-binding tag to the molecule was able to break the inherent pseudosymmetry of the N-glycan since different PCSs were observed for each arm under paramagnetic conditions (lanthanide-chelating unit loaded with Dy^{3+}). Individual assignment of all signals allowed a detailed binding epitope analysis of the N-glycan bound to the galactose binding protein, galectin-3.

The bioactive conformation of a ligand can be derived from the proton–proton NOEs observed in NOESY experiments carried out in the presence of the protein.[48,92–94] Transferred NOESY (trNOESY) experiments were one of the first NMR methodologies to be proposed for use in screening[47] and, well before that, NOE experiments had been used to determine the bound conformation of a ligand in the binding site of a protein.[95] Although the translation of NOE intensities into proton–proton distances is a routine task for small molecules, this conversion is more complicated in trNOESY experiments because the NOE intensity can be contaminated with indirect effects mediated by a third proton. One of these effects, known as spin diffusion, is very efficient in proteins. In trNOESY experiments, the mixing time should be sufficiently short to minimize spin diffusion and reduce the contribution from the free ligand, but long enough to detect NOEs from the bound state.

In proteins that contain a paramagnetic nucleus as a cofactor, identification of binding sites can be derived from paramagnetic relaxation experiments.[96] This source of relaxation outweighs other mechanisms and the relaxation rate of each ligand proton is determined by its proximity to the paramagnetic center. This allows the distance and orientation of the ligand with respect to the nucleus to be established. When the protein does not contain a paramagnetic metal, a specific residue, usually a cysteine, can be labeled with a TEMPO group to enable this approach.[41]

When two ligands bind simultaneously to a protein, they can be linked together to provide a ligand of much higher affinity. This concept was first put forward by Jencks[97] and fragment assembly has since been used to provide higher affinity ligands for several proteins.[98] Selection of a suitable linker to connect the two fragments is key to achieving the expected energy gain; the linker should be flexible enough to allow interaction of each fragment with the receptor to be optimized but not so flexible that there is an entropic cost arising from loss of mobility in the bound state. Information about the distance between the two fragments and their relative orientation is critical for the rational design of an appropriate linker. Co-crystallization of the two fragments or protein-detected NMR studies are the best options for obtaining this information, but are not always feasible. In the absence of structural information, the observation of intermolecular NOEs (ILOEs) between the two ligands in a NOESY spectrum can be extremely helpful in working out how to link the two fragments.[99] As with the trNOESY experiments, short mixing times should be used to reduce spin diffusion mediated by the protein. A NOESY experiment for the two ligands in solution, but without the protein, should be conducted as a negative control to rule out the possibility of interactions between unbound fragments. If there is no inter-ligand NOE in the absence of protein, this confirms that a NOE between the two fragments is observed only when they are brought into proximity by binding to the protein. The concept of SAR by ILOEs has led to the design of potential bidentate derivatives using compound mixtures.[100–102] Although spin diffusion can complicate interpretation of ILOE data, a technique known as INPHARMA (protein-mediated inter-ligand NOEs for pharmacophore mapping) takes advantage of this indirect effect to determine the relative orientation of two competing ligands.[103–105] The magnetization of the first ligand is transferred to the protein before it dissociates from the receptor and is then picked up by a second ligand that occupies the same binding site, giving rise to a correlation peak between the two ligands. A combination of INPHARMA, trNOEs and STD-based ligand epitope mapping was used to apply restraints for determining the binding pose of a ligand bound to protein kinase A.[106] For this protein, the docking score derived from a combination of these three ligand-detected NMR experiments was able to predict the binding mode observed in the crystal structure, whereas scoring using INPHARMA or STD data alone failed to reproduce the crystallographic results.

2.6 Example of Application to Medicinal Chemistry Projects

Although fragment screening could be carried out using just one of the ligand-based NMR methods described above, a combination of two (or more) NMR approaches is often preferred to verify hits. The different NMR methods can be used in parallel screens or can be used in a stepwise manner, where the hits identified in the initial NMR-based screen are validated using a second NMR method. The NMR screening of 5′-cytosolic nucleotidase II (cN-II) is an example of parallel screening.[107] A fragment library composed of 300 fragments was screened by STD and water-LOGSY, and only fragments that showed binding in both experiments were considered to be ligands, leading to a 10% hit rate. Most of the NMR hits were able to inhibit the nucleotidase activity of cN-II. Subsequent IN-PHARMA experiments showed NOEs between two fragments and a known anthraquinone-like cN-II inhibitor (AdiS), indicating that these fragments bind to the inosine 5′-monophosphate binding site of the enzyme. An example of consecutive NMR methods is provided by the screening of a library of 1152 diverse fragments against choline kinase α (ChoKα1).[108] STD was used for the primary binding assay, giving 55 hits (4.8% hit rate). The structural diversity of these hits suggested that they were likely to bind to different sites on the protein. ChoKα1 catalyzes the first step in phosphatidylcholine biosynthesis and recognizes both ATP and choline. Competitive ^{19}F-NMR experiments, using 2-fluoroadenosine or fluorocholine chloride as fluorinated spy compounds, were performed to monitor binding to the ATP and choline sites, respectively. These experiments showed that 21 compounds were competitive with the spy for the ATP site and 13 compounds were competitive with the spy for the choline-binding pocket. The remaining compounds were deprioritized because they were regarded as weak competitors, allosteric binders or artefacts. Competitive ^{19}F-NMR experiments on the STD hits thus allowed hit validation and identification of potential binding sites. The majority of the fragment hits showed weak K_d values (0.5–2 mM), highlighting the unique ability of NMR methods to detect weak affinity ligands. One fragment hit showed stronger affinity ($K_d = 132$ μM) and a crystal structure showed this fragment located in the choline-binding site, in agreement with the competitive ^{19}F-NMR results.

Bromodomains are protein domains that recognize acetylated lysine residues, such as those present on histones, and function as epigenetic "readers" of lysine acetylation. Several bromodomains have been subjected to fragment screening using ligand-based NMR methods, efforts that were complemented with protein-observed NMR or other biophysical techniques. For instance, Zhang and coworkers screened a library of 539 soluble, commercially available fragments against BRD4, a member of the BET (bromodomain and extra terminal domain) family.[109] The compounds were clustered into pools of 8–10 compounds with minimum NMR signal overlap and screened using STD and $T_1\rho$ methods. The combination of NMR experiments yielded ten validated hits, of which four showed BRD4 inhibition in fluorescence anisotropy binding assays. The binding regions of five of the hit compounds were identified by HSQC NMR experiments using a ^{15}N-labeled protein, and the interaction between three of them and BRD4 was characterized in more detail by X-ray crystallography. In a second example, three ligand-detected NMR methods, STD, Water-LOGSY and ^{1}H-CPMG, were used for fragment screening against BRD7, another BET family bromodomain.[110] A primary hit rate of 1.1% yielded ten plausible ligands, five of which were confirmed when tested individually. Protein-based chemical shift perturbation experiments were then performed for these hits to confirm interaction with the acetyl-lysine recognition site and to measure the binding affinities. In a third example, a fragment screening campaign against the C-terminal bromodomain of P300/CBP-associated factor (PCAF, also known as histone acetyltransferase KAT2B) was performed to identify a chemical probe that would allow unravelling of the function of this domain and evaluation of its therapeutic potential.[111] A fragment library composed of approximately 950 commercially available fragments was screened by target immobilized NMR, using mixtures of four compounds. A total of 63 hits were identified (6.6% hit rate), confirming the high ligandability that had been predicted for this bromodomain. NMR hits were further validated using surface plasmon resonance and thermal shift assays, leading to seven crystal structures. These three examples of the application of NMR screening to bromodomain targets highlight the versatility of NMR methods for fragment screening and illustrate how different methods can be used in parallel.

As well as playing an important role in fragment screening and hit validation, ligand-based NMR methods can also be used to provide information about binding modes, as in the case of bacterial RNA polymerase (RNAP), a validated but largely underexploited target for the development of antibiotics.[112] A virtual screen using a 3D pharmacophore model that combined protein features from the RNAP switch region and ligand features from myxopyronin A (Myx), a well-known switch region binder, identified a hit compound containing a ureidothiophene-2-carboxylic acid core that showed antibacterial activity against Gram-positive bacteria. Confirmation that the hit compound was binding to the RNAP switch region was obtained through competitive STD experiments. Stepwise addition of the hit compound to the RNAP/Myx complex led to a decrease in the intensity of STD signals for Myx, accompanied by an increase in the intensity of STD signals for the hit, suggesting that the hit competes with Myx for the same binding site in the RNAP switch region. INPHARMA experiments were then carried out to exclude allosteric effects on the displacement of Myx and to gain insights into the relative binding orientation. Using this combination of NMR experiments, the authors were able to propose a plausible binding mode, in which the hit compound occupies the same area as the 2-pyrone moiety of Myx.

Protein-based NMR methods can be successfully used for fragment linking, as exemplified by the recently approved drug, venetoclax. Identification of second site binders can also be accomplished using ligand-based methods, as described by Amgen scientists for β-secretase (BACE).[113] The researchers designed a ^{19}F-NMR screen to identify fragments that would occupy the S3 subpocket and could give selectivity against related aspartic proteases. The ^{19}F-NMR experiments were conducted in the presence of a high-affinity compound that filled a large portion of the active site, without blocking the S3 subpocket. Seven fragments were identified and these were then subjected to competition experiments with a compound that occupies the entire pocket, including the S3 subpocket, to rule out binding to other sites on the protein. The most potent fragment was selected for linking to an analogue of the primary ligand used in the initial screen. Information about the proximity and relative orientation of the two ligands obtained by ILOE experiments was used to guide the linking strategy. Four compounds were prepared based on the ILOE results and all of them showed higher affinity than the two individual fragments. Affinity was increased by as much as 350-fold and selectivity relative to cathepsin D was increased by up to 2000-fold.

2.7 Conclusions and Future Perspectives

The ligand-based NMR methods described in this chapter monitor the changes in an NMR parameter (relaxation, NOE) that occur when a small molecule (the ligand) binds to a large molecule (the protein target). NMR methods are able to identify much lower affinity ligands than can be detected using enzymatic assays or other biophysical techniques. NMR methods can be carried out in a competitive manner to assess the specificity of an interaction and to establish an affinity ranking order for a set of ligands that compete with the same probe molecule. During the drug discovery process, initial hits are transformed into more potent and selective compounds that have suitable molecular properties for further development. The optimization process can also be guided by ligand-based NMR approaches, such as trNOESY, INPHARMA or ILOE, which provide partial structural information to characterize the protein–ligand interaction. Over the last 20 years, ligand-based NMR techniques have been successfully applied to a variety of soluble targets and they are now considered to be versatile and mature tools that offer key benefits over conventional functional assays or other biophysical techniques.

The application of NMR techniques to membrane proteins, which represent more than 50% of current drug targets, remains challenging because of the difficulties in protein expression and purification and the selection of a proper membrane-mimicking system. Ligand-observed NMR methods are able to provide binding information for these difficult targets using intact membranes or receptors expressed on the surface of living cells. As examples, intact membranes have been used to characterize the interactions of GPCR40 with a family of agonists[114] and the acquisition of STD and trNOESY data using living cancer cells was used to study recognition of a family of cyclic peptides by integrin $\alpha_{\upsilon}\beta_{3}$.[115] In cells that have a strong tendency to aggregate and settle, it may be necessary to acquire STD data under high resolution magic angle spinning conditions. This approach has already been used to study the interaction of the sodium-glucose transport protein SGLT1 with its ligands in cells derived from solid tissues.[116] Ligand-observed NMR methods are thus promising tools for the characterization of membrane protein binding events that allow intact living cells to be studied in an NMR tube.

References

1. M. Pellecchia, *Chem. Biol.*, 2005, **12**, 961–971.
2. T. Carlomagno, *Annu. Rev. Biophys. Biomol. Struct.*, 2005, **34**, 245–266.
3. A. L. Skinner and J. S. Laurence, *J. Pharm. Sci.*, 2008, **97**, 4670–4695.
4. M. Goldflam, T. Tarrago, M. Gairi and E. Giralt, *Methods Mol. Biol.*, 2012, **831**, 233–259.
5. P. J. Hajduk, R. P. Meadows and S. W. Fesik, *Q. Rev. Biophys.*, 1999, **32**, 211–240.
6. G. C. K. Roberts, *Drug Discovery Today*, 2000, **5**, 230–240.
7. T. Diercks, M. Coles and H. Kessler, *Curr. Opin. Chem. Biol.*, 2001, **5**, 285–291.
8. M. Heller and H. Kessler, *Pure Appl. Chem.*, 2001, **73**, 1429.
9. M. Pellecchia, D. S. Sem and K. Wuthrich, *Nat. Rev. Drug Discovery*, 2002, **1**, 211–219.
10. B. J. Stockman and C. Dalvit, *Prog. Nucl. Magn. Reson. Spectrosc.*, 2002, **41**, 187–231.
11. F. Jasna, L. Christopher and X. Xiaoling, *Curr. Top. Med. Chem.*, 2003, **3**, 81–97.
12. B. Meyer and T. Peters, *Angew. Chem., Int. Ed.*, 2003, **42**, 864–890.
13. C. Fernandez and W. Jahnke, *Drug Discovery Today Technol.*, 2004, **1**, 277–283.
14. J. W. Peng, J. Moore and N. Abdul-Manan, *Prog. Nucl. Magn. Reson. Spectrosc.*, 2004, **44**, 225–256.
15. C. A. Lepre, J. M. Moore and J. W. Peng, *Chem. Rev.*, 2004, **104**, 3641–3676.
16. S. W. Homans, *Angew. Chem., Int. Ed.*, 2004, **43**, 290–300.
17. M. Pellecchia, B. Becattini, K. J. Crowell, R. Fattorusso, M. Forino, M. Fragai, D. Jung, T. Mustelin and L. Tautz, *Expert Opin. Ther. Targets*, 2004, **8**, 597–611.
18. J. R. Huth, C. Sun, D. R. Sauer and P. J. Hajduk, *Methods in Enzymology, Academic Press*, 2005, vol. 394, pp. 549–571.
19. X. Zhang, H. Tang, C. Ye and M. Liu, *Drug Discovery Today Technol.*, 2006, **3**, 241–245.
20. M. Pellecchia, I. Bertini, D. Cowburn, C. Dalvit, E. Giralt, W. Jahnke, T. L. James, S. W. Homans, H. Kessler, C. Luchinat, B. Meyer, H. Oschkinat, J. Peng, H. Schwalbe and G. Siegal, *Nat. Rev. Drug Discovery*, 2008, **7**, 738–745.
21. A. M. Kelly, D. S. Matthew and P. Robert, *Combin. Chem. High Throughput Screening*, 2009, **12**, 285–295.
22. R. Powers, *Expert Opin. Drug Discovery*, 2009, **4**, 1077–1098.
23. C.-O. Ramon, *Curr. Top. Med. Chem.*, 2011, **11**, 43–67.
24. T. Maurer, in *Methods in Enzymology*, ed. C. K. Lawrence, Academic Press, 2011, vol. 493, pp. 469–485.
25. C. A. Lepre, *Methods Enzymol.*, 2011, **493**, 219–239.
26. H. L. Eaton and D. F. Wyss, in *Methods in Enzymology*, ed. C. K. Lawrence, Academic Press, 2011, vol. 493, pp. 447–468.

27. M. J. Harner, A. O. Frank and S. W. Fesik, *J. Biomol. NMR*, 2013, **56**, 65–75.
28. D. M. Dias and A. Ciulli, *Prog. Biophys. Mol. Biol.*, 2014, **116**, 101–112.
29. S. B. Shuker, P. J. Hajduk, R. P. Meadows and S. W. Fesik, *Science*, 1996, **274**, 1531–1534.
30. P. J. Hajduk, R. P. Meadows and S. W. Fesik, *Science*, 1997, **278**, 497–499.
31. H. Widmer and W. Jahnke, *Cell Mol. Life Sci.*, 2004, **61**, 580–599.
32. T. Oltersdorf, S. W. Elmore, A. R. Shoemaker, R. C. Armstrong, D. J. Augeri, B. A. Belli, M. Bruncko, T. L. Deckwerth, J. Dinges, P. J. Hajduk, M. K. Joseph, S. Kitada, S. J. Korsmeyer, A. R. Kunzer, A. Letai, C. Li, M. J. Mitten, D. G. Nettesheim, S. Ng, P. M. Nimmer, J. M. O'Connor, A. Oleksijew, A. M. Petros, J. C. Reed, W. Shen, S. K. Tahir, C. B. Thompson, K. J. Tomaselli, B. Wang, M. D. Wendt, H. Zhang, S. W. Fesik and S. H. Rosenberg, *Nature*, 2005, **435**, 677–681.
33. R. Z. Edward, Y. Jiangli, M. Huaping, D. K. Allen and J. S. Michael, *Curr. Top. Med. Chem.*, 2003, **3**, 25–37.
34. U. L. G. Christian Ludwig, *Front. Biosci.*, 2009, **14**, 4565–4574.
35. C. A. Pawel Sledz and A. Ciulli, *NMR Biomol.*, 2012, **15**, 265–280.
36. J. L. W. M. L. Rowe and M. J. Howard, *Modern NMR Tech. Synth. Chem.*, 2015, 63–123.
37. G. Roberts, *Protein NMR Spectroscopy: Practical Techniques and Application*, 2011, pp. 221–267.
38. P. J. Hajduk, E. T. Olejniczak and S. W. Fesik, *J. Am. Chem. Soc.*, 1997, **119**, 12257–12261.
39. S. Vanwetswinkel, R. J. Heetebrij, J. van Duynhoven, J. G. Hollander, D. V. Filippov, P. J. Hajduk and G. Siegal, *Chem. Biol.*, 2005, **12**, 207–216.
40. M. Kobayashi, K. Retra, F. Figaroa, J. G. Hollander, E. Ab, R. J. Heetebrij, H. Irth and G. Siegal, *J. Biomol. Screening*, 2010, **15**, 978–989.
41. W. Jahnke, S. Rüdisser and M. Zurini, *J. Am. Chem. Soc.*, 2001, **123**, 3149–3150.
42. W. Jahnke, *ChemBioChem*, 2002, **3**, 167–173.
43. C. Dalvit, P. E. Fagerness, D. T. A. Hadden, R. W. Sarver and B. J. Stockman, *J. Am. Chem. Soc.*, 2003, **125**, 7696–7703.
44. T. Tengel, T. Fex, H. Emtenas, F. Almqvist, I. Sethson and J. Kihlberg, *Org. Biomol. Chem.*, 2004, **2**, 725–731.
45. C. Dalvit, *Prog. Nucl. Magn. Reson. Spectrosc.*, 2007, **51**, 243–271.
46. P. Balaram, A. A. Bothner-By and E. Breslow, *J. Am. Chem. Soc.*, 1972, **94**, 4017–4018.
47. B. Meyer, T. Weimar and T. Peters, *Eur. J. Biochem.*, 1997, **246**, 705–709.
48. L. Marilisa, H. F. Hudson, C. Chui Sien and P. Maurizio, *Curr. Drug Discovery Technol.*, 2006, **3**, 91–100.
49. M. Mayer and B. Meyer, *Angew. Chem., Int. Ed.*, 1999, **38**, 1784–1788.
50. J. Klein, R. Meinecke, M. Mayer and B. Meyer, *J. Am. Chem. Soc.*, 1999, **121**, 5336–5337.

51. V. V. Krishnan, *Curr. Anal. Chem.*, 2005, **1**, 307–320.
52. D. W. Begley, S. O. Moen, P. G. Pierce and E. R. Zartler, in *Current Protocols in Chemical Biology*, ed. J. W. Sons, John Wiley & Sons, Inc., 2013, pp. 251–268.
53. A. Viegas, J. Manso, F. L. Nobrega and E. J. Cabrita, *J. Chem. Educ.*, 2011, **88**, 990–994.
54. A. Bhunia, S. Bhattacharjya and S. Chatterjee, *Drug Discovery Today*, 2012, **17**, 505–513.
55. V. Jayalakshmi and N. R. Krishna, *J. Magn. Reson.*, 2002, **155**, 106–118.
56. C. Dalvit, G. Fogliatto, A. Stewart, M. Veronesi and B. Stockman, *J. Biomol. NMR*, 2001, **21**, 349–359.
57. C. Dalvit, P. Pevarello, M. Tatò, M. Veronesi, A. Vulpetti and M. Sundström, *J. Biomol. NMR*, 2000, **18**, 65–68.
58. A. H. Siriwardena, F. Tian, S. Noble and J. H. Prestegard, *Angew. Chem., Int. Ed.*, 2002, **41**, 3454–3457.
59. W. Jahnke, P. Floersheim, C. Ostermeier, X. Zhang, R. Hemmig, K. Hurth and D. P. Uzunov, *Angew. Chem., Int. Ed.*, 2002, **41**, 3420–3423.
60. C. Dalvit, M. Flocco, S. Knapp, M. Mostardini, R. Perego, B. J. Stockman, M. Veronesi and M. Varasi, *J. Am. Chem. Soc.*, 2002, **124**, 7702–7709.
61. D. Claudio, F. Maria, V. Marina and J. S. Brian, *Combin. Chem. High Throughput Screening*, 2002, **5**, 605–611.
62. O. Cala and I. Krimm, *J. Med. Chem.*, 2015, **58**, 8739–8742.
63. F. Lee, *Curr. Top. Med. Chem.*, 2003, **3**, 39–53.
64. B. W. Dubois and A. S. Evers, *Biochemistry*, 1992, **31**, 7069–7076.
65. L. Fielding, *Prog. Nucl. Magn. Reson. Spectrosc.*, 2007, **51**, 219–242.
66. J. Angulo, P. M. Enríquez-Navas and P. M. Nieto, *Chem. – A Eur. J.*, 2010, **16**, 7803–7812.
67. X. Zhang, A. Sänger, R. Hemmig and W. Jahnke, *Angew. Chem., Int. Ed.*, 2009, **48**, 6691–6694.
68. M. D. Shortridge, D. S. Hage, G. S. Harbison and R. Powers, *J. Comb. Chem.*, 2008, **10**, 948–958.
69. R. J. Hall, P. N. Mortenson and C. W. Murray, *Prog. Biophys. Mol. Biol.*, 2014, **116**, 82–91.
70. D. Joseph-McCarthy, A. J. Campbell, G. Kern and D. Moustakas, *J. Chem. Inf. Model.*, 2014, **54**, 693–704.
71. D. A. Erlanson, S. W. Fesik, R. E. Hubbard, W. Jahnke and H. Jhoti, *Nat Rev Drug Discovery*, 2016, **15**, 605–619.
72. P. J. Hajduk and J. Greer, *Nat Rev Drug Discovery*, 2007, **6**, 211–219.
73. V. M. Consonni Roberto, *Structure-Activity Relationship in Drug Development by NMR Spectroscopy*, 2011, vol 1, pp. 67–101.
74. R. Z. Edward and M. Huaping, *Curr. Top. Med. Chem.*, 2007, **7**, 1592–1599.
75. M. N. Schulz, J. Landstrom, K. Bright and R. E. Hubbard, *J. Comput.-Aided Mol. Des.*, 2011, **25**, 611–620.
76. J. Fejzo, C. A. Lepre, J. W. Peng, G. W. Bemis, Ajay, M. A. Murcko and J. M. Moore, *Chem. Biol.*, 1999, **6**, 755–769.

77. J. Edgar, D. John and J. J. B. Marcel, *Curr. Top. Med. Chem.*, 2003, **3**, 11–23.
78. G. W. Bemis and M. A. Murcko, *J. Med. Chem.*, 1996, **39**, 2887–2893.
79. A. Vulpetti, U. Hommel, G. Landrum, R. Lewis and C. Dalvit, *J. Am. Chem. Soc.*, 2009, **131**, 12949–12959.
80. J. Wolfgang, F. Andreas, J. J. B. Marcel, C. G. Paris, H. Jutta, M. N. Carlo and B. P. Lawrence, *Curr. Top. Med. Chem.*, 2003, **3**, 69–80.
81. W. Jahnke, L. B. Perez, C. G. Paris, A. Strauss, G. Fendrich and C. M. Nalin, *J. Am. Chem. Soc.*, 2000, **122**, 7394–7395.
82. W. Jahnke, M. J. Blommers, C. Fernandez, C. Zwingelstein and R. Amstutz, *ChemBioChem*, 2005, **6**, 1607–1610.
83. J. Vazquez, S. K. De, L.-H. Chen, M. Riel-Mehan, A. Emdadi, J. Cellitti, J. L. Stebbins, M. F. Rega and M. Pellecchia, *J. Med. Chem.*, 2008, **51**, 3460–3465.
84. G. K. Jarori, S. B. Iyer, S. R. Kasturi and U. W. Kenkare, *Eur. J. Biochem.*, 1990, **188**, 9–14.
85. J. B. Baell and G. A. Holloway, *J. Med. Chem.*, 2010, **53**, 2719–2740.
86. R. H. Folmer, *Drug Discovery Today*, 2016, **21**, 491–498.
87. M. Mayer and B. Meyer, *J. Am. Chem. Soc.*, 2001, **123**, 6108–6117.
88. S. Kemper, M. K. Patel, J. C. Errey, B. G. Davis, J. A. Jones and T. D. Claridge, *J. Magn. Reson.*, 2010, **203**, 1–10.
89. Y. Mizukoshi, A. Abe, T. Takizawa, H. Hanzawa, Y. Fukunishi, I. Shimada and H. Takahashi, *Angew. Chem., Int. Ed.*, 2012, **51**, 1362–1365.
90. C. Ludwig, P. J. A. Michiels, X. Wu, K. L. Kavanagh, E. Pilka, A. Jansson, U. Oppermann and U. L. Günther, *J. Med. Chem.*, 2008, **51**, 1–3.
91. A. Canales, A. Mallagaray, J. Pérez-Castells, I. Boos, C. Unverzagt, S. André, H.-J. Gabius, F. J. Cañada and J. Jiménez-Barbero, *Angew. Chem., Int. Ed.*, 2013, **52**, 13789–13793.
92. C. Post, *Curr. Opin. Struct. Biol.*, 2003, **13**, 581–588.
93. M. P. Williamson, Annual Reports on NMR Spectroscopy, Academic Press, 2009, vol. 65, pp. 77–109.
94. J. Jiménez-Barbero, J. L. Asensio, F. J. Cañada and A. Poveda, *Curr. Opin. Struct. Biol.*, 1999, **9**, 549–555.
95. P. Balaram, A. A. Bothner-By and E. Breslow, *Biochemistry*, 1973, **12**, 4695–4704.
96. I. Bertini, M. Fragai, Y. M. Lee, C. Luchinat and B. Terni, *Angew. Chem., Int. Ed.*, 2004, **43**, 2254–2256.
97. W. P. Jencks, *Proc. Natl. Acad. Sci.*, 1981, **78**, 4046–4050.
98. O. Ichihara, J. Barker, R. J. Law and M. Whittaker, *Mol. Inf.*, 2011, **30**, 298–306.
99. D. Li, E. F. DeRose and R. E. London, *J. Biomol. NMR*, 1999, **15**, 71–76.
100. B. Becattini and M. Pellecchia, *Chemistry*, 2006, **12**, 2658–2662.
101. M. F. Rega, B. Wu, J. Wei, Z. Zhang, J. F. Cellitti and M. Pellecchia, *J. Med. Chem.*, 2011, **54**, 6000–6013.
102. P. Sledz, H. L. Silvestre, A. W. Hung, A. Ciulli, T. L. Blundell and C. Abell, *J. Am. Chem. Soc.*, 2010, **132**, 4544–4545.

103. J. Orts, C. Griesinger and T. Carlomagno, *J. Magn. Reson.*, 2009, **200**, 64–73.
104. V. M. Sanchez-Pedregal, M. Reese, J. Meiler, M. J. Blommers, C. Griesinger and T. Carlomagno, *Angew. Chem., Int. Ed.*, 2005, **44**, 4172–4175.
105. I. Krimm, *MedChemComm*, 2012, **3**, 605–610.
106. J. Pilger, A. Mazur, P. Monecke, H. Schreuder, B. Elshorst, S. Bartoschek, T. Langer, A. Schiffer, I. Krimm, M. Wegstroth, D. Lee, G. Hessler, K. U. Wendt, S. Becker and C. Griesinger, *Angew. Chem., Int. Ed.*, 2015, **54**, 6511–6515.
107. Z. Marton, R. Guillon, I. Krimm, Preeti, R. Rahimova, D. Egron, L. P. Jordheim, N. Aghajari, C. Dumontet, C. Périgaud, C. Lionne, S. Peyrottes and L. Chaloin, *J. Med. Chem.*, 2015, **58**, 9680–9696.
108. S. G. Zech, A. Kohlmann, T. Zhou, F. Li, R. M. Squillace, L. E. Parillon, M. T. Greenfield, D. P. Miller, J. Qi, R. M. Thomas, Y. Wang, Y. Xu, J. J. Miret, W. C. Shakespeare, X. Zhu and D. C. Dalgarno, *J. Med. Chem.*, 2016, **59**, 671–686.
109. J.-l. Yu, T.-t. Chen, C. Zhou, F.-l. Lian, X.-l. Tang, Y. Wen, J.-k. Shen, Y.-c. Xu, B. Xiong and N.-X. Zhang, *Acta Pharmacol. Sin.*, 2016, **37**, 984–993.
110. N. Wang, F. Li, H. Bao, J. Li, J. Wu and K. Ruan, *ChemBioChem*, 2016, **17**, 1456–1463.
111. A. Chaikuad, S. Lang, P. E. Brennan, C. Temperini, O. Fedorov, J. Hollander, R. Nachane, C. Abell, S. Müller, G. Siegal and S. Knapp, *J. Med. Chem.*, 2016, **59**, 1648–1653.
112. M. Fruth, A. Plaza, S. Hinsberger, J. H. Sahner, J. Haupenthal, M. Bischoff, R. Jansen, R. Müller and R. W. Hartmann, *ACS Chem. Biol.*, 2014, **9**, 2656–2663.
113. J. B. Jordan, D. A. Whittington, M. D. Bartberger, E. A. Sickmier, K. Chen, Y. Cheng and T. Judd, *J. Med. Chem.*, 2016, **59**, 3732–3749.
114. S. Bartoschek, T. Klabunde, E. Defossa, V. Dietrich, S. Stengelin, C. Griesinger, T. Carlomagno, I. Focken and K. U. Wendt, *Angew. Chem., Int. Ed.*, 2010, **49**, 1426–1429.
115. D. Potenza, F. Vasile, L. Belvisi, M. Civera and E. M. V. Araldi, *ChemBioChem*, 2011, **12**, 695–699.
116. C. Airoldi, S. Giovannardi, B. La Ferla, J. Jiménez-Barbero and F. Nicotra, *Chem. – Eur. J.*, 2011, **17**, 13395–13399.

CHAPTER 3

Receptor-based NMR Techniques in Drug Discovery

V. N. SIVANANDAM, GANEKO BERNARDO-SEISDEDOS AND OSCAR MILLET*

Protein Stability and Inherited Diseases Laboratory, CIC bioGUNE, Parque Tecnológico de Vizcaya, Ed. 800, 48160 Derio, Spain
*Email: omillet@cicbiogune.es

3.1 Introduction

In the absence of an estimation of the affinity constant, NMR is undoubtedly the most suitable technique to investigate protein–ligand binding since it is sensitive to multiple timescales, from the weak non-specific "millimolar" interactions to the covalent-like "sub-nanomolar" associations. Furthermore, it can be efficiently applied to characterize the binding event by monitoring both the target protein and the ligand. Such versatility has resulted in a plethora of NMR-based experiments, ingeniously developed to exploit the different physical properties that become altered upon ligand binding. The goal for this chapter is to review these methods, from the most recent published contributions back to the gold-standard protocols, always complementing the procedural aspects with representative examples. We will only focus on the receptor-based methods and this chapter has to be considered as highly complementary to the one covering ligand-based NMR methods.

Nuclear chemical shift is exquisitely sensitive to changes in its chemical environment, resulting in it being the most powerful NMR-based reporter for the protein–ligand interaction event. Due to its importance, this chapter thoroughly reviews the theoretical basis of the chemical shift perturbation and the technical achievements that have expanded its realm to investigating larger systems (*i.e.*, TROSY, segmental labeling, *etc.*). In this context, we will discuss chemical shift perturbation as a case study of chemical exchange since ligand binding is always a dynamic event. Other properties like the scalar or dipolar couplings and, more importantly, paramagnetic relaxation effects have also been extensively exploited and they are also reviewed to a certain detail. Finally, special attention will be paid to membrane proteins due to the methodological and technical challenges of the system and the enormous importance that such receptors have in biology, biomedicine and drug discovery.

3.2 Protein–ligand Binding Understood as an Exchange Process

In the cellular environment, to exert their function, proteins and enzymes are able to selectively recognize a ligand within a myriad of metabolites and small molecules. Cellular metabolism is fast and changing and protein shape must evolve to produce non-covalent, moderate-to-weak interactions with substrates. The reversible nature of such protein–ligand intermolecular associations requires their consideration as an exchange process, even when the binding mechanism does not entail any conformational rearrangement.

The phenomenon of chemical exchange effectively describes the dynamic process that governs the time-dependent exposure of a given nucleus to two or more chemical environments.[1] A two-state conversion is the simplest equilibrium (Figure 3.1), with the nucleus exchanging between two different chemical environments (*i.e.*, protein–ligand ↔ apo-protein):

$$P + L \xrightarrow{K_D} P \cdot L \tag{3.1}$$

where P represents the protein, L the ligand, $P \cdot L$ the complex and K_D is the dissociation constant that regulates the process:

$$K_D = \frac{[P] \cdot [L]}{[P \cdot L]} \tag{3.2}$$

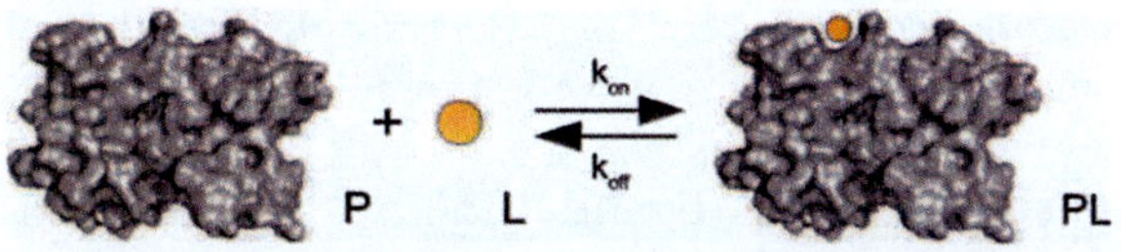

Figure 3.1 The protein–ligand binding process. A protein named "P" interacts with a ligand "L", which could eventually be another protein, a small molecule, an ion, a DNA molecule, *etc.* The kinetic constants are represented and the dissociation constant (K_D) is defined in eqn (3.2) and related to the kinetic rates in eqn (3.3).

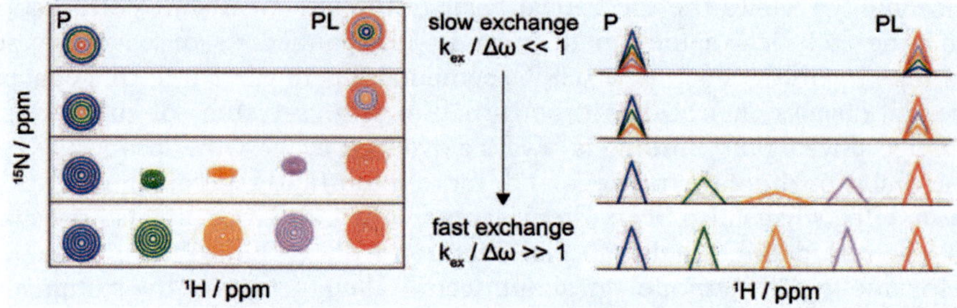

Figure 3.2 The binding process, as observed by NMR spectroscopy. Left: simulation of a single peak in the free (P) and ligand-bound (PL) states, at different saturation ratios and for the different timescales of line broadening. Right: projection of the signals from the spectrum. Each signal is colored according to the titration point, where blue corresponds to the free form and red corresponds to the fully saturated signal.

The brackets account for the concentrations. Chemical exchange results in chemical shift and line broadening (Figure 3.2) and it can be roughly divided into three regimes based on the chemical shift difference ($\Delta\omega = \delta_A - \delta_B$, in Hz) and the exchange rate between states (k_{ex}). The ratio of the two magnitudes ($k_{ex}/\Delta\omega$) ultimately determines the timescale of the conformational exchange line broadening.[2] The three regimes are:

(i) *Slow exchange* ($k_{ex}/\Delta\omega \ll 1$). Assuming that the association event (on-rate) is controlled by the diffusion limit, this regime normally corresponds to strong binding, with K_D values between nanomolar and low micromolar. From the spectroscopic point of view, well-differentiated peaks are observed, corresponding to the different states of the protein (Figure 3.2). In this scenario, the related intensities vary according to the free/bound protein ratio.

(ii) *Fast exchange* ($k_{ex}/\Delta\omega \gg 1$). Only one signal is observed at a chemical shift that varies from the free state value towards the complex chemical shift, proportionally to the free/bound protein ratio. For the same reasons as in the slow exchange, and under the same considerations, fast exchange is often related to weak binding.

(iii) *Intermediate exchange* ($k_{ex}/\Delta\omega \cong 1$). A single signal is observed, often with a non-lorentzian shape and showing maximum line broadening.

The relationship between ligand binding and the NMR signal and the chemical exchange-induced associated line broadening is subtle. Not considering external factors (*i.e.*, temperature, buffer, pH, *etc.*), the chemical shift change ($\Delta\omega$) is solely dependent on the structural features of the exchanging species while the exchange rate, k_{ex}, also depends on the concentrations of the participating molecules. In a canonical titration experiment, the protein remains at a fixed concentration and the ligand concentration is varied within experiments. Under this experimental set-up, k_{ex} can be formulated as the sum of a unimolecular *off*-rate (ligand independent) plus a *pseudo*-bimolecular *on*-rate (ligand dependent):

$$k_{ex} = k_{on} \cdot [L] + k_{off} \tag{3.3}$$

where k_{on} and k_{off} represent the true kinetic (unimolecular) rates for the binding process. Eqn (3.3) underlines the dependence of k_{ex} (and the exchange timescale) on the ligand concentration. The constant k_{on} is normally limited by the diffusion rate, with values ranging between 10^8–10^9 s^{-1}. However, this rate can decrease if the number of productive collisions is reduced. This is the case, for example, for HIV protease, where k_{on} decreases down to 10^4–10^5 s^{-1} because flap motions are required for the binding pocket to be accessible.[3] Alternatively, the k_{on} rate can also increase, for instance when electrostatic forces accelerate the formation of the encounter complex.[4] Nonetheless, these cases are rather exceptional and we will assume a standard value of k_{on} rate for the rest of the discussion.

When purely in the slow regime, k_{off} is very small and k_{ex} normally falls outside of the microsecond–millisecond (μs–ms) time window at almost every ligand concentration, so the exchange line broadening is negligible. Moreover, assuming that the ligand is much smaller in size, the target protein will suffer no additional line broadening due to changes in the rotational diffusion tensor and the ratio of intensities can be reliably used to estimate the K_D (see Section 3.1).

In the intermediate and fast exchange regimes, k_{ex} falls in the μs–ms time frame and exchange becomes an efficient relaxation mechanism for the involved nuclei. In the fast exchange limit, the signal line broadening is proportional to the product of the molar fractions for the free and bound protein ($X_P \cdot X_{PL}$), thus implying that the maximum exchange will be observed exactly when 50% of the protein is saturated with ligand (Figure 3.2, right). In this regime, the K_D of the process can be obtained by monitoring the chemical shift perturbation (CSP) as a function of ligand concentration (see Section 3.1).

Finally, the exchange regime of the binding process, as monitored by NMR, may also be modulated by any external factors also affecting the K_D. Assuming a positive change in heat capacity upon binding, temperature will shift the timescale towards fast exchange if the process is enthalpically driven and the opposite applies if binding is dominated by entropic factors. When the binding coordinate implies changes in solvent accessible area and/or total molecular volume, crowders and/or osmotic pressure may significantly alter the timescale and the thermodynamics of the association event.

3.3 Chemical Shift Perturbation Monitoring the Receptor

3.3.1 General Considerations

Chemical shift perturbation (CSP) of the protein spectrum signals, also termed chemical shift mapping, constitutes the most widely used NMR-based experiment to characterize protein–ligand binding. It provides information at the atomic level because it monitors the entire protein and binding perturbs *only* the chemical shift of the residues directly involved in the association event. Thus, the spectrum will not change at all if the potential ligand does not bind to the target protein. The CSP experiment is widely used in drug discovery because it removes the false positives observed using other techniques such as spectrophotometry, calorimetry or enzymatic assay.

Specifically, CSP is very helpful in all of the steps of the drug discovery process as it always demonstrates direct association. For instance, protein–protein interactions often become interesting targets, susceptible to influence by drugs.[5] In this scenario, CSP is particularly suitable as it can demonstrate that the compound (*i.e.*, small molecules or mimicking peptides) binds to the appropriate protein region, negating any inter-protein interactions.[6,7]

A non-trivial requirement for this method is that the chemical assignment of the spectral signals must be obtained. Nowadays, this is a standard procedure that works routinely for proteins below 20 kDa, and can also be achieved, at least partially, for much larger systems (up to 40 kDa). Chemical shift assignment requires an appropriate isotopic labeling scheme for the protein, followed by extensive use of triple-resonance multidimensional experiments. In practice, this procedure is tedious and time limiting and can span several weeks (if not months). Nevertheless, even when the assignments are not accessible, the titration curves (chemical shift *vs.* concentration of ligand) can still be adjusted to obtain the dissociation constant of the ligand.

The high-resolution structure of the protein constitutes another important asset to adequately interpret the information obtained from the CSP experiment. Interestingly, such a structure may be obtained from other approaches (*i.e.*, X-ray crystallography or electron cryo-microscopy) since the link to the primary structure is already provided by the chemical shift assignment. Mapping the affected residues in the structure will typically show them clustered in specific regions, ultimately characterizing the ligand's binding site.

In the most widespread implementation, the protein under consideration[8] is usually ^{15}N labeled, with a concentration ranging between 50 and 200 μM and the unlabeled ligand/drug is titrated against the protein. Provided that the chemical shift assignment of the protein is available, a simple comparison of the protein's 2D ^{15}N–^{1}H-HSQC (**H**eteronuclear **S**ingle **Q**uantum **C**orrelation) spectra[9,10] with and without the ligand is enough to pinpoint the residues that are interacting with the ligand. In the case where the high-resolution structure of the protein is also known, the affected region of the tertiary structure becomes immediately identified.

The change in the chemical shift reflects the type of exchange happening between the protein and substrate. The timescale can be readily determined depending on whether signals in the NMR spectra displace (fast), broaden (intermediate) or disappear (slow). In slow exchange, the identification of a coupled pair of signals (*i.e.,* belonging to the same nucleus) should be enough to determine the molar fractions of the free and bound forms and, by extension, the K_D of the process. However, it is more accurate to obtain a set of experimental data at different protein saturation fractions and to fit the dataset to the following hyperbolic equation:

$$\frac{[P \cdot L]}{P_T} = \frac{[L]}{[L] + K_D} \tag{3.4}$$

where P_T is the total protein concentration and $[P \cdot L]/PT$ is the fraction of bound protein. Additionally, in the slow exchange regime, a NOESY (**N**uclear **O**verhauser **E**nhancement **S**pectroscop**Y**) experiment will connect the exchanging coupled signals with a cross-peak (EXSY (**EX**change **S**pectroscop**Y**)),[11] whose intensity contains information about the kinetic rates.[12] The problem here is that 2D-NOESY often suffers from severe signal overlap and a more sophisticated version of EXSY, the zz-exchange experiment, allows the same information to be obtained but based on 2D-HSQC.[13]

Slowly exchanging association often offers the possibility of the structural elucidation of the complex, by NMR or other biophysical methods, since it is associated with strong binding and the complex exhibits long residence times. This is of the utmost importance for the drug discovery process, since only high-resolution structures of the protein–lead compounds can provide the atomic detail required for further refinement of the molecule on the path to a new drug.

In the fast exchange regime, CSPs are proportional to the protein saturation ratio and they are calculated using the following general formula (for ^{1}H and ^{15}N chemical shift changes):

$$\Delta\delta = \sqrt{\frac{1}{2}\left[\delta_H^2 + (\alpha \cdot \delta_N^2)\right]} \tag{3.5}$$

where $\Delta\delta$ is the average Euclidean distance moved between the free and bound resonance. The factor α takes into account the different sensitivity of the ^{1}H and ^{15}N nuclei to the environment and often the giromagnetic constant ($\gamma(^1H)/\gamma(^{15}N)$) ratio is used.[14] One consequence of the extreme sensitivity of the chemical shift to changes in the environment is that CSP is able to detect very weak associations, with K_D values of millimolar and higher. This feature is particularly useful in the early stages of the drug discovery process (*i.e.*, screening of compounds) since it provides valuable information about the binding epitope, even at such low affinities.

3.3.2 Expanding the Molecular Weight Available for a Protein Target

In theory, every natural protein is a putative target for drug discovery since it exerts a function that can be chemically modulated. In practice though, some protein families that possess well-established roles in cellular signaling and metabolism and are more druggable (*i.e.*, kinases, GPCR, ion channels, *etc.*) are most often finally selected as pharmaceutical targets. Importantly, the physicochemical properties of the targeted protein are solely decided by the functional properties of the molecule and they often do not suit the requirements of NMR experiments, particularly regarding the molecular weight limit. Fortunately, novel NMR experiments have been devised to expand the molecular weight of systems that can be addressed by this technique. In this section we discuss advancements like TROSY, methyl-TROSY and specific labeling strategies, trying to complement the discussion with representative examples from the literature.

TROSY (**T**ransverse **R**elaxation **O**ptimized **S**pectroscop**Y**) is perhaps the most relevant recent advancement for the study of large molecules by NMR.[15] TROSY makes use of the partial cancellation of two independent relaxation mechanisms that become entangled through a cross-correlated mechanism (Figure 3.3). In a non-decoupled (^{1}H,^{15}N) spectrum, TROSY selects only the line that (partially) cancels out the two relaxation mechanisms, thus producing a sharper line. Even though this results in a loss of half the signal,[15] that signal can easily be reduced to less than half,[16,17] and in large systems (MW > 50 kDa) this has proven to be the only way to detect a signal at all.

Other implementations like CRIPT or CRINEPT[18] acknowledge the fact that, for very large systems, the shorter the pulse sequence, the better, and the HMQC (**H**eteronuclear **M**ultiple **Q**uantum **S**pectroscopy) scheme substitutes the HSQC for polarization transfer. Using these novel experiments, systems as large as the tobacco mosaic virus ($\approx$40.8 MDa) can be observed by NMR, but no experiments for the assignment of these very large systems are available.

TROSY combines well with deuteration or perdeuteration of the sample.[19,20] In the fingerprint experiment for proteins, the ^{15}N–^{1}H-HSQC–TROSY implementation leaves the dipolar contributions to other protons as the remaining dominant relaxation mechanism. Pervasive deuteration of the sample drastically reduces proton density and the line shape of the amide peak (the TROSY component) becomes much sharper. The only problem of this technique is that, after heterologous bacterial expression of the protein, (per)deuteration will also eliminate the amide protons, which have to be exchanged back during purification, normally through a refolding protocol.

TROSY has been extensively used in combination with CSP to characterize protein–ligand interactions as well as macromolecular assemblies. For instance, the N-terminal nucleotide binding domain (NBD) of the 70 kDa heat shock protein (HSP70) is known to allosterically regulate the C-terminal substrate-binding domain (SBD).[21] Backbone CSPs (^{1}H, ^{15}N and ^{13}CO) were determined for different conformational states of HSP70 to unravel the

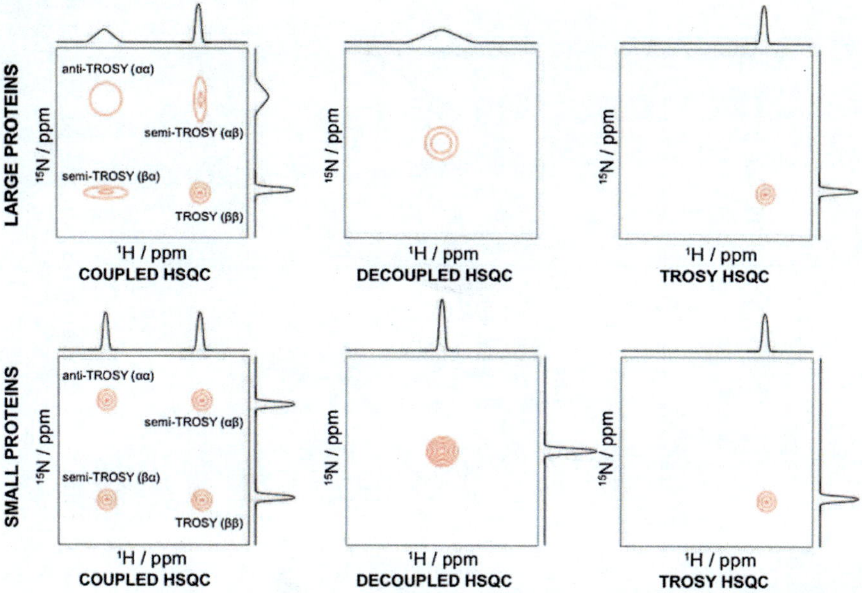

Figure 3.3 Transverse Relaxation-Optimized Spectroscopy (TROSY) for small and large proteins. TROSY effect in the peak intensity of the fully decoupled and fully coupled ^{1}H–^{15}N-HSQC spectrum. TROSY–HSQC selectively cancels the transverse relaxation component caused by DD and CSA, and the signal line width reflects the different partial cancelation of the relaxation mechanisms in each component of the ^{15}N–^{1}H correlation. This effect is only noticeable for very large molecules (*i.e.*, MW > 35 kDa).

nucleotide binding coordinates and how they modulate the segmental dynamics of the complex. In another example, Park and coworkers[22] used a different reporter (backbone amide signal disappearance) to demonstrate that the p53 DNA-binding domain (DBD) undergoes conformational changes to adopt a molten globule-like state in the presence of the 90 kDa chaperone heat shock protein HSP90.

3.3.3 Methyl TROSY-based Approaches

The selective labeling of methyl groups surrounded by a high deuteration background was pioneered in the last decade by Lewis Kay and coworkers and has allowed the study by NMR of biomolecules of unprecedented size (much larger than 100 kDa).[23–26] Specifically, the methyl labeling strategy selectively targets the Ile-δ1 or Leu-δ and Val-γ positions in proteins expressed in *Escherichia coli* and subsequently uses TROSY–HMQC[27–29] to exploit the 3-fold symmetry in the CH_3 moiety and the fast rotation of methyl groups around the C–C bond (Figure 3.4). Methyl labeling is not only restricted to ILV residues but it can also be introduced in a residue-specific and stereo-specific way to other amino acids such as Thr, Ile-γ2, Ala and Met.[23,26,29]

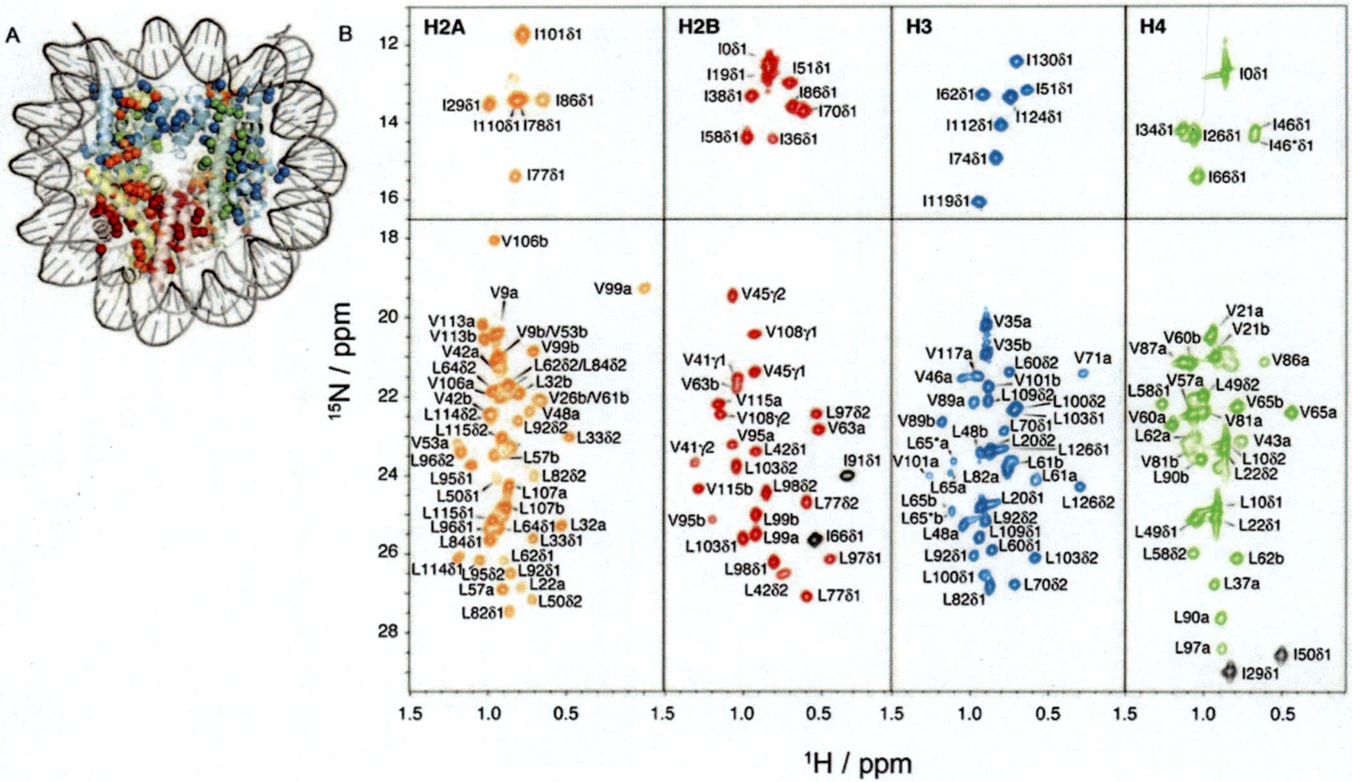

Figure 3.4 The methyl TROSY strategy applied to the nucleosome. (A) Distribution of the ILV methyl groups in the nucleosome (PDB 2PYO) where histones have been colored as: H2A, orange; H2B, red; H3, blue; H4, green; and DNA, gray. (B) Methyl-TROSY spectra with labeled ILV signals in each histone. On the top, δ1 methyl groups of Ile residues. On the bottom, the Leu-δ1/δ2 and Val-γ1/γ2 methyl groups.
Adapted with permission from H. Kato, H. van Ingen, B.-R. Zhou, H. Feng, M. Bustin, L. E. Kay and Y. Bai, *Proceedings of the National Academy of Sciences*, 2011, **108**, 12283–12288.

Chemical shift perturbation monitoring in isolated methyl-labeled residues is able to report on biomolecular events, eliminating resonance crowding, and thus providing a reliable and powerful option to investigate ligand binding to very large proteins. For instance, in an impressive undertaking, the interaction between an entire proteasome (360 kDa) and reversible inhibitors was investigated using methyl-TROSY technology.[30,31] The NMR results provided a picture in which gate exchange proceeds through many steps involving only very small segment sizes of the proteasome. In another recent example, shown in Figure 3.4, Kato and coworkers used CSP in combination with paramagnetic relaxation enhancement (PRE, see below) to assign up to 90% of the methyl groups in a high mobility group nucleosomal (HMGN, 230 kDa).[32,33] Finally, Cabrita and coworkers[34] were able to follow protein folding in the ribosome-nascent chain complex during cotranslational folding events. To that end, the ribosome was uniformly labeled (^{2}H, isoleucine-δ1 (Ile-δ1) or U-^{15}N-labeled under high perdeuteration) to access the chemical shift perturbations from the ^{1}H, ^{13}C (Ile-δ1) methyl groups and ^{1}H–^{15}N backbone signals *via* methyl TROSY and SOFAST–HMQC.[35]

3.4 Paramagnetic Spin-labels for Lead Discovery and Optimization

3.4.1 General Considerations

Spin-labels or paramagnetic tags are small molecules/metal-chelates with unpaired electrons. Most lanthanides and a significant number of metallic elements of the periodic table are paramagnetic and the same is valid for stable, free radical organic molecules. The presence of an unpaired electron induces large effects on the relaxation properties of the neighboring protons that can be quantified.[36] Therefore, they have been extensively applied in lead discovery and optimization in the pharmaceutical industry.[37]

Unpaired electrons increase the relaxation rates of surrounding protons *via* an electron–nuclear dipolar mechanism, which leads to a so-called paramagnetic relaxation enhancement (PRE) in the protein. This can be measured up to 35 Å from the paramagnetic center in a precise way[38] and affects the transverse relaxation of the nuclei:

$$R_2^{\mathrm{tot}} = R_2^{\mathrm{dia}} + R_2^{\mathrm{para}} \tag{3.6}$$

where R_2^{dia} accounts for diamagnetic effects and R_2^{para} accounts for the paramagnetic contribution to line broadening. Three distinct mechanisms, elsewhere reviewed,[38–41] drive the dependence between PRE and the relaxation time of the unpaired electron (τ_e):

(i) When the molecule possesses an unpaired electron with a long τ_e, dipolar relaxation between the electron and nuclear spins becomes the major source of proton relaxation.

(ii) When the paramagnetic center is characterized by a short τ_e, the thermally averaged Curie-spin mechanism will dominate the paramagnetic relaxation contribution to line broadening.
(iii) Cross-correlation effects between nuclear dipole–dipole and electron curie-spin must also be considered and are dependent on the overall molecular tumbling τ_c.

When the magnitude of the magnetic moment of the unpaired electron depends on the molecular orientation with respect to the magnetic field vector, paramagnetism also results in a chemical shift variation (pseudo contact shift, PCS). PCS is usually much larger than CSP since its magnitude is proportional to r^{-3} (equation defined in Chapter 2). Such a large influence area provides useful long-term restraints, widely used in protein structure determination and segmental motion analysis. Interestingly, PCS is proportional to the anisotropy of the magnetic susceptibility tensor while all paramagnetic species exhibit dipolar PREs, so the two properties can be independently tuned, for instance by carefully choosing the lanthanide ion.

From the experimental point of view, PREs can be estimated from the peak intensity ratio of the paramagnetic and diamagnetic protein spectra. Normally, for distances below 10 Å, the effect is large enough to produce signal disappearance due to intense line broadening. In the case where the paramagnetic center also induces PCS, chemical shifts in the spectrum will become considerably displaced, normally requiring full reassignment of the spectral peaks. The detailed mechanism of how to extract distances and dynamic information has been extensively described by Clore and coworkers.[38]

3.4.2 Accessorizing Proteins with Spin Labels

There are several ways to attach paramagnetic spin-labels to a target protein.[41,42] The simplest one consists of replacing calcium with a lanthanide in a calcium binding protein, as first demonstrated with synagtotagmyn[43] and also used in many other natural systems like calmodulin[44] and others.[41] Even though this is an elegant and easy entry for the spin label, the strategy is obviously limited by the natural availability of a divalent ion-binding site, so incorporation of the spin label most often requires protein engineering. In this context, a short three-residue sequence, Gly–Ser–His, known as the 'ATCUN motif' (amino terminal Cu^{2+} (Ni^{2+}) binding motif) to bind Cu^{2+} at the N-terminal was developed by Lewis Kay and coworkers.[45] The sequence can be easily inserted into the protein *via* a one-step recombinant cloning method. PREs measured on a ubiquitin system bound with the ATCUN motif resulted in signal broadening with no associated PCS because Cu^{2+} magnetic susceptibility anisotropy was modest. In an alternative approach, Kato and coworkers[33] used spin-labels (Mn^{2+} attached to a cysteine residue) at certain positions to confirm the binding sites in a methyl-TROSY experiment of the HMGN2–nucleosome complex. Results showed that the resonances of the methyl groups were affected depending on their distance from the spin-label, thus rationalizing CSPs.

3.4.3 The SLAPSTIC Experiment

Chemical modification of a protein also allows conjugation of spin labels to certain amino acid side chains.[42,46] Wolfgang Jahnke and coworkers (Novartis) were the first to exploit this idea to develop a novel screening method known as SLAPSTIC (**S**pin **L**abels **A**ttached to **P**rotein **S**ide chains as a **T**ool to identify **I**nteracting **C**ompounds).[37,46,47] The SLAPSTIC technique relies on PRE effects exerted by the protein molecule towards the substrate. The method involves attaching a spin label at one or two sites close to the possible binding pocket and monitoring the changes in ligand resonances (Figure 3.5). In the absence of a spin label, the ligands are defined by sharp lines in the NMR spectrum. Yet, in the presence of the paramagnetic agent, the electron gyromagnetic ratio is 658 times that of the proton gyromagnetic ratio and the relaxation experienced due to the unpaired electron is 658^2 times larger as compared to the diamagnetic relaxation induced by protons. Thus, when the spin labels are located close to the active site or the binding pocket, the ligand experiences strong PRE (and eventually PCS) and ligand resonances are severely affected. One of the greatest advantages of using paramagnetic spin labels is the identification of low-affinity binding molecules. The rapid exchange between the loose and bound form of weak ligands can still be detected by SLAPSTIC, even at highly reduced concentrations of proteins, due to the very efficient relaxation mechanism.

The first application of this strategy was shown for FKBP,[46] which has multiple lysine residues within the radius of 15–20 Å from the binding pocket of the pipecolinic acid moiety (one of its putative substrates). The protein spin label with an organic nitroxide radical, also known as

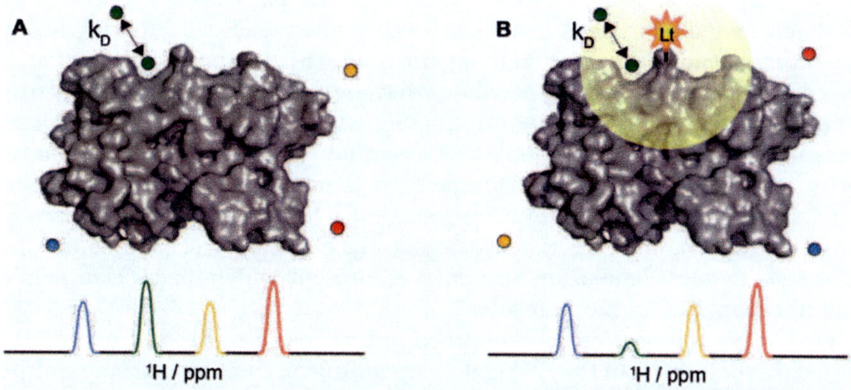

Figure 3.5 The SLAPSTIC experiment. (A) In the absence of any paramagnetic label, a simulated monodimensional spectrum shows a single peak for each chemical compound (colored spheres). (B) Once the paramagnetic group is placed close to a binding site, the monodimensional spectrum shows a decrease in the green peak, due to its association to the protein and as a consequence of a direct paramagnetic relaxation.

TEMPO (2,26,6,-tetramethyl-piperidin-1-oxyl), induced the disappearance of *p*-hydroxybenzanilid within a mixture of non-binding aromatic compounds.

Finding a second-site ligand neighboring the main binding site is often desirable in the drug discovery process. Both ligands can then be linked to produce a single compound with much higher affinity (SAR by NMR, see Section 6). However, the challenge here is to ensure saturation at the primary ligand-binding site, as often ligands suffer from weak affinity and/or low water solubility. To that end, the ligand can also be paramagnetically tagged to benefit from the PRE-induced disappearance of the signals from the ligand.[47] Moreover, the mixing time can also be tuned to best distinguish the new binding site.

3.5 Residual Dipolar Couplings (RDCs)

Dipolar couplings constitute a physical response originating from the interaction between two magnetically active nuclei that are close in space.[43,48,49] The resultant dipolar couplings (DCs) are generally in the order of kHz and are partially responsible for the broader lines in the solid-state NMR spectrum. In (isotropic) solution, though, Brownian motion averages out the dipolar coupling line broadening and a sharp line shape is obtained. Yet, by placing the biomolecule in an anisotropic medium that creates a degree of anisotropy, it adopts a preferred orientation towards the magnetic field B_0 and the dipolar couplings are reintroduced to a certain degree (residual dipolar coupling, RDC). Thus, under alignment conditions, RDCs manifest themselves as an apparent addition of a few Hz to the scalar coupling.

Weak alignment conditions can be obtained in different ways. Some molecules intrinsically show a large magnetic susceptibility anisotropy, already adopting a preferred orientation towards B_0. However, most target proteins are globular and an anisotropic alignment tensor is required to align the biomolecule to the external magnetic field. This can be created artificially by adding liquid crystals or polymers that are soluble and uniformly anisotropic over the entire sample, and stable at different pH, ionic and temperature conditions (Figure 3.6). It is essential for the alignment media not to interact with the protein. Bicelles,[50,51] filamentous phages pf1,[52] polyacrylamide gels and synthetic reagents such as DHPC/DMPC (hexanoyl-phosphatidyl-choline/dimyristoyl-phophatidyl-choline)[53] or C12E5/hexanol (dodecyl-pentaethylene-oxide/hexanol)[54] represent widely used examples of alignment media for biomolecules.

Protein–ligand or protein–protein complexes can be analysed using RDC-based methodology. In the free state, depending on their anisotropies, either the protein or the ligand may be under the influence of the alignment media but due to their difference in size, shape, orientation and τ_c, they will experience a different type of ordering. In the bound state, though, they will be subjected to the same degree of alignment tensor. Therefore, individual calculation of the alignment tensors of the protein and the ligand should only coincide if the ligand is interacting with the protein. It is important to

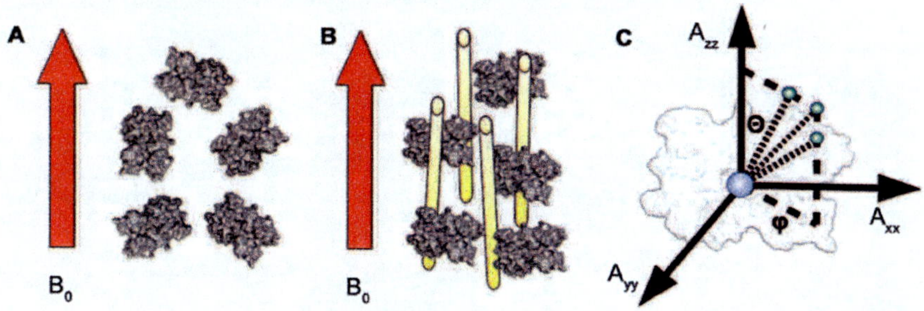

Figure 3.6 Reintroducing dipolar coupling in liquid-state NMR. (A) In isotropic media, protein tumbles freely and the dipolar coupling contribution cancels because all of the protein orientations towards the magnetic field have equal probability. (B) In anisotropic media, proteins acquire preferential orientation towards the magnetic field. The incomplete averaging of all of the protein's possible orientations leads to RDCs. (C) Protein preferential orientation can be described by an alignment tensor. For example, measured RDC between ^{15}N–^{1}H depends on the orientation of the internuclear vector with respect to the alignment tensor, which is described by the polar coordinates θ and Φ.

emphasize that observed RDCs can lead to K_D estimations since the contributions of the free state and the bound state forms are dependent on the affinity. Finally, the RDC dataset can be easily introduced in protein–ligand docking programs such as HADDOCK[55] to define the relative orientation of both molecules within the complex.

Due to their large sensitivity to complex geometry, RDCs can also be applied to test the efficiency of a compound in dissociating a protein–protein interaction (a highly recurring pharmaceutical target). For instance, Brüschweiler and coworkers used RDCs as a reporter of conformational changes produced by Cu^+ and ATP ligands on the N- and P-domains of Cu^+-ATPase ion transporter,[56] showing that the N-/P- interdomain orientation leads to the relative openness of the transporter and modulates the traffic of Cu^+ in different steps along the translocational cycle. Molecular allosteric modulators of this process could be readily monitored using this technique.

3.6 Structure–Activity Relationship (SAR) by NMR

SAR by NMR is a strategy that aims to identify and optimize drug candidates by combining (covalently linking) small organic molecules with low to moderate affinities that bind at proximal binding sites of a validated target (protein, DNA, *etc.*) to give rise to a high affinity and specific drug candidate.[57] The SAR concept is straightforward and very intuitive, as shown in Figure 3.7. It is based on the receptor's CSP upon ligand binding and the availability of the protein structure, elucidated either by NMR spectroscopy, X-ray crystallography or electron cryo-microscopy.

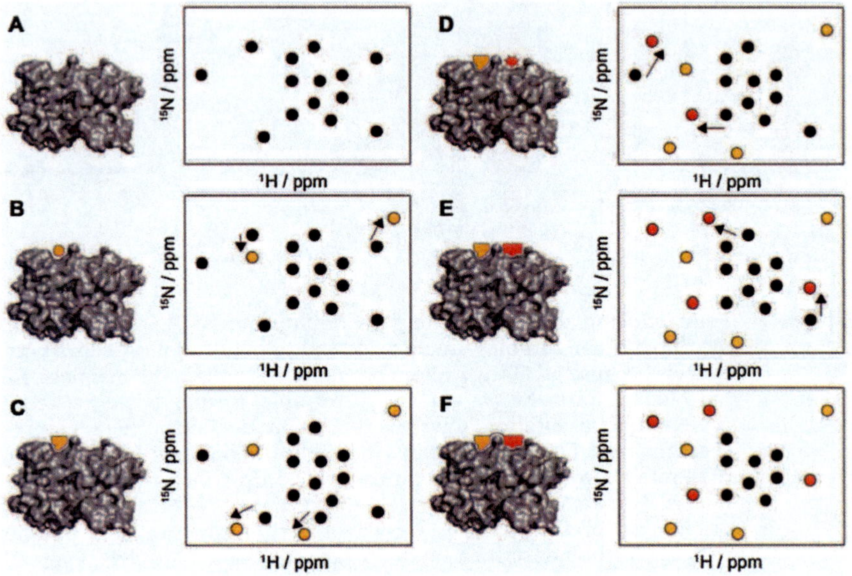

Figure 3.7 The SAR by NMR method. (A) In the absence of any ligand, a protein with two putative binding sites shows a ^{15}N–^{1}H-HSQC fingerprint. (B) After screening a library of compounds, a hit is detected, as revealed by localized peak displacements. (C) Based on the chemical structure of the molecule, this hit is optimized for binding. (D) A second screening of compounds is performed at the second binding site, in this case in the presence of the previous compound. (E) After the new hit detection, the molecule affinity is again optimized. (F) Finally, both compounds are chemically linked to improve the overall affinity.

From the implementation point of view, proteins must be isotopically labeled, normally ^{15}N enriched, although ^{13}C labeling is more suitable for larger proteins due to its higher sensitivity and favourable relaxation. Then, a library of compounds is first screened over the main binding site in the target macromolecule using CSP (Figure 3.7A–B). The dissociation constant can then be obtained from the following expression,

$$K_D = \frac{(P_T - [P])(L_T - [P \cdot L]))}{[P \cdot L]} \tag{3.7}$$

where P_T and L_T correspond to the total protein and ligand concentrations added and $[P \cdot L]$ is the concentration of the formed complex. $[P \cdot L]$ can be estimated from the chemical shift perturbation experiment:

$$[P \cdot L] = P_T \cdot \frac{\delta_{obs} - \delta_{free}}{\Delta\delta_{max}} \tag{3.8}$$

where $\Delta\delta_{max}$ corresponds to the maximum chemical shift perturbation obtained after saturating the protein with the ligand.

Once a hit is identified, analogues of the ligand with optimized affinities are screened at the main binding site (Figure 3.7C). A library of compounds is subsequently scrutinized for secondary binding sites, either in the presence or in the absence of the first optimized compound (Figure 3.7D–E). Only those compounds that produce significant CSP in proximal binding sites are selected and further modified. After the identification of the two compounds, their relative location and orientation in the tertiary complex are determined by NMR (NOEs, RDCs, *etc.*) or X-ray crystallography. Both compounds are then chemically connected, trying to maintain the spatial orientation. Afterwards, the hybrid compound is tested for efficacy (Figure 3.7F). The theoretical affinity for the tethered compound should match the product of the K_D values plus the linker stabilization of the complex.

SAR is a reliable technique and has several advantages. Since the screening is receptor-based, there is a lack of background signals coming from non-binding ligands. The only requirement for the compound is to be soluble in water at around 1 mM and there is no need for functional assays, as CSP analysis rapidly locates the binding sites over the protein structure. Furthermore, SAR by NMR is fast and combines well with High Throughput Screening. On the other hand, considering the limiting factors, the protein is limited in size (MW$<$50 kDa) and, more importantly, chemical shift assignment and the tertiary structure of the protein structure are required.

SAR methodology has been applied to the discovery of selective Bcl-2 inhibitors.[58] Hajduk and coworkers identified a binding site for diphenylmethanes and another proximal binding site for biaryl acids. After optimizing analogue compounds for the first binding site (best K_D of 20 μM) and for the second site (best K_D of 400 μM), they combined both compounds. The final K_D value reached 40 nM (as compared to a theoretical value of 8 nM) and showed promising results in follicular lymphoma cell lines. Another interesting example targeted EPAC, a multidomain cAMP mediator in the regulation involved in several human diseases such as diabetes, cognitive impairment, renal and heart failure, (pulmonary) hypertension, asthma, and chronic obstructive pulmonary disease.[59] Zhou and coworkers employed SAR by NMR to optimize EPAC antagonists, based on a previously identified EPAC competitor, 2-(isoxazol-3-yl)-2-oxo-N′-phenyl-acetohydrazonoyl cyanide (with an IC_{50} of 8.9 μM).[59] Chemical modifications included several group substitutions and optimization of the linker between the aromatic rings. The strategy succeeded in lowering the IC_{50} down to 0.9 μM.

3.7 NMR-based Drug Discovery in Membrane Proteins

Up to 30% of proteins encoded by eukaryotic cells are membrane proteins (MP) but they constitute less than 3% of the Protein Data Bank deposited structures (out of 123 800). Such a low number reflects the challenge of membrane protein structural biology, while it says little about the functional relevance of these systems. In fact, 60% of pharmaceutical drug targets are membrane proteins.[60] In this context, NMR represents a robust alternative for structure-based drug screening in membrane proteins as it bypasses the need for crystallization of a detergent-soluble sample and it can monitor conformational changes without the need for structure determination, thus providing high-quality information with atomic resolution.

3.7.1 Solution *versus* Solid State NMR

Solution NMR has proven useful in the elucidation of high-resolution structures of membrane proteins of small size, mostly showing a β-barrel fold. For instance, the bacterial integral membrane protein OmpX has been structurally characterized in detergent[61] and, more recently, in phospholipid nanodiscs as well.[62] Furthermore, it allows the study of dynamics, as shown by the conformational landscape of the membrane protein PagP, unraveled by solution NMR spectroscopy.[63] However, line broadening in solution is only acceptable for membrane assemblies (protein plus lipids or lipid analogues) with $\tau_c < 100$ ns (*i.e.*, MW < 40 kDa), and large MP proteins with α-helical topology, which constitute the most attractive targets of the pharmaceutical industry (*i.e.*, GPCRs, *etc.*), are very often beyond the limit of applicability of liquid-state NMR.

As the rotational correlation time and molecular size increase, solid-state NMR (ssNMR) becomes the method of choice. In general, the dipolar coupling contribution to line broadening is in the order of kHz, which complicates the ssNMR spectra. Moreover, the powder pattern adds a layer of complexity that also affects the signal's line shape. However, both magnitudes are dependent on the second order Legendre term ($3\cos^2\theta - 1$) and they can easily be cancelled out by fast rotation (10–00 kHz) at the magic angle ($\theta = 54.7°$). The resulting experiment (**M**agic **A**ngle **S**pinning ssNMR, MAS-ssNMR) allows the protein samples to be studied without the need for obtaining a monodisperse population and is more tolerant to sample heterogeneity.

This strategy has been successfully exploited for the structural determination of attractive drug targets like the epidermal growth factor receptor (EGFR),[64] the amiloidogenic Aβ peptide in the fibril conformation,[65] and potassium channels,[66] among many others. In another striking example applied to pharmacology, Mei Hong and coworkers investigated the M2 protein of the influenza A virus, which forms a tetrameric proton channel targeted by the antiviral drugs rimantadine and amantadine.[67] They used MAS-ssNMR techniques, especially rotational-echo double-resonance (REDOR) experiments and ^{2}H NMR spectra, to detect the sub-nanometer inter-atomic distances and the drugs' relative orientation.[68] Two amantadine-binding sites in M2 protein across the phospholipid bilayer were reported, one localized in the lumen of the channel surrounded by the residues mutated in the amantadine-resistant viruses and another one situated in the C-terminal of the protein and only saturated at high drug concentrations.

3.7.2 Isotope Labeling in Membrane Proteins

Choosing a proton as the nucleus to characterize MP–ligand interactions by NMR has the advantage of sensitivity since its ^{1}H abundance is 100%. However, a major caveat is that background signals from the detergents will be present and dominate the spectrum at the protein:detergent ratio normally used (about 1 : 100). Deuterated detergents are an attractive alternative to circumvent this problem, but they are expensive and not always commercially available. In the case where non-deuterated detergents are used, there are available multiple solvent suppression experiments such as sculpting suppression,[69] WET suppression[70] or coherence selection.[71] For instance, Klein-Seetharaman and coworkers investigated canidin-3-glucoside binding to the full-length membrane receptor in dodecyl-maltose detergent micelles using one dimensional selective excitation of ^{1}H sculpting experiments.[72]

Protein expression and backbone labeling (^{15}N or ^{13}C and eventually ^{2}H) are feasible in bacteria and yeast but prohibitively expensive in mammalian cell lines. Fluorine (^{19}F) NMR-based observation becomes a very interesting alternative for MP proteins (Figure 3.8). In general, ^{19}F is not common in

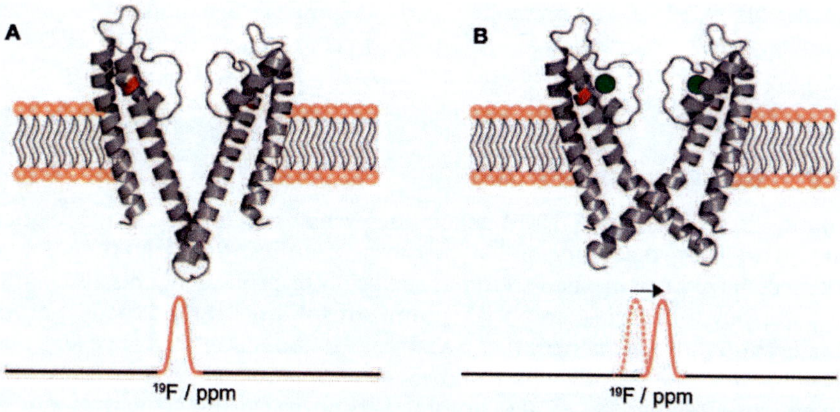

Figure 3.8 Selective labeling in a membrane protein. In KcsA, a single cysteine is located close to the channel pore. In the absence of any ligand, the ^{19}F spectrum of the ^{19}F label KcsA (at the cysteine) shows a single peak (Left). After ligand binding (Right), the ^{19}F peak shifts due to a change in the chemical environment and a plausible conformational change. For clarity, only two subunits have been represented from the KcsA channel (PDB 1R3J).

nature (proteins, lipids or detergents) so ^{19}F-NMR does not suffer from background signals. It has 100% abundance and 83% sensitivity, as compared to protons. Moreover, ^{19}F contains nine surrounding electrons, which establish a perfect sensor for environmental changes such as Van der Waals or electrostatic interactions. The chemical shift anisotropy is very large, resulting in well-resolved peaks. Finally, ^{19}F compounds are often commercially accessible.

There are two major ways of introducing fluorine into proteins.[73] Isotope-labeled amino acids can be incorporated into the cell expression media. ^{19}F-Trp, -Phe and -Tyr analogues are often used and the fluorine insertion does not usually perturb the final 3D structure. Alternatively, fluorine can be introduced by a chemical reaction through activated cysteines (Figure 3.8). 2,2,2-trifluoroethanethiol (TET)[74] and 3-bromo-1,1,1-trifluoroacetone (BTFA)[75] are widely used reagents for this purpose. In this approach, only solvent-exposed cysteines are labeled. When fluorine incorporation is done to more than one residue, ^{19}F can be assigned either by mutagenesis or by complementation with a paramagnetic ion.

In membrane proteins, ^{19}F labeling has been used to localize ligand-binding sites and also to monitor the different folding transitions, conformational changes and oligomerization states induced upon ligand binding.[76,77] For instance, the β_2-adrenergic receptor (β_2AR) is a G protein-coupled receptor that is responsible for the signal transduction of different extracellular ligands. Wüthrich and coworkers characterized the receptor response to various ligands using ^{19}F-NMR.[78,79] Collecting 1D spectra for ^{19}F, they were able to observe CSP and to cluster the receptor response mechanisms into two groups: ligands shifting ^{TET}Cys265 to the most active conformation possible are transmitting the signal *via* a canonical G protein pathway while the chemical shift perturbation observed for ^{TET}Cys327 is transmitted through a β-arresting pathway.

3.8 Concluding Remarks

Thanks to its versatility, NMR spectroscopy has developed a fluid relationship with drug discovery and a plethora of conventional NMR pulse sequences have been applied to the observation of protein–ligand association at multiple timescales and with atomic resolution. Brand new methodologies and pulse sequences (*i.e.*, SAR by NMR and SLAPSTIC) have also been developed to specifically assist in drug development. The set of NMR properties exploited can be easily combined to optimize the structural and dynamic information that can be obtained. The NMR methods described in this chapter combine well with high throughput strategies and they are extremely complementary to the ligand-monitored methods, described elsewhere in this book. It is expected that drug discovery will continue to benefit from the continuous expansion of protein targets that can be investigated by NMR spectroscopy.

References

1. J. Cavanagh, W. J. Fairbrother, A. G. Palmer, N. J. Skelton and M. Rance, *Protein NMR Spectroscopy: Principles and Practice*, Elsevier Science, 2010.
2. O. Millet, J. P. Loria, C. D. Kroenke, M. Pons and A. G. Palmer, *J. Am. Chem. Soc.*, 2000, **122**, 2867–2877.
3. R. Ishima, D. I. Freedberg, Y.-X. Wang, J. M. Louis and D. A. Torchia, *Structure*, 1999, **7**, S1–S12.
4. E. Meneses and A. Mittermaier, *J. Biol. Chem.*, 2014, **289**, 27911–27923.
5. C. Corbi-Verge and P. M. Kim, *Cell Commun. Signal.*, 2016, **14**, 8.
6. E. Barile and M. Pellecchia, *Chem. Rev.*, 2014, **114**, 4749–4763.
7. M. Pellecchia, D. S. Sem and K. Wuthrich, *Nat. Rev. Drug Discovery*, 2002, **1**, 211–219.
8. E. R. P. Zuiderweg, *Biochemistry*, 2002, **41**, 1–7.
9. G. Bodenhausen and D. J. Ruben, *Chem. Phys. Lett.*, 1980, **69**, 185–189.
10. D. M. Dias and A. Ciulli, *Prog. Biophys. Mol. Biol.*, 2014, **116**, 101–112.
11. J. Jeener, B. H. Meier, P. Bachmann and R. R. Ernst, *J. Chem. Phys.*, 1979, **71**, 4546–4553.
12. G. Ortega, T. Diercks and O. Millet, *Chem. Biol.*, 2015, **22**, 1597–1607.
13. N. A. Farrow, O. Zhang, J. D. Forman-Kay and L. E. Kay, *J. Biomol. NMR*, 1994, **4**, 727–734.
14. M. P. Williamson, *Prog. Nucl. Magn. Reson. Spectrosc.*, 2013, **73**, 1–16.
15. K. Pervushin, R. Riek, G. Wider and K. Wüthrich, *Proc. Natl. Acad. Sci.*, 1997, **94**, 12366–12371.
16. J. P. Loria, M. Rance and A. G. Palmer, *J. Magn. Reson.*, 1999, **141**, 180–184.
17. K. Pervushin, *Q. Rev. Biophys.*, 2000, **33**, 161–197.
18. R. Riek, G. Wider, K. Pervushin and K. Wüthrich, *Proc. Natl. Acad. Sci.*, 1999, **96**, 4918–4923.
19. A. Eletsky, A. Kienhöfer and K. Pervushin, *J. Biomol. NMR*, 2001, **20**, 177–180.
20. M. Salzmann, K. Pervushin, G. Wider, H. Senn and K. Wüthrich, *Proc. Natl. Acad. Sci.*, 1998, **95**, 13585–13590.
21. A. Zhuravleva and L. M. Gierasch, *Proc. Natl. Acad. Sci.*, 2011, **108**, 6987–6992.
22. S. J. Park, B. N. Borin, M. A. Martinez-Yamout and H. J. Dyson, *Nat. Struct. Mol. Biol.*, 2011, **18**, 537–541.
23. D. P. Frueh, *Prog. Nucl. Magn. Reson. Spectrosc.*, 2014, **78**, 47–75.
24. V. Tugarinov, V. Kanelis and L. E. Kay, *Nat. Protoc.*, 2006, **1**, 749–754.
25. V. Tugarinov and L. E. Kay, *J. Am. Chem. Soc.*, 2003, **125**, 13868–13878.
26. S. Wiesner and R. Sprangers, *Curr. Opin. Struct. Biol.*, 2015, **35**, 60–67.
27. L. E. Kay, *J. Magn. Reson.*, 2011, **210**, 159–170.
28. J. E. Ollerenshaw, V. Tugarinov and L. E. Kay, *Magn. Reson. Chem.*, 2003, **41**, 843–852.
29. R. Rosenzweig and L. E. Kay, *Annu. Rev. Biochem.*, 2014, **83**, 291–315.
30. M. P. Latham, A. Sekhar and L. E. Kay, *Proc. Natl. Acad. Sci.*, 2014, **111**, 5532–5537.

31. A. M. Ruschak, T. L. Religa, S. Breuer, S. Witt and L. E. Kay, *Nature*, 2010, **467**, 868–871.
32. H. Kato, J. Jiang, B.-R. Zhou, M. Rozendaal, H. Feng, R. Ghirlando, T. S. Xiao, A. F. Straight and Y. Bai, *Science*, 2013, **340**, 1110.
33. H. Kato, H. van Ingen, B.-R. Zhou, H. Feng, M. Bustin, L. E. Kay and Y. Bai, *Proc. Natl. Acad. Sci.*, 2011, **108**, 12283–12288.
34. L. D. Cabrita, A. M. E. Cassaignau, H. M. M. Launay, C. A. Waudby, T. Wlodarski, C. Camilloni, M.-E. Karyadi, A. L. Robertson, X. Wang, A. S. Wentink, L. S. Goodsell, C. A. Woolhead, M. Vendruscolo, C. M. Dobson and J. Christodoulou, *Nat. Struct. Mol. Biol.*, 2016, **23**, 278–285.
35. P. Schanda, Ē. Kupče and B. Brutscher, *J. Biomol. NMR*, 2005, **33**, 199–211.
36. I. Bertini, C. Luchinat and G. Parigi, *Solution NMR of Paramagnetic Molecules: Applications to Metallobiomolecules and Models*, Elsevier Science, 2001.
37. W. Jahnke, *BioNMR in Drug Research*, Wiley-VCH Verlag GmbH & Co. KGaA, 2003, pp. 341–354.
38. G. M. Clore and J. Iwahara, *Chem. Rev.*, 2009, **109**, 4108–4139.
39. I. Bertini, C. Luchinat and G. Parigi, *NMR of Biomolecules*, Wiley-VCH Verlag GmbH & Co. KGaA, 2012, pp. 154–171.
40. J. Koehler and J. Meiler, *Prog. Nucl. Magn. Reson. Spectrosc.*, 2011, **59**, 360–389.
41. G. Otting, *Annu. Rev. Biophys.*, 2010, **39**, 387–405.
42. F. Rodriguez-Castañeda, P. Haberz, A. Leonov and C. Griesinger, *Magn. Reson. Chem.*, 2006, **44**, S10–S16.
43. M. À. Contreras, J. Ubach, Ò. Millet, J. Rizo and M. Pons, *J. Am. Chem. Soc.*, 1999, **121**, 8947–8948.
44. I. Bertini, C. Del Bianco, I. Gelis, N. Katsaros, C. Luchinat, G. Parigi, M. Peana, A. Provenzani and M. A. Zoroddu, *Proc. Natl. Acad. Sci. U. S. A.*, 2004, **101**, 6841–6846.
45. L. W. Donaldson, N. R. Skrynnikov, W.-Y. Choy, D. R. Muhandiram, B. Sarkar, J. D. Forman-Kay and L. E. Kay, *J. Am. Chem. Soc.*, 2001, **123**, 9843–9847.
46. W. Jahnke, S. Rüdisser and M. Zurini, *J. Am. Chem. Soc.*, 2001, **123**, 3149–3150.
47. W. Jahnke, *ChemBioChem*, 2002, **3**, 167–173.
48. H. C. Kung, K. Y. Wang, I. Goljer and P. H. Bolton, *J. Magn. Reson., Series B*, 1995, **109**, 323–325.
49. J. R. Tolman, J. M. Flanagant, M. A. Kennedy and J. H. Prestegard, *Biophysics*, 1995, **92**, 9279–9283.
50. N. Tjandra and A. Bax, *Science*, 1997, **278**, 1111–1114.
51. A. Dvoretsky, E. M. Abusamhadneh, J. W. Howarth and P. R. Rosevear, *J. Biol. Chem.*, 2002, **277**, 38565–38570.
52. J. Ma, G. I. Goldberg and N. Tjandra, *J. Am. Chem. Soc.*, 2008, **130**, 16148–16149.
53. C. R. Sanders and J. P. Schwonek, *Biochemistry*, 1992, **31**, 8898–8905.

54. M. Rückert and G. Otting, *J. Am. Chem. Soc.*, 2000, **122**, 7793–7797.
55. S. J. de Vries, A. D. J. van Dijk, M. Krzeminski, M. van Dijk, A. Thureau, V. Hsu, T. Wassenaar and A. M. J. J. Bonvin, *Proteins*, 2007, **69**, 726–733.
56. D. Meng, L. Bruschweiler-Li, F. Zhang and R. Brüschweiler, *Biochemistry*, 2015, **54**, 5095–5102.
57. S. B. Shuker, P. J. Hajduk, R. P. Meadows and S. W. Fesik, *Science*, 1996, **274**, 1531–1534.
58. A. M. Petros, J. R. Huth, T. Oost, C.-M. Park, H. Ding, X. Wang, H. Zhang, P. Nimmer, R. Mendoza, C. Sun, J. Mack, K. Walter, S. Dorwin, E. Gramling, U. Ladror, S. H. Rosenberg, S. W. Elmore, S. W. Fesik and P. J. Hajduk, *Bioorg. Med. Chem. Lett.*, 2010, **20**, 6587–6591.
59. N. Ye, Y. Zhu, H. Chen, Z. Liu, F. C. Mei, C. Wild, H. Chen, X. Cheng and J. Zhou, *J. Med. Chem.*, 2015, **58**, 6033–6047.
60. L. Fagerberg, K. Jonasson, G. von Heijne, M. Uhlén and L. Berglund, *Proteomics*, 2010, **10**, 1141–1149.
61. C. Fernández, C. Hilty, G. Wider, P. Güntert and K. Wüthrich, *J. Mol. Biol.*, 2004, **336**, 1211–1221.
62. F. Hagn and G. Wagner, *J. Biomol. NMR*, 2015, **61**, 249–260.
63. P. M. Hwang, R. E. Bishop and L. E. Kay, *Proc. Natl. Acad. Sci. U. S. A.*, 2004, **101**, 9618–9623.
64. M. Kaplan, S. Narasimhan, C. de Heus, D. Mance, S. van Doorn, K. Houben, D. Popov-Čeleketić, R. Damman, E. A. Katrukha, P. Jain, W. J. C. Geerts, A. J. R. Heck, G. E. Folkers, L. C. Kapitein, S. Lemeer, P. M. P. van Bergen en Henegouwen and M. Baldus, *Cell*, 2016, **167**, 1241–1251, e1211.
65. A. K. Schütz, T. Vagt, M. Huber, O. Y. Ovchinnikova, R. Cadalbert, J. Wall, P. Güntert, A. Böckmann, R. Glockshuber and B. H. Meier, *Angew. Chem., Int. Ed.*, 2015, **54**, 331–335.
66. B. J. Wylie, M. P. Bhate and A. E. McDermott, *Proc. Natl. Acad. Sci.*, 2014, **111**, 185–190.
67. S. D. Cady, K. Schmidt-Rohr, J. Wang, C. S. Soto, W. F. DeGrado and M. Hong, *Nature*, 2010, **463**, 689–692.
68. T. Gullion and J. Schaefer, *J. Magn. Reson.*, 1969, **81**, 196–200.
69. T. L. Hwang and A. J. Shaka, *J. Magn. Reson., Ser. A*, 1995, **112**, 275–279.
70. S. H. Smallcombe, S. L. Patt and P. A. Keifer, *J. Magn. Reson. Ser. A*, 1995, **117**, 295–303.
71. G. Bodenhausen, H. Kogler and R. R. Ernst, *J. Magn. Reson.*, 1969, **1984**(58), 370–388.
72. K. C. Tirupula, F. Balem, N. Yanamala and J. Klein-Seetharaman, *Photochem. Photobiol.*, 2009, **85**, 463–470.
73. T. Didenko, J. J. Liu, R. Horst, R. C. Stevens and K. Wüthrich, *Curr. Opin. Struct. Biol.*, 2013, **23**, 740–747.
74. J. Klein-Seetharaman, E. V. Getmanova, M. C. Loewen, P. J. Reeves and H. G. Khorana, *Proc. Natl. Acad. Sci. U. S. A.*, 1999, **96**, 13744–13749.
75. P. A. Luchette, R. S. Prosser and C. R. Sanders, *J. Am. Chem. Soc.*, 2002, **124**, 1778–1781.

76. C. T. Gee, K. E. Arntson, A. K. Urick, N. K. Mishra, L. M. L. Hawk, A. J. Wisniewski and W. C. K. Pomerantz, *Nat. Protoc.*, 2016, **11**, 1414–1427.
77. M. A. Danielson and J. J. Falke, *Annu. Rev. Biophys. Biomol. Struct.*, 1996, **25**, 163–195.
78. R. Horst, J. J. Liu, R. C. Stevens and K. Wüthrich, *Angew. Chem., Int. Ed.*, 2013, **52**, 10762–10765.
79. J. J. Liu, R. Horst, V. Katritch, R. C. Stevens and K. Wuthrich, *Science*, 2012, **335**, 1106–1110.

CHAPTER 4

Molecular Mechanisms of Drug Action: X-ray Crystallography at the Basis of Structure-based and Ligand-based Drug Design

J. DE RUYCK,[a] G. ROOS,[a,b] E.-M. KRAMMER,[a,b] M. PRÉVOST,[b] M. F. LENSINK[a] AND J. BOUCKAERT*[a]

[a] Unité de Glycobiologie Structurale et Fonctionnelle, UMR 8576 of the Centre National de la Recherche Scientifique and the University of Lille, 50 Avenue de Halley, 59658 Villeneuve d'Ascq, France; [b] Université Libre de Bruxelles (ULB), Structure and Function of Biological Membranes, CP 206/2, Boulevard du Triomphe, 1050 Brussels, Belgium
*Email: julie.bouckaert@univ-lille1.fr

4.1 Introduction

Human health is continuously challenged by microbes and stress caused by toxic substances. Acute infections accompanied with a severe acute host response can become chronic, and continued inflammation can develop into dysplasia and cancer.[1,2] Rational drug designs are multidisciplinary approaches that exist to develop new drugs in order to cope with these challenges and block chronic inflammation,[3–8] and these can be started from two sides: the target and the ligand. Both approaches are based on structural knowledge of either the target (target-based drug design) or the ligands (ligand-based drug design). Usually, the target is a protein that is implicated in the disease and its progression. The interactions made by proteins are specific and selective because their three-dimensional (3-D) structures have been evolutionary optimized for a more or less dynamic fit. Influencing the activities of biomolecules by the mimicry or modulation of their structures can be potentially beneficial and thus present a therapy.

Developing a lead molecule (a chemical compound with good binding properties to the target) or an effective drug is challenging, even for known targets, and is performed by iterative experimental and computer-based approaches. Advances in genome sequencing and improvements in obtaining structural information by means of X-ray crystallography and nuclear magnetic resonance (NMR) spectroscopy have facilitated the design of novel therapeutic agents. Nowadays, thanks to increasingly powerful computational platforms, integrative approaches have minimized the time needed for the identification, characterization and optimization of drug candidates.[9] High-resolution 3-D protein structures are a particularly useful input in this development process, as they can give the molecular and atomic details of interactions made by the target. The crystallographic target is represented by a model, which is constructed to be congruent with the experimental electron density maps. This can be done with very high accuracy due to the atomic and sub-atomic resolution of X-ray diffraction. Although X-ray crystallography can be actively pursued using fragment-based lead discovery to guide hit-to-lead-to-drug evolution,[10] the finding of a binding partner is today still often serendipitous. These binders either co-purify with the target from the source materials, often indicative of a high affinity because no dissociation occurred during the purification process,[11] or they can be one of the components of one of the buffer solutions used for purification and crystallization. Also, during the soaking of the crystals with cryoprotectants, these small polar molecules like glycerol[12] and ethylene glycol[13,14] can be found in the crystal structure occupying potential binding pockets and thereby stabilizing the target.

Experimentally determined high-resolution 3-D structures, of the protein and/or protein–ligand complexes and/or ligand(s) alone, offer the molecular basis for the conceptualization of novel drug molecules. The 3-D structure of a target can be used to validate hits from biochemical and cellular assays, for high-throughput virtual screening of chemical compound databases by molecular docking[15–17] or for *de novo* ligand design starting from small fragment databases (target-based drug design).[18–20] For example, small organic molecules can bind to the active site of the target protein with a higher affinity than the natural substrate and thus block or repress their function. Structural knowledge of the target interaction site allows correlation with the ligand binding and its driving forces such as enthalpy, entropy, dynamics and solvation. Crystal structures are most frequently applied in the optimization of drug properties in the hit-to-lead phase, to design agonists and antagonists of molecular mechanisms in a structure-based approach. Crystal structures of target–ligand complexes can also form the basis of a ligand-based approach. Central in the latter approach is the determination of the ligand properties in order to link them to affinity. Crystallographic data on ligand interactions with the target may indicate the importance, and can guide the selection, of the structural features of the ligands, for the calculation of affinity. This has a predictive value, for example, prior to the synthesis of new compounds.[21] In this chapter, it will be documented using several examples how crystal structures form a valuable and comprehensive basis for drug discovery.

4.2 Applications

4.2.1 Structure/Dynamics/Affinity Relationships – Rational Drug-design

Nowadays, high-resolution structures solved by X-ray crystallography are often combined with binding affinity data, molecular dynamics (MD) simulations and quantum mechanical (QM) calculations to unravel the molecular mechanism of a target protein of interest. Curiously, the knowledge of the protein's mode of action is often obtained by employing a similar group of methods to those that are used to rationally design and analyze new chemical compounds, and some of the latter may represent future drugs against the protein of interest. We introduce the case of FimH, a mannose-specific lectin at the tip of the type-1 pili of *E. coli* and a major instigator of urinary tract infections, as well as having involvement in dysbiosis in Crohn's disease.[22,23] Targeting this protein through new mannoside-based anti-bacterial anti-adhesives represents a promising alternative to the standard antibiotic treatments of such infections. The FimH crystal structures evidence a small, deep and negatively charged cavity at the tip of the protein as the mannose-binding pocket, in which the sugar is tightly bound by several direct and water-mediated hydrogen bonds.[24,25] The amphipathic mannose is drawn by electrostatic attraction through a collar of hydrophobic residues into the pocket, where it latches through directional hydrogen-bond coordination onto the ligand residues. Tyr48 and Tyr137 form the so-called tyrosine gate (Figure 4.1) located on the hydrophobic outer rim of the mannose-binding pocket.[11] The tyrosine gate has recently attracted great interest because of its potential to generate nanomolar affinities for mannosides conjugated to hydrophobic aglycons through the making of favourable van der Waals and stacking interactions within the gate.[13,14,26–28] Mannopyranosides α-glycosidically linked to either biphenyl groups (BF) or with hydrophobic alkyl chain groups such as *n*-heptyl α-D-mannopyranoside have been proven to be particularly strong binders to FimH (Figure 4.1).[11,26,29] The combination of X-ray crystallography and affinity measurements has been successfully used to build structure–activity relationships for FimH antagonists as anti-adhesive drugs and the FimH adhesin as the drug target. Several of these studies compared the affinity obtained through the measurement of different parameters obtained using surface plasmon resonance (SPR) detection in a microfluidics system,

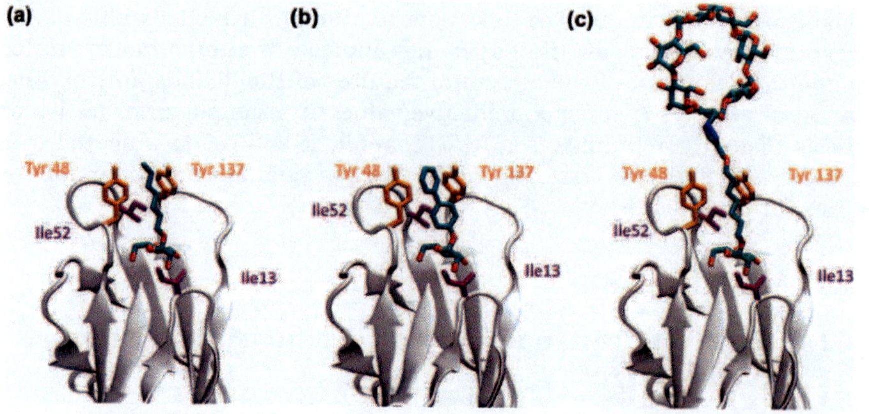

Figure 4.1 FimH in complex with different ligands. The binding position of (a) *n*-heptyl α-D-mannopyranoside (PDB entry 4buq), (b) *para*-biphenyl α-D-mannopyranoside (5fwr) and (c) βCD-1HM (5ab1) depicted in atom-wise colors. The protein backbone is shown in a cartoon presentation (white) with the tyrosine gate residues (48 and 137; orange) and gatekeeping isoleucine residues (13 and 52; purple).

isothermal microcalorimetry (ITC), glycan arrays and other molecular and cellular displacement and inhibition assays.[11,14,21,26–28,30–33] Other studies involved obtaining experimental data on the structure and dynamics of the drug–target interactions using NMR,[13,28] small-angle X-ray scattering (SAXS) and dynamic light scattering (DLS).[34,35] These techniques give very useful information when correlated with thermodynamic quantities that contribute to affinity other than the enthalpic change, namely the binding entropy and the stoichiometry of the interaction. This demonstrates that the organization and the particular shape of the binding site also dictate the 3-D structure of the most potent FimH antagonists and therefore constitute fundamental knowledge in structure-based and ligand-based design.

The major difficulties encountered in using crystal structures as the starting materials for rational design are (1) the quasi lack of dynamic information that can be derived from crystal structures and (2) the assessment of molecular affinities of the target protein that are intrinsic to the structure of the drug compound and independent of conditions such as pH, concentration or buffer composition. There exist solutions to overcome these hindrances towards optimized drug design. For example, in a combined crystallographic–thermodynamic study of human carbonic anhydrases, the determination of intrinsic affinities was achieved by the systematic addition of functionalities to benzenesulfonamide inhibitors.[36] Here, the subtraction of the pH-dependent buffer contribution from the enthalpy of binding gave access to the Gibbs free energies and enthalpies that corresponded to the modifications of the structures of the compounds and their ability to interact in the crystal structure. This permitted the construction of a compound-structure correlation map with the intrinsic Gibbs energy of binding.[36] A solution to overcome the static nature of crystal structures is to apply simulations. The crystal structures of proteins can be subjected to MD simulation, and also their complexes with drug ligands can be obtained by docking when a crystal structure of the complex is not available. For example, using a combination of MD simulations and QM calculations, it could be revealed why the FimH Tyr137Ala (Y137A) mutant shows such a dramatic loss of affinity without being in direct contact with its mannose ligand.[37] In the ligand-free state of Y137A FimH, several of the binding site residues (136, 137 and 48) exhibit backbone dihedral angles that are normally only found upon binding of the mannose. This is because the Y137A mutation disrupts a dynamic coupling between Tyr137 and Tyr48 *via* the inner Ile52 residue and holds the binding cavity in a conformation capable of highly energetic mannose binding.[37]

Another example of how crystal structures can contribute to the understanding of drug properties is captured by the complex of βCD-1HM with the FimH adhesin (Figure 4.1c).[38] In contrast to HM, βCD-1HM eliminated adherent-invasive *E. coli* (AIEC) and the accompanying inflammation in a Crohn's disease murine model, and moreover permitted normal healing of the colon tissue upon clearance of the infection.[39] The HM-part of compound βCD-1HM binds in an identical fashion to HM, with the mannose and heptyl moieties at the same positions. This was surprising because βCD-1HM displays a significantly lower binding affinity towards FimH in ITC measurements.[34] The β-cyclodextrin does not interact with FimH and its electron density is ill defined. These observations indicate that this part of the molecule is able to move freely in the solvent channel of the crystal, which was confirmed in MD simulations.[38] Thus, the β-cylodextrin moiety added to HM is likely to modulate the pharmacokinetics, but not the *in vivo* affinity of βCD-1HM for FimH. A multivalent version of this compound, βCD-7HM, did not deliver any significant advantage over the monovalent βCD-1HM in the treatment of the AIEC-infected Crohn's disease mice.[39] However, it made a marked 100-fold difference in the effective dose (only 2 μg per mouse, instead of 200 μg) needed to clear uropathogenic *E. coli* from the mouse bladder.[34] The high potency of βCD-7HM to treat urinary tract infections can be explained by the ability of the heptavalent compound to interact with different FimH molecules simultaneously and to induce FimH aggregation. Using a combination of ITC, SAXS and DLS, βCD-7HM was shown to bind firmly and with the highest affinity to three FimH adhesin domains. At higher concentrations of FimH, dimers of these trimers were formed, which initiated protein precipitation.[34] Fimbrial entanglement leads to large bacterial aggregates that are less prone to adhering to the uroepithelial cells.[40,41] Intravenously injected βCD-7HM reached the mouse bladder within minutes, where it was retained for over 24 h, thus long enough to cover multiple pathogenic life cycles of the invasive *E. coli* and to prevent fluxing bacteria from rebinding.[42] This highlighted βCD-7HM as a potential candidate for a one-shot non-antibiotic treatment of urinary tract infections.[34]

Sometimes crystal structures give structural information (far) beyond our expectations. This can have far-reaching consequences, like, for example, the serendipitous discovery of a lead molecule for drug design. The determination of the crystal structure of ligand-free Y137A FimH led to the discovery of ethylenediaminetetraacetic acid (EDTA) bound in several orientations near Thr53, Asn136 and Glu50 (Figure 4.2). These are all amino acid residues that are important for shear-force enhanced *E. coli* adhesion to vascular epithelium.[43] EDTA was present in the protein extraction buffer.[37] This is the second time that a ligand, bound to the FimH adhesin, was found serendipitously: the first one, butyl α-D-mannose, originating from the culture medium containing yeast extract, indicated the high affinity of FimH for alkylated mannosides (Figure 4.2).[11] A third discovery was a bound Ni^{2+} ion, present at 10 mM concentration in the crystallization condition, which interconnects with the mannose-binding pocket *via* Asp47.[26] This ion binds at the lateral side of FimH, rather than being involved in head-to-tail associations. The EDTA molecule was not found in the mannoside-occupied wild type FimH, or the Y48A or Y137A mutants, in spite of the identical extraction, purification and crystallization protocols.[37] The binding of EDTA is most likely enabled by the relaxed state of the mannose-binding pocket in Y137A FimH prior to mannose binding. This mutant also allowed the insertion of the *C*-terminus of a neighbouring molecule in the crystal, giving rise to a mini-fimbrius (Figure 4.2). The EDTA-binding site marks the beginning of a belt of positively charged amino acids (blue-coloured in Figure 4.2), similar to what is found in fibronectin-binding polyadhesins.[44–46] Whereas these residues are completely conserved within the *E. coli* species, arginines 92, 98 and 132 are variant in the otherwise highly identical *Klebsiella pneumoniae* and *K. oxytoca*.[31] The belt may thus indicate a protein-docking site in the continuation of the high-mannose-binding site on FimH, a site that determines species-specific tissue tropism.

The results presented here considerably improve our understanding of the FimH–mannoside interaction and provide a guide for optimised design strategies of FimH antagonists. To further improve the usability of the drug molecules under physiological conditions, we chose glycosidic linkages that are not enzymatically degradable, by changing the O-linker atom to N, S, or C of the substituted mannosides. Both *N*-linked[33] and *C*-linked[47] mannosides were co-crystallized with FimH (Figure 4.3). The crystallization of the less soluble *C*-linked mannosides led to the discovery of a lipophilic docking site other than the tyrosine gate: the napthyl structure on *C*-linked mannosides was shown to interact dynamically near Ile13 and Phe142.[47] This site is also a secondary docking site for biantennary mannosides.[48] The Ile13 residue is positioned in the middle of the clamp loop, which is thus called because it closes over the mannose-binding site under shear force-induced conformational changes in FimH.[12] Such inhibitors could thus potentially further facilitate the breaking of bonds under shear.[49,50]

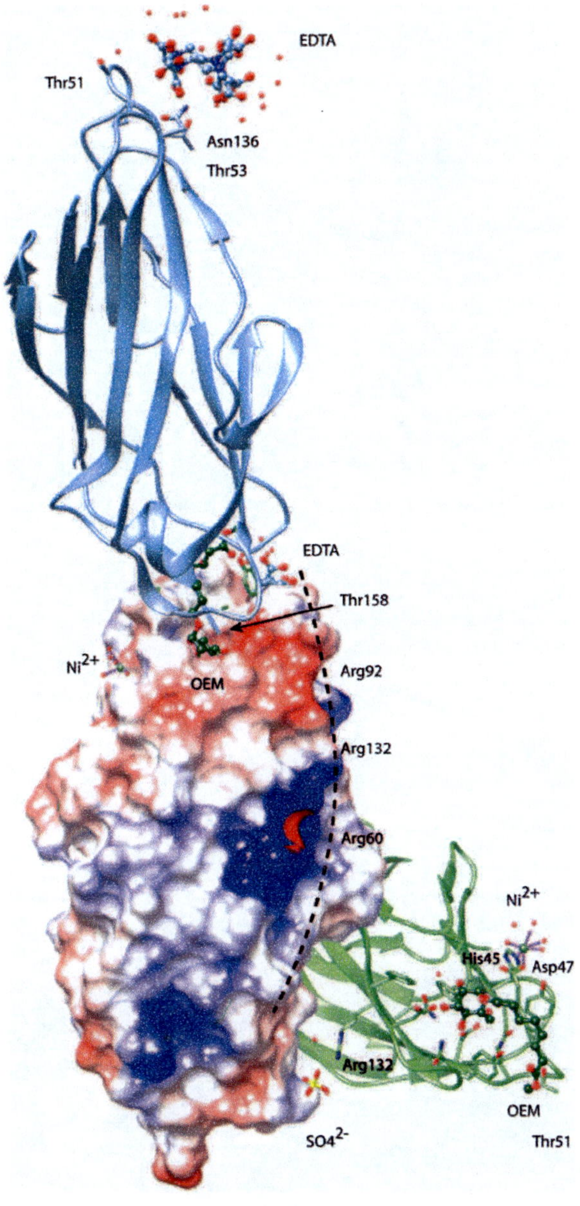

Figure 4.2 Presentation of serendipitously-found binding sites on the FimH adhesin. The figure is a superposition of the Y137A mutant FimH (coulombic surface of PDB entry 5fx3, dashed line indicates the belt of positive charges) on chain A of the crystal structure of FimH in complex with octyl ester mannoside (OEM, moss green ball-and-stick presentation) (PDB entry 4avi), showing both nickel and sulfate binding. Chain B of PDB entry 4avi is depicted in green, the intrusive neighbour in the crystal of the Y137A mutant FimH is shown in blue.

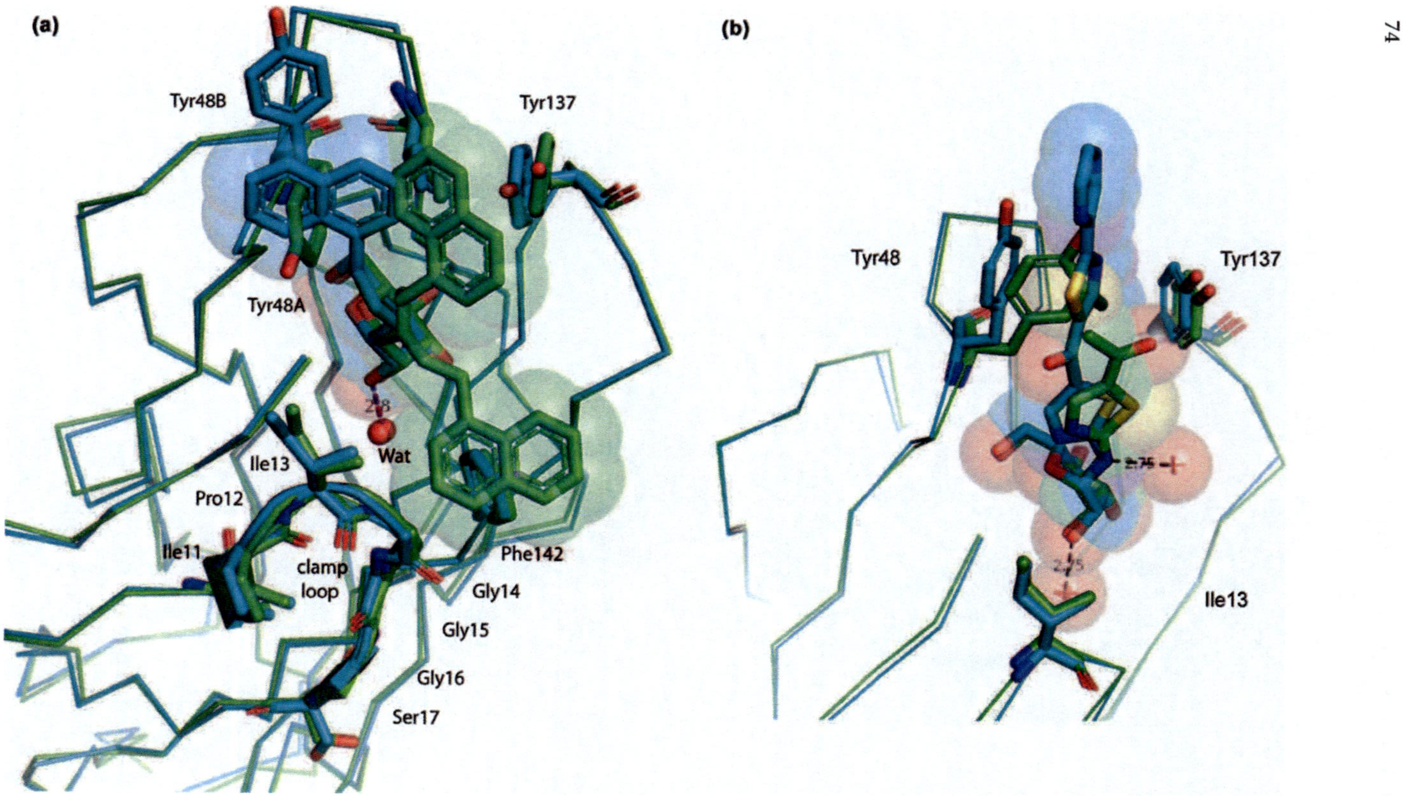

Figure 4.3 *C*- and *N*-Glycosidally linked mannosides as FimH inhibitors. The van der Waals spheres of the ligand are shown to illustrate its aromatic stacking interactions within the tyrosine gate. (a) Naphthalene ethyl α-D-mannopyranoside bound in at least two conformations in chain A (green) of the crystal, and one conformation in chain B (blue) (PDB entry 5abz) making hydrophobic van der Waals interactions with the clamp loop containing Ile13. (b) Superimposition of FimH structures bound to heterocyclic thiazolylamine-containing α-D-mannosides (PDB entries 3zl2 (green) and 5mts (blue)).

4.2.2 Crystal Structures as an Input for *In silico* Drug-design: Docking and Scoring

Existing crystal structures can form the input to obtain new structural information, without the necessity of obtaining new crystals, by the use of docking procedures. Molecular docking is nowadays often used to obtain the structural information of a ligand–protein complex and as a predictive tool to identify critical residues of a protein that could be involved in enzymatic mechanisms or biological assemblies. These so-called hot spots can then be targeted and used as a starting point for the design of new putative drugs against the protein of interest. Basically, the aim of molecular docking is to give information on the ligand–receptor complex structure and on how well a particular ligand binds to its receptor using computational methods. Initial docking protocols kept the protein immobile, giving rise to the term rigid-body docking. In contrast, induced-fit (IF) docking allows for extra flexibility of the receptor molecule by including a minimization step of a particular receptor region. Independent of which protocol is used, the interaction energy is estimated using scoring functions, which probe intermolecular interactions and desolvation in terms of enthalpy and entropy.[51,52]

Whereas most of the docking programs are able to successfully predict the conformation of the ligand within the binding site, they usually fail at reproducing the absolute binding free energy of the ligand–receptor. This challenge has opened up the field of research to new scoring functions able to overcome these limitations.[53,54] Efforts have been undertaken to improve the quality of scoring by, for example, introducing a semi-empirical quantum mechanical (SQM) approach.[55] In the SQM approach, interaction energies of the ligand and relevant parts of the protein are calculated at the (semi-empirical) QM level, whereas the rest of the protein is treated using a classical-mechanical force field. Nevertheless, the correlation between the calculated docking binding energies and the inhibitory capacity (IC) as defined by the IC50 (concentration of the compound that causes 50% inhibition) values is often not always evident. This is because IC50 values depend on the type of assay used; for instance IC50 values in a kinetic assay could depend on the conditions under which binding was measured and not only on the intrinsic affinity.[56]

Structure-based design has also been applied to conceive potent inhibitors of FimH. The lead molecules were based on the HM, namely α-D-mannopyranoside moieties equipped with an alkyl tail. These were fine-tuned to improve their affinity towards the FimH protein.[38] HM analogs featuring an exocyclic oxygen, sulfur, carbon or nitrogen at the anomeric centre were docked in three different rigid receptor structures (PDB entries 4avi, 2vco and 3zl2). For all docked HM analogs, the position of the mannose moiety always nicely matched that of the original crystal ligand, but the position of the end of their alkyl chain varied with the receptor structure used. This suggested that the positions and orientations of the residue side chains in

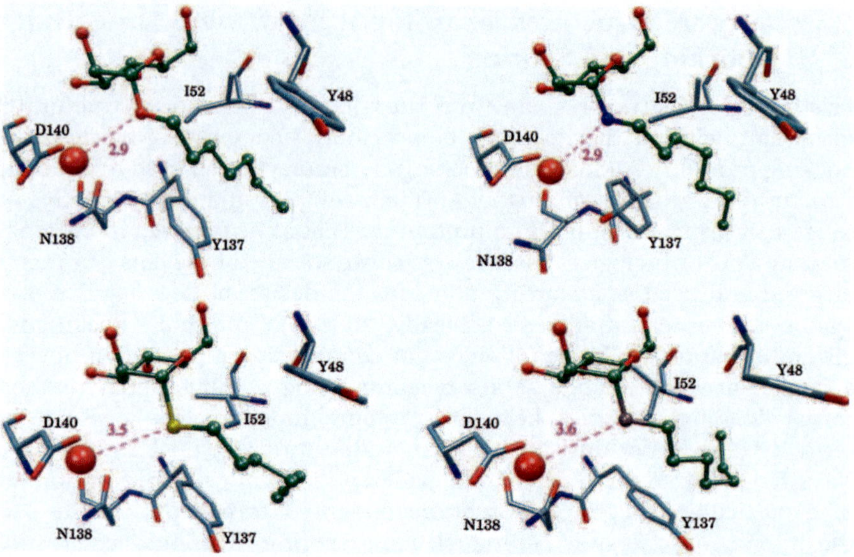

Figure 4.4 Comparison of the best Glide scored poses of induced-fit docked HM compounds with an oxygen, nitrogen, sulfur (yellow) or carbon (grey) atom at the anomeric centre in the binding site of the receptor protein 4AVI. Overall, the four compounds adopt similar positions in the receptor structure. The nature of the exocyclic atom at the anomeric center influenced the distance to the closest crystal water molecule: with the *C*- and *S*-glycosidic atoms, the water molecule was found at a larger distance (shown in Å) from the exocyclic atom. The ligand is depicted by ball-and-stick, the neighboring residues as sticks and the water molecule as a sphere.

the crystal structure impacted the conformations of the ligand terminal part and pointed to a concomitant movement of these moieties with their neighboring residues. To take this into account, IF docking calculations were used. A region of 5 Å around the ligand was minimized, allowing extra flexibility of the receptor. All four compounds adopted a position similar to each other. In addition, their position was independent of the receptor molecule used. However, the water molecule close to the exocyclic atom at the anomeric center was located farther away in compounds with carbon or sulfur compared to those with an oxygen or nitrogen glycosidic linker (Figure 4.4).

A very recent example of a rational drug design approach was to counter the rapid emergence of antibiotic-resistant bacteria by targeting a family of antibiotic-modifying enzymes, the aminoglycoside phosphotransferases (APH).[57] A combination of ITC, electrostatic potential calculations, MD simulations and X-ray crystallography was applied to study the interactions between the enzyme and different aminoglycosides. An allosteric site in these proteins was discovered by normal mode analysis (Figure 4.5). This site was subsequently used as a binding pocket receptor for the docking of a

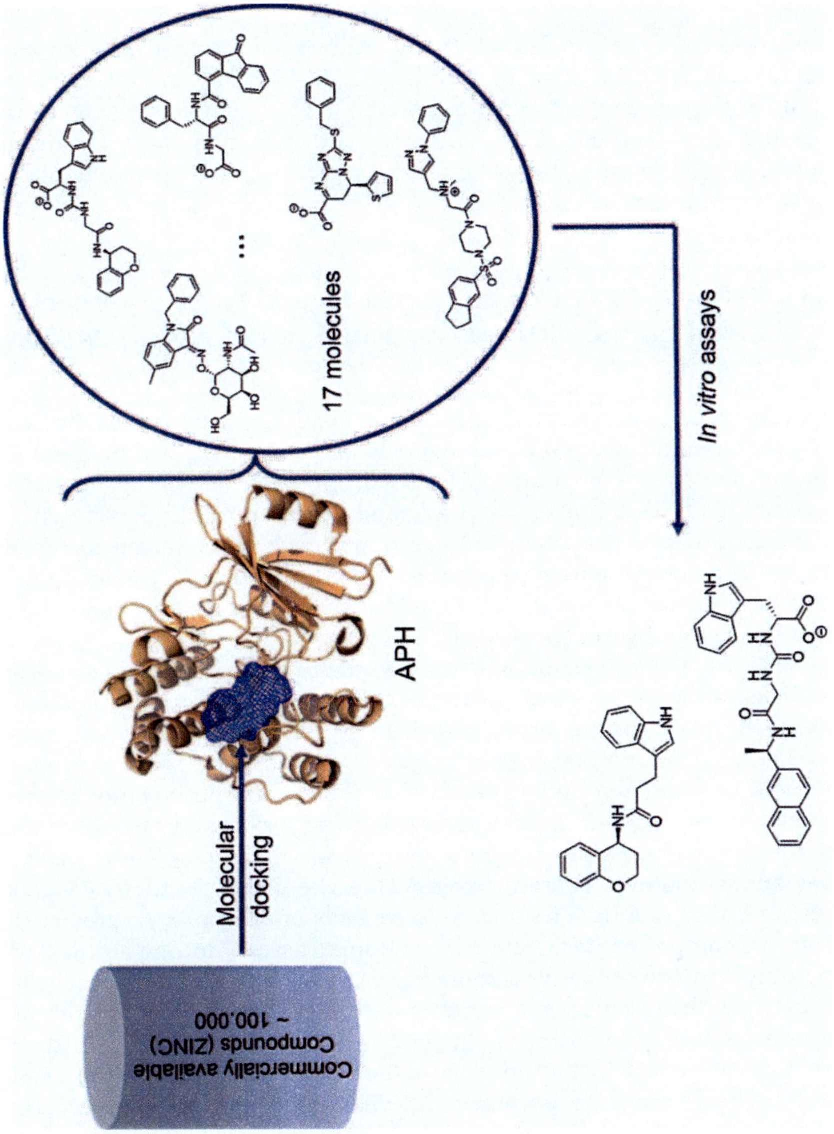

Figure 4.5 Structure-based drug-design on aminoglycoside phosphotransferases (PDB entry 1l8t).

large library of commercially available compounds. From this screening, seventeen of the highest ranked compounds were tested *in vitro* in kinetic experiments in order to evaluate their ability to inhibit APHs.[58] Two hits were of special interest and provided a structural basis for the development of combination therapies consisting of antibiotics plus APH allosteric inhibitors.

Numerous other examples of the identification of pharmacologically relevant leads, resulting from the combination of docking and experimental analysis of the selected hits, can be mentioned, as demonstrated with the rhinovirus capsid protein VP1,[59] with the inflammation indicator myeloperoxidase[60,61] and with the design of new putative anti-infectious agents against tuberculosis and malaria.[15,62,63]

4.2.3 Rational Drug Design from a Ligand-based Approach Based on Properties Formulated in Chemical Reactivity Theory

The mechanistic understanding of protein–ligand interactions is at the heart of new therapeutics. Insight into the mechanism of activity can be obtained following two strategies: a structure-based approach, largely discussed in other sections of this chapter, and a ligand-based approach, which will be discussed in this section. A ligand-based approach can be used in combination with a structure-based approach to optimize the pharmacomodulation process, *e.g.*, hits from high throughput screenings can be analyzed and selected based on ligand properties.[64]

Insight into the molecular structure and properties of ligands can be given by concepts formulated in chemical reactivity theory, such as atomic charges, electrostatic potential, polarizability, electronegativity, chemical potential, hardness and electrophilicity.[65–72] Despite their success in many applications, it is important to note that these concepts are not sharply defined[73] as they are not directly associated with experimentally observable values and often multiple schemes derived for their quantification exist (a well-known example is atomic charges).[74] The use of reactivity concepts can be useful to identify chemically active sites on a ligand, to predict the activity of a compound in a chemical reaction, to assess binding affinity, and to quantify structure–activity relationships. Often, an integrated approach is used in which the calculation of molecular properties is an essential part of a complete toolbox used to rationally design new molecules.[75–79] Molecular properties can be calculated on isolated ligand molecules or on the ligand bound to its receptor to obtain a detailed description of the receptor-binding site and of the principles of ligand binding. In the latter case, linear scaling approaches can make it possible to treat solvated macromolecules at the (semi-empirical) quantum-mechanical level.[80]

Among all of the reactivity descriptors defined today, the Fukui function $f(r)$ together with atomic charges is by far the most popular reactivity descriptor.

The Fukui function indicates the regions where the molecule preferentially reacts (regioselectivity). For example, plots of the Fukui function were used to identify the preferential oxidation sites of novel bis-arylalkylamines.[81] These compounds were designed from high-throughput virtual screening and activity studies as new human myeloperoxidase (MPO) inhibitors based on the detected bis-2,20-[(dihydro-1,3(2*H*,4*H*)pyrimidinediyl)bis(-methylene)]phenol lead compound. Human MPO is a key enzyme in the immune defence system that catalyzes the production of strong oxidizing and halogenating antimicrobial products such as hypochlorous acid (HOCl) from hydrogen peroxide (H_2O_2) and chlorine anions (Cl^-).[82] In the search for new lead MPO-inhibiting compounds, a structure-based screening of the public pharmaceutical database ASINEX was performed.[83] Three successive protocols with enhanced precision in the docking and scoring process were used to filter the molecules. From this procedure, lead compounds were selected based on their scored affinities and on their predicted interactions with the residues of the MPO catalytic pocket. In the next step, these compounds were evaluated by measuring their IC_{50} value for inhibition of chlorination activity and MPO-mediated LDL oxidation. The bis-arylalkylamines feature two aromatic phenol rings (ring A and B, Figure 4.6). Structural information obtained from docking calculations showed that ring B forms a stacking interaction with the MPO porphyrin ring. This structural information is congruent with the spots of highest reactivity towards oxidation, as shown by the Fukui function. From the Fukui function, it can be seen that oxidation preferentially happens in ring B and not in ring A, in agreement with the structural information obtained *via* docking experiments (Figure 4.6). Also, the IC_{50} values and redox potential measurements could be coupled to the importance of that particular ring B.[61]

4.3 Future Perspectives

As we have seen throughout this chapter, structural and biochemical data are crucial in a structure-based drug design process. As this strategy is applied more and more to discover new potential drugs, efforts have been made to significantly improve ways of obtaining and resolving crystallographic structures of ligand–target complexes. However, there is still room for improvement: the results from a recent fragment-based lead discovery study on endothiapepsin were alarming because only half the number of possible X-ray hits of endothiapepsin complexes had been predicted by precrystallization *in vitro* screening.[84] One of the most promising approaches in structure- and function-based drug design lies in fragment-based screening, which works independently of prior high-throughput screenings in other biochemical assays.[10,85] This shotgun approach is oriented towards the annotation of functions of the proteins, and speeding up and increasing the impact of selective (countering off-target interactions) drugs.[86] Its success strongly depends on obtaining the crystals of the protein and their subsequent high-throughput refinement. Improvement can, for example, be established by reducing the required manual handling steps, such as

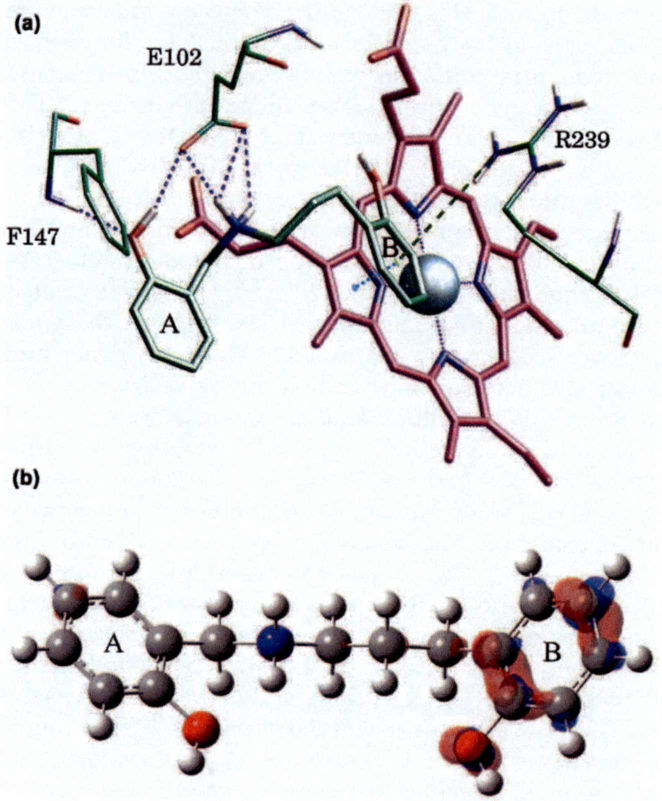

Figure 4.6 (a) Docked pose of bis-arylalkylamine compound in the X-ray MPO structure 1dnw. (b) Non-condensed nucleophilic Fukui function maps calculated as $\rho(\underline{r})\ (N_0) - \rho(\underline{r})\ (N_0 - 1)$ shown on the 0.005 isodensity surface (red), indicating ring B as the site that is preferentially oxidized.

in the soaking of crystals with ligands[87,88] and in harvesting them. Co-crystallization screening with the fragment or metabolite library of choice can be accomplished using dry crystallization and X-ray diffraction data collection *in situ*.[85] In dry crystallization, the drug molecules are dried on the plates before co-crystallization with the target protein. This is a significant advantage compared to soaking, where drug molecules need dimethyl sulfoxide (DMSO) as solvent, and opens up new opportunities for the screening of most chemical libraries.

The thermal shift assay, but also microscale thermophoresis[89] and differential scanning calorimetry, are suitable techniques to evaluate the influence of the fragment, compound or metabolite on protein stability and activity before setting up crystallization assays.[90] Co-crystallization permits complex formation with the target protein, even when the fragments or metabolites have very low affinities. This significantly improves the discovery rate of new binders over other biophysical screening methods. Also, it can lead to the discovery of fragments of allosteric regulators at previously unexplored binding sites on the target.[84] For example, the binding affinity of the FimH adhesin to mannosylated receptors is upregulated by circulating antibodies that wedge in between the lectin and pilin of FimH by binding to the amino-acid variable interdomain region.[91]

The recent development of high-throughput refinement pipelines using ultra-rapid detectors and high-brilliance X-ray beams has expanded the data sets that one can collect from crystals of a target protein. Collective data deposits of target protein crystal structures in complex with a series of drug molecules are being enabled at the PDB.[9] Another aspect of drug design, which has not yet been included in the standard procedure, is the inherent, functionally important dynamics of protein molecules. Nowadays, several molecular modelling packages such as Phenix[92] or CCP4[93] can generate ensembles of structures where local molecular vibrations are sampled by MD simulations. Also, improvements are being made in the definition of the electrostatics in the force fields, *e.g.*, by introducing multipole moments, dipolar polarization and machine learning protocols,[94–96] making the MD simulations more reliable.

In this way, ensemble refinement extracts dynamical details from the X-ray data that allow a more comprehensive understanding of structure–dynamics–function relationships. This method was successfully applied on human factor D, a thrombin-like serine protease involved in immune response, exhibiting conformational dynamics that locked the unbound native state into an ordered inactive conformation *via* the self-inhibitory loop.[97]

Acknowledgements

We acknowledge support from the Federation of Research for Structural & Functional Biochemistry of Biomolecular Assemblies, FRABio. JdR acknowledges funding from the Nord-Pas-de-Calais Regional Council. EMK is a postdoctoral researcher and MP is a maître de recherches of the Fonds de la Recherche Scientifique de Belgique (F.R.S-FNRS) in Belgium.

References

1. I. U. Mysorekar, M. A. Mulvey, S. J. Hultgren and J. I. Gordon, *J. Biol. Chem.*, 2002, **277**, 7412–7419.
2. T. J. Hannan, I. U. Mysorekar, C. S. Hung, M. L. Isaacson-Schmid and S. J. Hultgren, *PLoS Pathog.*, 2010, **6**, e1001042.
3. J. A. Jones, K. G. Virga, G. Gumina and K. E. Hevener, *MedChemComm*, 2016, **7**, 1694–1715.
4. M. Bassetto, A. Massarotti, A. Coluccia and A. Brancale, *Curr. Opin. Pharmacol.*, 2016, **30**, 116–130.
5. S. Choi, S. J. Macalino, M. Cui and S. Basith, *Curr. Med. Chem.*, 2016, **23**, DOI: 10.2174/0929867323666160902160535.

6. S. K. Gill, A. F. Christopher, V. Gupta and P. Bansal, *Perspect. Clin. Res.*, 2016, 7, 115–122.
7. M. Gore and N. S. Desai, *Methods Mol. Biol.*, 2014, **1168**, 313–321.
8. A. Abdolmaleki, J. Ghasemi and F. Ghasemi, *Curr. Drug Targets*, 2017, **18**, 556–575.
9. R. E. Amaro, *Structure*, 2016, **24**, 1225–1226.
10. J. Schiebel, S. G. Krimmer, K. Rower, A. Knorlein, X. Wang, A. Y. Park, M. Stieler, F. R. Ehrmann, K. Fu, N. Radeva, M. Krug, F. U. Huschmann, S. Glockner, M. S. Weiss, U. Mueller, G. Klebe and A. Heine, *Structure*, 2016, **24**, 1398–1409.
11. J. Bouckaert, J. Berglund, M. Schembri, E. De Genst, L. Cools, M. Wuhrer, C. S. Hung, J. Pinkner, R. Slattegard, A. Zavialov, D. Choudhury, S. Langermann, S. J. Hultgren, L. Wyns, P. Klemm, S. Oscarson, S. D. Knight and H. De Greve, *Mol. Microbiol.*, 2005, **55**, 441–455.
12. I. Le Trong, P. Aprikian, B. A. Kidd, M. Forero-Shelton, V. Tchesnokova, P. Rajagopal, V. Rodriguez, G. Interlandi, R. Klevit, V. Vogel, R. E. Stenkamp, E. V. Sokurenko and W. E. Thomas, *Cell*, 2010, **141**, 645–655.
13. S. Vanwetswinkel, A. N. Volkov, Y. G. Sterckx, A. Garcia-Pino, L. Buts, W. F. Vranken, J. Bouckaert, R. Roy, L. Wyns and N. A. van Nuland, *J. Med. Chem.*, 2014, **57**, 1416–1427.
14. A. Wellens, M. Lahmann, M. Touaibia, J. Vaucher, S. Oscarson, R. Roy, H. Remaut and J. Bouckaert, *Biochemistry*, 2012, **51**, 4790–4799.
15. J. de Ruyck, G. Brysbaert, R. Blossey and M. F. Lensink, *Adv. Appl. Bioinf. Chem.*, 2016, **9**, 1–11.
16. M. Sturlese, M. Bellanda and S. Moro, *Mol. Inf.*, 2015, **34**, 513–525.
17. D. Kilburg and E. Gallicchio, *Adv. Protein Chem. Struct. Biol.*, 2016, **105**, 27–57.
18. B. C. Doak, R. S. Norton and M. J. Scanlon, *Pharmacol. Ther.*, 2016, DOI: 10.1016/j.pharmthera.2016.07.003.
19. S. Kuhne, A. J. Kooistra, R. Bosma, A. Bortolato, M. Wijtmans, H. F. Vischer, J. S. Mason, C. de Graaf, I. J. De Esch and R. Leurs, *J. Med. Chem.*, 2016, DOI: 10.1021/acs.jmedchem.6b00981.
20. C. F. Chang, W. H. Lin, Y. Y. Ke, Y. S. Lin, W. C. Wang, C. H. Chen, P. C. Kuo, J. T. Hsu, B. J. Uang and H. P. Hsieh, *Eur. J. Med. Chem.*, 2016, **124**, 186–199.
21. G. Roos, A. Wellens, M. Touaibia, N. Yamakawa, P. Geerlings, R. Roy, L. Wyns and J. Bouckaert, *ACS Med. Chem. Lett.*, 2013, **4**, 1085–1090.
22. L. K. Mydock-McGrane, Z. T. Cusumano and J. W. Janetka, *Expert Opin. Ther. Pat.*, 2016, **26**, 175–197.
23. S. G. Gouin, *Future Med. Chem.*, 2016, **8**, 1903–1906.
24. C. S. Hung, J. Bouckaert, D. Hung, J. Pinkner, C. Widberg, A. DeFusco, C. G. Auguste, R. Strouse, S. Langermann, G. Waksman and S. J. Hultgren, *Mol. Microbiol.*, 2002, **44**, 903–915.
25. S. D. Knight and J. Bouckaert, *Top. Curr. Chem.*, 2009, **288**, 67–107.

26. A. Wellens, C. Garofalo, H. Nguyen, N. Van Gerven, R. Slattegard, J. P. Hernalsteens, L. Wyns, S. Oscarson, H. De Greve, S. Hultgren and J. Bouckaert, *PLoS One*, 2008, **3**, e2040.
27. Z. Han, J. S. Pinkner, B. Ford, R. Obermann, W. Nolan, S. A. Wildman, D. Hobbs, T. Ellenberger, C. K. Cusumano, S. J. Hultgren and J. W. Janetka, *J. Med. Chem.*, 2010, **53**, 4779–4792.
28. B. Fiege, S. Rabbani, R. C. Preston, R. P. Jakob, P. Zihlmann, O. Schwardt, X. Jiang, T. Maier and B. Ernst, *ChemBioChem*, 2015, **16**, 1235–1246.
29. O. Sperling, A. Fuchs and T. K. Lindhorst, *Org. Biomol. Chem.*, 2006, **4**, 3913–3922.
30. J. Bouckaert, J. Mackenzie, J. L. de Paz, B. Chipwaza, D. Choudhury, A. Zavialov, K. Mannerstedt, J. Anderson, D. Pierard, L. Wyns, P. H. Seeberger, S. Oscarson, H. De Greve and S. D. Knight, *Mol. Microbiol.*, 2006, **61**, 1556–1568.
31. E. Lonardi, K. Moonens, L. Buts, A. R. de Boer, J. D. Olsson, M. S. Weiss, E. Fabre, Y. Guerardel, A. M. Deelder, S. Oscarson, M. Wuhrer and J. Bouckaert, *Biology*, 2013, 2, 894–917.
32. D. Eris, R. C. Preston, M. Scharenberg, F. Hulliger, D. Abgottspon, L. Pang, X. Jiang, O. Schwardt and B. Ernst, *ChemBioChem*, 2016, **17**, 1012–1020.
33. S. Brument, A. Sivignon, T. I. Dumych, N. Moreau, G. Roos, Y. Guerardel, T. Chalopin, D. Deniaud, R. O. Bilyy, A. Darfeuille-Michaud, J. Bouckaert and S. G. Gouin, *J. Med. Chem.*, 2013, **56**, 5395–5406.
34. J. Bouckaert, Z. Li, C. Xavier, M. Almant, V. Caveliers, T. Lahoutte, S. D. Weeks, J. Kovensky and S. G. Gouin, *Chem. - Eur. J.*, 2013, **19**, 7847–7855.
35. S. D. Weeks and J. Bouckaert, *Methods Mol. Biol.*, 2014, **1200**, 511–526.
36. A. Zubriene, A. Smirnov, V. Dudutiene, D. D. Timm, J. Matuliene, V. Michailoviene, A. Zaksauskas, E. Manakova, S. Grazulis and D. Matulis, *ChemMedChem*, 2016, DOI: 10.1002/cmdc.201600509.
37. S. Rabbani, E.-M. Krammer, G. Roos, A. Zalewski, R. Preston, S. Eid, P. Zihlmann, M. Prevost, M. F. Lensink, A. Thompson, B. Ernst and J. Bouckaert, *IUCrJ*, 2017, **4**, 7–23.
38. D. A. Dorta, A. Sivignon, T. Chalopin, T. Dumych, G. Roos, R. Bilyy, D. Deniaud, E. M. Krammer, J. De Ruyck, M. Lensink, J. Bouckaert, N. Barnich and S. G. Gouin, *ChemBioChem*, 2016, DOI: 10.1002/cbic.201600018.
39. A. Sivignon, X. Yan, D. Alvarez Dorta, R. Bonnet, J. Bouckaert, E. Fleury, J. Bernard, S. G. Gouin, A. Darfeuille-Michaud and N. Barnich, *MBio*, 2015, **6**, e01298–15.
40. M. Almant, V. Moreau, J. Kovensky, J. Bouckaert and S. G. Gouin, *Chemistry*, 2011, **17**, 10029–10038.
41. S. G. R. Gouin, G. Roos and J. Bouckaert, *Top. Med. Chem.*, 2014, **12**, 123–168.

42. S. S. Justice, C. Hung, J. A. Theriot, D. A. Fletcher, G. G. Anderson, M. J. Footer and S. J. Hultgren, *Proc. Natl. Acad. Sci. U. S. A.*, 2004, **101**, 1333–1338.
43. T. Feenstra, M. S. Thogersen, E. Wieser, A. Peschel, M. J. Ball, R. Brandes, S. C. Satchell, T. Stockner, F. M. Aarestrup, A. J. Rees and R. Kain, *Eur. J. Clin. Microbiol. Infect. Dis.*, 2017, **36**, 467–478.
44. S. Rego, T. J. Heal, G. R. Pidwill, M. Till, A. Robson, R. J. Lamont, R. B. Sessions, H. F. Jenkinson, P. R. Race and A. H. Nobbs, *J. Biol. Chem.*, 2016, **291**, 15985–16000.
45. A. A. Berry, Y. Yang, N. Pakharukova, J. A. Garnett, W. C. Lee, E. Cota, J. Marchant, S. Roy, M. Tuittila, B. Liu, K. G. Inman, F. Ruiz-Perez, I. Mandomando, J. P. Nataro, A. V. Zavialov and S. Matthews, *PLoS Pathog.*, 2014, **10**, e1004404.
46. N. Pakharukova, M. Tuittila, S. Paavilainen and A. Zavialov, *Acta Crystallogr. Sect. F: Struct. Biol. Commun.*, 2017, **73**, 450–454.
47. J. De Ruyck, M. F. Lensink and J. Bouckaert, *IUCrJ.*, 2016, **3**, 163–167.
48. T. Tomasic, S. Rabbani, M. Gobec, I. M. Raščan, C. Podlipnik, B. Ernst and M. Anderluh, *MedChemComm*, 2014, **5**, 1247–1253.
49. O. Zagorodko, J. Bouckaert, T. Dumych, R. Bilyy, I. Larroulet, A. Y. Serrano, D. A. Dorta, S. G. Gouin, S. O. Dima, F. Oancea, R. Boukherroub and S. Szunerits, *Biosensors*, 2015, 5, 276-287.
50. S. Szunerits, O. Zagorodko, V. Cogez, T. Dumych, T. Chalopin, D. Alvarez Dorta, A. Sivignon, N. Barnich, A. Harduin-Lepers, I. Larroulet, A. Yanguas Serrano, A. Siriwardena, A. Pesquera, A. Zurutuza, S. G. Gouin, R. Boukherroub and J. Bouckaert, *Biology*, 2016, 5.
51. X. Y. Meng, H. X. Zhang, M. Mezei and M. Cui, *Curr. Comput.-Aided Drug Des.*, 2011, 7, 146–157.
52. S.-Y. Huang, S. Z. Grinter and X. Zou, *Phys. Chem. Chem. Phys.*, 2010, **12**, 12899–12908.
53. K. M. Merz Jr., *J. Chem. Theory Comput.*, 2010, **6**, 1018–1027.
54. Y. C. Chen, *Trends Pharmacol. Sci.*, 2015, **36**, 78–95.
55. J. Fanfrlik, A. K. Bronowska, J. Rezac, O. Prenosil, J. Konvalinka and P. Hobza, *J. Phys. Chem. B*, 2010, **114**, 12666–12678.
56. J. Soubhye, M. Prevost, P. Van Antwerpen, K. Zouaoui Boudjeltia, A. Rousseau, P. G. Furtmuller, C. Obinger, M. Vanhaeverbeek, J. Ducobu, J. Neve, M. Gelbcke and F. O. Dufrasne, *J. Med. Chem.*, 2010, **53**, 8747–8759.
57. E. Kaplan, J. F. Guichou, L. Chaloin, S. Kunzelmann, N. Leban, E. H. Serpersu and C. Lionne, *Biochim. Biophys. Acta*, 2016, **1860**, 802–813.
58. N. Leban, E. Kaplan, L. Chaloin, S. Godreuil and C. Lionne, *Biochim. Biophys. Acta*, 2017, **1861**, 3464–3473.
59. L. Da Costa, M. Roche, E. Scheers, A. Coluccia, J. Neyts, T. Terme, P. Leyssen, R. Silvestri and P. Vanelle, *Eur. J. Med. Chem.*, 2016, **115**, 453–462.
60. J. Soubhye, I. Aldib, M. Prevost, B. Elfving, M. Gelbcke, M. Podrecca, R. Conotte, J. M. Colet, P. G. Furtmuller, C. Delporte, A. Rousseau,

M. Vanhaeverbeek, J. Neve, C. Obinger, K. Zouaoui-Boudjeltia, P. Van Antwerpen and F. Dufrasne, *J. Pharm. Pharmacol.*, 2014, **66**, 1122–1132.
61. I. Aldib, M. Gelbcke, J. Soubhye, M. Prevost, P. G. Furtmuller, C. Obinger, B. Elfving, I. C. Alard, G. Roos, C. Delporte, G. Berger, D. Dufour, K. Zouaoui Boudjeltia, J. Neve, F. Dufrasne and P. Van Antwerpen, *Eur. J. Med. Chem.*, 2016, **123**, 746–762.
62. J. de Ruyck and J. Wouters, *Curr. Protein Pept. Sci.*, 2008, **9**, 117–137.
63. J. de Ruyck, J. Wouters and C. D. Poulter, *Curr. Enzyme Inhib.*, 2011, **7**, DOI: 10.2174/157340811796575317.
64. F. Shah, P. Mukherjee, J. Gut, J. Legac, P. J. Rosenthal, B. L. Tekwani and M. A. Avery, *J. Chem. Inf. Model.*, 2011, **51**, 852–864.
65. G. Roos, P. Geerlings and J. Messens, *J. Phys. Chem. B*, 2009, **113**, 13465–13475.
66. G. te Velde, F. M. Bickelhaupt, E. J. Baerends, C. Fonseca Guerra, S. J. A. Van Gisbergen, J. G. Snijders and T. Ziegler, *J. Comput. Chem.*, 2001, **22**, 931–967.
67. P. Geerlings, F. De Proft and W. Langenaeker, *Chem. Rev.*, 2003, **103**, 1793–1873.
68. R. G. Pearson and R. G. Parr, *J. Am. Chem. Soc.*, 1983, **105**, 7512–7516.
69. R. G. Parr and W. Yang, *J. Am. Chem. Soc.*, 1984, **106**, 4049–4050.
70. R. G. Parr, L. Von Szentpaly and S. B. Liu, *J. Am. Chem. Soc.*, 1999, **121**, 1922–1924.
71. R. G. Parr and R. G. Pearson, *J. Am. Chem. Soc.*, 1983, **105**, 7512–7516.
72. C. Morell, B. Herrera, S. Gutierrez-Oliva, M. L. Ceron, A. Grand and A. Toro-Labbe, *J. Phys. Chem. A*, 2012, **116**, 7074–7081.
73. S. Liu, in *Chemical Reactivity Theory: A Density Functional View*, ed. P. K. Chattaraj, Taylor and Francis, Boca Raton, FL, 2009.
74. J. F. Gonthier, S. N. Steinmann, M. D. Wodrich and C. Corminboeuf, *Chem. Soc. Rev.*, 2012, **41**, 4671–4687.
75. O. Coskuner, D. E. Bergeron, L. Rincon, J. W. Hudgens and C. A. Gonzalez, *J. Phys. Chem. A*, 2008, **112**, 2940–2947.
76. O. Coskuner, D. E. Bergeron, L. Rincon, J. W. Hudgens and C. A. Gonzalez, *J. Phys. Chem. A*, 2009, **113**, 2491–2499.
77. G. Roos, N. Foloppe, K. Van Laer, L. Wyns, L. Nilsson, P. Geerlings and J. Messens, *PLoS Comput. Biol.*, 2009, **5**, e1000461.
78. S. Brument, A. Sivignon, T. I. Dumych, N. Moreau, G. Roos, Y. Guerardel, T. Chalopin, D. Deniaud, R. O. Bilyy, A. Darfeuille-Michaud, J. Bouckaert and S. G. Gouin, *J. Med. Chem.*, 2013, **56**, 5395–5406.
79. D. H. Lopez, M. A. Fiol-deRoque, M. A. Noguera-Salva, S. Teres, F. Campana, S. Piotto, J. A. Castro, R. J. Mohaibes, P. V. Escriba and X. Busquets, *PLoS One*, 2013, **8**, e72052.
80. J. Khandogin and D. M. York, *Proteins*, 2004, **56**, 724–737.
81. I. Aldib, M. Gelbcke, J. Soubhye, M. Prevost, P. G. Furtmuller, C. Obinger, B. Elfving, I. C. Alard, G. Roos, C. Delporte, G. Berger, D. Dufour, K. Zouaoui Boudjeltia, J. Neve, F. Dufrasne and P. Van Antwerpen, *Eur. J. Med. Chem.*, 2016, **123**, 746–762.

82. S. J. Klebanoff, *J. Leukocyte Biol.*, 2005, **77**, 598–625.
83. I. Aldib, J. Soubhye, K. Zouaoui Boudjeltia, M. Vanhaeverbeek, A. Rousseau, P. G. Furtmuller, C. Obinger, F. Dufrasne, J. Neve, P. Van Antwerpen and M. Prevost, *J. Med. Chem.*, 2012, **55**, 7208–7218.
84. J. Schiebel, N. Radeva, S. G. Krimmer, X. Wang, M. Stieler, F. R. Ehrmann, K. Fu, A. Metz, F. U. Huschmann, M. S. Weiss, U. Mueller, A. Heine and G. Klebe, *ACS Chem. Biol.*, 2016, **11**, 1693–1701.
85. M. Gelin, V. Delfosse, F. Allemand, F. Hoh, Y. Sallaz-Damaz, M. Pirocchi, W. Bourguet, J. L. Ferrer, G. Labesse and J. F. Guichou, *Acta Crystallogr., Sect. D: Biol. Crystallogr.*, 2015, **71**, 1777–1787.
86. J. F. Guichou and G. Labesse, *Structure*, 2012, **20**, 1617–1619.
87. A. le Maire, M. Gelin, S. Pochet, F. Hoh, M. Pirocchi, J. F. Guichou, J. L. Ferrer and G. Labesse, *Acta Crystallogr. Sect. D: Biol. Crystallogr.*, 2011, **67**, 747–755.
88. F. U. Huschmann, J. Linnik, K. Sparta, M. Uhlein, X. Wang, A. Metz, J. Schiebel, A. Heine, G. Klebe, M. S. Weiss and U. Mueller, *Acta Crystallogr., Sect. F: Struct. Biol. Commun.*, 2016, **72**, 346–355.
89. P. Linke, K. Amaning, M. Maschberger, F. Vallee, V. Steier, P. Baaske, S. Duhr, D. Breitsprecher and A. Rak, *J. Biomol. Screening.*, 2016, **21**, 414–421.
90. J. P. Renaud, C. W. Chung, U. H. Danielson, U. Egner, M. Hennig, R. E. Hubbard and H. Nar, *Nat. Rev. Drug Discovery*, 2016, **15**, 679–698.
91. V. B. Rodriguez, B. A. Kidd, G. Interlandi, V. Tchesnokova, E. V. Sokurenko and W. E. Thomas, *J. Biol. Chem.*, 2013, **288**, 24128–24139.
92. B. T. Burnley, P. V. Afonine, P. D. Adams and P. Gros, *eLife*, 2012, **1**, e00311.
93. R. A. Nicholls, F. Long and G. N. Murshudov, *Acta Crystallogr., Sect. D: Biol. Crystallogr.*, 2012, **68**, 404–417.
94. Y. Shi, Z. Xia, J. Zhang, R. Best, C. Wu, J. W. Ponder and P. Ren, *J. Chem. Theory Comput.*, 2013, **9**, 4046–4063.
95. S. J. Davie, N. Di Pasquale and P. L. A. Popelier, *J. Chem. Phys.*,**145**, DOI: 10.1063/1.4962197.
96. P. L. A. Popelier, *Phys. Scr.*, 2016, **91**, 033007.
97. F. Forneris, B. T. Burnley and P. Gros, *Acta Crystallogr., Sect. D: Biol. Crystallogr.*, 2014, **70**, 733–743.

CHAPTER 5

Mass Spectrometry in Biophysics: from High Throughput Screening to Structural Biology

T. GHEYI*[a] AND M. MOLINA-MARTIN*[b]

[a] Lilly Biotechnology Center, 10290 Campus Point Dr., San Diego 92121 California, USA; [b] Analytical Technologies Department, Centro de Investigación Lilly S.A., Alcobendas 28108, Madrid, Spain
*Email: gheyi_tarun@lilly.com; molina_manuel@lilly.com

5.1 Introduction

Mass spectrometry (MS) as a concept was introduced more than a century ago by J. J. Thomson, who measured the electron charge-to-mass ratio back in 1897.[1] Since this pioneering work, the technique has matured to play a prominent role in science and technology. Within the health and life sciences and pharmaceutical sector, MS has become an indispensable tool that is applied from early drug discovery to clinical diagnosis.[2,3] Its widespread application in pharmacokinetic and pharmacodynamic analyses and Phase I and II drug metabolism studies,[4–7] together with its universal use in early drug discovery,[8,9] testify to its impact in drug discovery and development processes.

In a mass spectrometer, atoms or molecules from the sample in the gas phase are ionized immediately before their introduction into the detector.

The instrument separates the ions in the mass analyser according to their mass-to-charge (*m*/*z*) ratios and the detector accurately measures the relative abundance of these ions. A typical mass spectrometer contains the following elements: (i) a sample introduction device (a probe or an interface), (ii) an ionization source to generate ions, (iii) at least one mass analyser to separate ions based on their *m*/*z*, (iv) a detector to quantify the abundance of ions and (v) a data analysis system (Figure 5.1a). The typical output from this instrument is a series of MS spectra, exemplified in Figure 5.1b.

A more detailed description of the fundamentals of MS and the types of detectors more frequently used nowadays is out of the scope of this chapter and can be found elsewhere.[6,8,9]

As an analytical technique, modern MS not only has outstanding sensitivity but also exhibits very high resolving power, which results in exceptional selectivity. These features make MS a very popular analytical tool both for qualitative and quantitative purposes. This technique is the method of choice to confirm the identity of newly synthesised small molecules in drug discovery and to identify and characterize biomolecules in the biopharmaceutical industry. One of the limitations of MS is that successful generation of ions from the sample is a prerequisite to detection and thus the initial phase of the sample and the ionization efficiency may limit the applicability of the technique. However, progress in recent decades on interfaces and ionization sources has largely resolved this issue and today

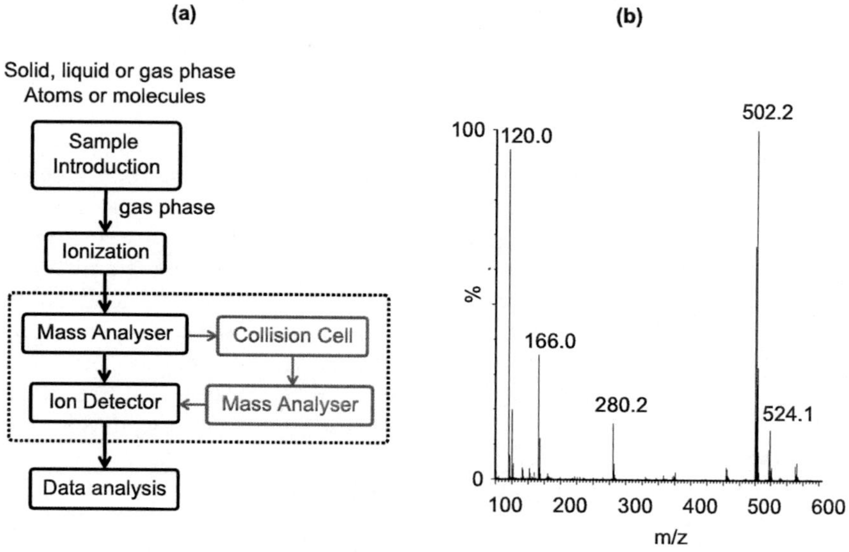

Figure 5.1 Schematic block diagram of a mass spectrometer (a); components shown in grey are used for tandem MS. Following data analysis, a mass spectrum is obtained (b) with *m*/*z* in the *x*-axis and ion intensities in the *y*-axis (given as relative or absolute intensities). The most intense peak in the spectrum (*m*/*z* of 502.2) is known as the base peak.

MS is routinely employed for diverse applications analysing a wide variety of samples. Another MS limitation has been the inability to distinguish between different isomers. This limitation is partially circumvented using tandem MS provided that the isomers have different fragmentation patterns. Furthermore, hyphenation of MS to separation techniques such as HPLC or GC significantly reduces this problem by adding a complementary dimension to the analysis. In the early days of MS, its application to thermolabile or high molecular weight compounds was seriously restrained, however the landmark development of soft ionization techniques during the end of last century, followed by atmospheric pressure ionization techniques such as matrix assisted laser desorption ionization (MALDI) or electrospray ionization (ESI), allowed the straightforward use of MS in this particular area.[10,11]

In this chapter we review the application of MS within the biophysical sciences. One advantage of MS as a biophysical tool is that it typically does not require chemical modifications or immobilization of the target or ligands, and as a result, protein structure, dynamics and interactions can be studied under near-physiological conditions in solution. The fundamentals of the introduction of the chemical species of interest into the MS detector, its transfer to the gas phase and ionization followed by its mass analysis are well understood. As a result, MS technology can be utilized in diverse ways to answer specific questions. Native MS of biomolecules to study their structure and dynamics is now a critical theme in structural biology.[9,12,13] Covalent approaches to interrogate proteins throw light on the three dimensional structure of the biomolecule and are used to explore the surface and topology of proteins and their complexes. This approach is also able to identify and characterize covalent inhibitors.[9,14,15] Hydrogen–deuterium exchange mass spectrometry (HDX-MS) has come of age in the study of protein structure and dynamics and protein–ligand interactions[8,9,16–18] and thus will be covered thoroughly. Finally, different MS screening approaches (based on changes in physicochemical properties of the biomolecules or the ligands upon binding) will be reviewed with special emphasis on the analysis of intact target–ligand complexes[19,20] and affinity selection mass spectrometry (AS-MS).[8,20–23] The line between conventional bioassays and biophysical work with MS is ambiguous, but for the sake of brevity, we have deliberately excluded the indirect MS-based screening of enzymatic targets. The intrinsic character of most of these bioassays is a functional assay, so reading of the substrate or product concentrations bears no direct biophysical implications.

5.2 Applications

Progress in Next-Generation Sequencing (NGS) has revived interest in understanding the biomolecular structure and dynamics of macromolecules at high resolution, particularly in order to diagnose several diseases. Studies are now centred towards the composition, stoichiometry, and spatial arrangement of biomolecules for their functional and conformational analysis in high-throughput mode.[24,25] Conscientious efforts are being made in utilizing genomic sequences and cloning resources in order to obtain high resolution structures for most protein domains.[26] However, structures of proteins in isolation cannot decipher biological functions in eukaryotic systems that involve multi-domain proteins networked in large complexes.[27,28] Understanding the interactions between proteins and with other molecules is often critical when elucidating their function.[29] Various techniques, including X-ray crystallography, nuclear magnetic resonance (NMR), mass spectrometry, circular dichroism (CD), Fourier-transform (FT) infrared (IR), Raman, ultraviolet (UV)-visible absorption and fluorescence spectroscopies, are used for protein analyses.[30] Among these, X-ray crystallography and NMR have outshone all other techniques in revealing high resolution structural information for macromolecules,[31,32] thereby contributing the majority of protein structures to the Protein Data Bank (www.pdb.org).[33] Although productivity in this field continues to accelerate, overall failure rates in determining the structure of a given target remain high[34–36] due to inherent size, stability and solubility issues.[37] Besides, membrane proteins bring additional challenges due to their poor performance in over-expression systems and the need for solubilisation prior to crystallization.[38,39] Lately, MS methods have been utilized successfully wherever classical structural biology approaches have failed. The application of MS to explain these unresolved structure–function relationships has guided the development of several methodologies. In particular, native mass spectrometry and covalent approaches have been widely applied within this context to study biomolecular structure and dynamics. HDX-MS has a prominent, dual role in revealing medium resolution structural details of proteins in solution and characterizing their interactions with other macromolecules and small molecules. Finally, MS-based screening approaches are typically used to identify and validate small molecules or fragments binding the target of interest.

5.2.1 Native Mass Spectrometry of Biomolecules to Study Structure and Dynamics

Non-covalent interactions, for example electrostatic, van der Waals, hydrophobic and hydrogen bonding, aid in molecular recognition processes, forming physical associations between proteins and other binding partners, *e.g.*, DNA, RNA, nucleic acids, lipids, carbohydrates and small molecules.[40,41] These complexes play a key role in performing cellular functions. Classical MS techniques are not amenable to protecting these associations due to the use of denaturing conditions to benefit the ionization process. Native MS in conjunction with ion mobility spectrometry (IMS) is a rapidly growing technology for systematic investigation of these complex biological samples. It has shown great promise in characterizing large groups of proteins interacting in molecular assemblies and diverse pathways in physiologically relevant conditions. ESI and MALDI are two soft ionization techniques that are utilized to gently ionize the complexes with minimal perturbation of the quaternary state. Among these two methods, ESI is the preferred technique to ionize and volatilize these assemblies as it preserves the non-covalent interactions relatively efficiently. Thus, this approach has been termed "native MS" because it tends to investigate the native structure and dynamics of macromolecular complexes. MALDI has not been very popular in studying these complexes because of its tendency to dissociate weak non-covalent interactions during sample preparation and/or ionization.[40] Even though native MS does not yield structures at atomic resolutions, its strength lies in investigating several co-existing biological molecules simultaneously and rapidly using very low sample amounts, thus overcoming certain limitations of conventional structural biology approaches, specifically X-ray crystallography and NMR spectroscopy. Larger complexes of up to several mega-daltons have been investigated by native MS.[42–45] Obtaining the mass of the intact complex together with the masses of the individual subunits provides knowledge of the stoichiometry of the subunits, relative binding strengths, the topology of the complex, and epitope maps.[46–48] Additional information on subunit replacements in solution can also be obtained with native MS.[49,50] Recent advancements in native MS together with data interpretation and visualization tools have significantly transformed studies on membrane proteins[51] (Figure 5.2). Despite these advances, native MS has not yet been applied for routine applications[52] and requires specialized skills. However, the information obtained about the structure, stoichiometry, stability and spatial arrangements of molecular assemblies through native MS cannot be obtained with alternative techniques. Investigators frequently couple native MS with IMS to allow greater understanding of individual proteins and the assembled complexes.[53–56] The progress within the field of IMS has been significant too. Drift-time ion mobility spectrometry (DTIMS), aspiration ion mobility spectrometry (AIMS), differential-mobility spectrometry (DMS) [also called high-field asymmetric waveform ion mobility spectrometry (FAIMS)] and traveling-wave ion mobility spectrometry (TWIMS) are the four ion mobility (IM) approaches that can be used in association with mass spectrometry.[57–65] DTIMS and FAIMS are the most common methods utilized for the structural investigation of individual proteins and complexes.[66] Traditional DTIMS permits the separation of analyte ions based on their ability to travel through a gas-filled enclosure in the presence of a weak electric field, thus providing an additional dimension for separation of the molecules under study. The drift time (t_d) in IM, therefore, not only depends on ion size and charge, but also on the shape of the ion itself, as the rotationally averaged collision cross section (CCS).[67] In DTIMS, direct measurement of CCS can be performed using the Mason–Schamp equation.[68,69] Alternatively, in FAIMS and TWIMS, there is no direct correlation between t_d and CCS because of the non-linear

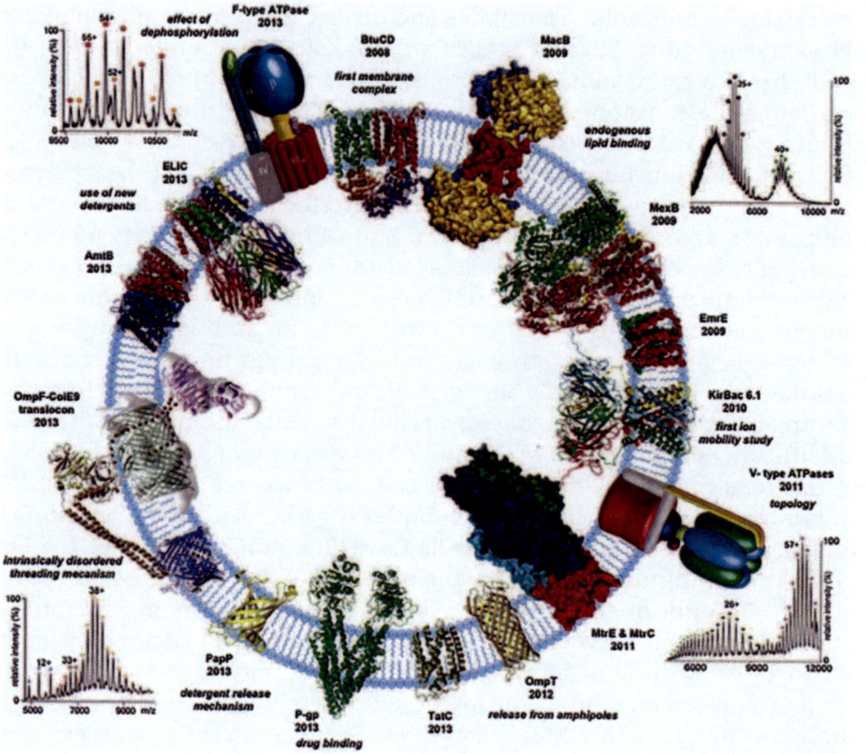

Figure 5.2 Development in the study of membrane–protein complexes by native MS. Reprinted from Structure, Volume 21, J. Marcoux and C. V. Robinson, Twenty Years of Gas Phase Structural Biology, 1541–1550, Copyright 2013, with permission from Elsevier.[51]

relationship between voltage and time. Therefore, well-characterized calibrants with known CCS values (from DTIMS) are needed as standards to relate t_d values of unknowns and estimate their CCS. In general, globular or compact ions exhibit lower values than the compared unfolded structures. Thus, careful control of experimental parameters is needed to preserve the natural structural features of the ions to prevent gas phase compaction and unfolding.[70] Several applications of IMS-MS have been reported, such as observing ligand-induced conformational differences[71–74] followed by kinase/ligand screening,[75] monitoring intermediate protein complexes during binding events,[76] revealing architectures of protein assemblies[77,78] and estimating differences in the packing and stability of membrane protein complexes.[79] The results obtained from IMS-MS are often integrated with computational approaches to extract meaningful and complementary information on macromolecular assemblies and binding orientations.

5.2.2 Covalent Approaches to Interrogate Proteins

MS-based covalent approaches have recently shown compelling evidence that they can provide valuable information on higher order structures of proteins. These approaches not only aid in quantifying the solvent accessibility of protein molecules but also monitor subtle conformational changes upon complex formation with their binding partners. In the following section we cover useful MS-based technological innovations that require covalent adduct formation to study higher order structure and their creative applications in solving complex scientific problems.

5.2.2.1 Chemical Cross-Linking

Over the years, chemical cross-linking in conjunction with MS (XL-MS) analysis has emerged as a powerful tool to examine the tertiary structure of proteins.[80–86] The general approach in protein XL-MS is the formation of covalent bonds between functional groups of adjacent amino acid residues of the same protein or adjacent protein molecules in their native or native-like state. This technique facilitates detection of the physical proximity between functional groups in a protein for its structural analysis[80,87] as cross-linked residues produce distance restraints in the native substrate that are equal to or shorter than the extended length of the cross-linker.[88] In addition, it is also used to identify and confirm spatially close residues involved in protein–protein interactions.[85,89] Obtaining these distance restraints benefits molecular modelling approaches in elucidating the structural features of protein complexes.[90,91] Two separate MS strategies can be utilized to confirm the identity of these cross-linked residues *vis-à-vis* "top-down" and "bottom-up". MS analysis of intact cross-linked proteins followed by fragmentation to peptides for detailed information is encompassed in the "top-down" approach. This approach has the advantage of eliminating the sample preparation step before MS analysis but is limited by high molecular weight (MW) and complex behaviour. The schematic representation of a typical "bottom-up" cross-linking workflow is shown in Figure 5.3. This shotgun proteomic technique is performed by enzymatically digesting cross-linked samples followed by MS analysis to confirm the identity of the involved peptides. However, this latter approach has its own challenges. These include separating the mixture of enzymatically-digested peptides before MS analysis along with the presence of a relatively low number of chemically cross-linked peptides as compared to free peptides. The ongoing advancement in MS over the last decade is providing ample opportunities to overcome these drawbacks, particularly by increasing sensitivity and mass accuracy whilst reducing the mass analysis time with improved data processing tools.[92]

Because of its ability to provide important structural information on proteins and their complexes, applications of XL-MS have exploded in the field of modern structural biology in the last few years. The strength of

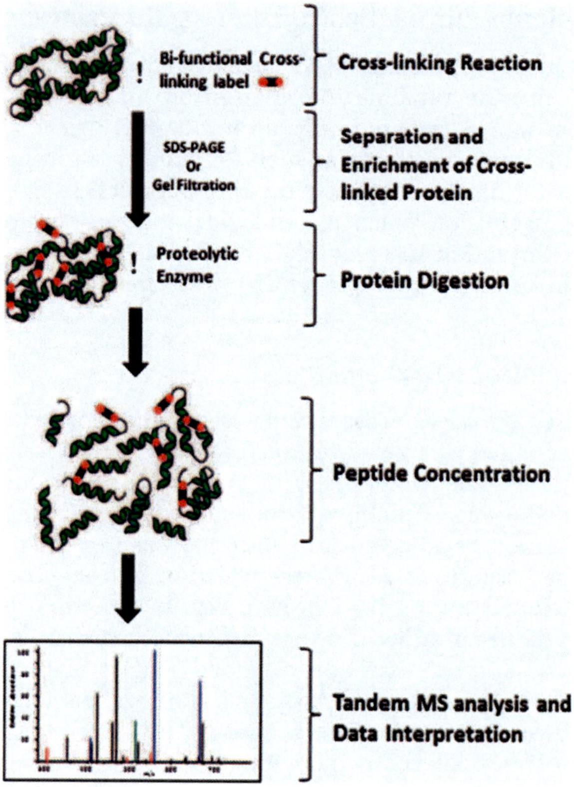

Figure 5.3 General scheme of a chemical cross-linking experiment.

XL-MS is that it can capture unstable or momentary protein interactions that cannot be otherwise preserved during routine protein isolation and purification processes. Initially, the XL-MS approach was confined to the analysis of single proteins or small complexes as reported in the literature over a decade ago.[91,93–95] This was due to challenges in the detection and identification of the cross-linked peptides from complex mixtures of peptides such as those derived from large protein complexes. In recent years, advances in mass spectrometry instrumentation and capabilities have aided the increased use of XL-MS to answer real biological questions. In addition, developments in novel cross-linking chemistries that help in enriching and detecting cross-linked peptides have provided various alternatives for researchers.[96,97] Moreover, the development of bioinformatics tools has significantly contributed to providing diverse detection, identification and visualization strategies.[98–108] Since these developments, XL-MS has made significant progress in elucidating protein 3D structures and protein–protein interactions. Presently, this technique is not limited to studying single proteins or simple complexes but is utilized for macro protein assemblies, core particles and RNA polymerase II complex.[84,85,109–111] It is also used to monitor protein dynamics and characterize different functional states of proteins,[88] protein–RNA binding sites,[112] and ligand–protein interactions in living cells and tissues.[113] Distance constraints identified by XL-MS have also been utilized to validate the substrate binding interface in molecular chaperones.[114] In addition to these applications, XL-MS has been integrated with other structural biology techniques (particularly Cryo-EM) to decode the structures of macro-molecular assemblies.[25,85] Observing the current use and evolution of XL-MS and informatics tools, we expect the scope of their utilization to further expand towards high-throughput applications and *in vivo* protein–protein interactions to reveal the functions of macro-molecular assemblies.

5.2.2.2 Hydroxyl Radical-Mediated Footprinting

Hydroxyl radical footprinting (HRF) coupled with MS has proven to be a reliable ally in investigating protein structures and mapping conformational changes. HRF assay determines the solvent accessibility of a macromolecule by covalently attaching stable labels using highly reactive hydroxyl radicals (•OH) on side chains of solvent-exposed amino acids.[115] As this oxidative modification occurs in multiple residues in a nonselective manner and the modified residues are easily detected,[116] HRF has been demonstrated to be an ideal tool for the structural characterization of proteins.[117–123] Furthermore, •OH is accepted as an optimal labelling reagent because of its size similarity to water molecules, and thus the extent of labelling is directly correlated with solvent accessibility.[123] In addition, the relative reactivity of the side chains with •OH has already been documented: Cys > Met > Trp > Tyr > Phe > Cystine > His > Leu ∼ Ile > Arg ∼ Lys ∼ Val > Ser ∼ Thr ∼ Pro > Gln ∼ Glu > Asp ∼ Asn > Ala > Gly.[124] Out of the 20 amino acids, Asp, Asn, Ala, Gly, Ser and Thr are the residues that either have low reactivity with •OH or their oxidation products cannot be easily detected. Nevertheless, the remaining 14 amino acids encompass ∼65% of the sequence of a typical protein, thus they can be utilized to investigate protein structure with reasonable resolution.[124] There are several methods of producing •OH in solution, including electron pulse radiolysis,[125–128] X-ray radiolysis,[129–131] photolysis of H_2O_2,[132–134] Fe(II)-EDTA Fenton chemistry[135] and electrochemical flow cells.[136] Each of these methods has advantages and disadvantages[115] that have to be assessed in the context of each particular experiment. Followed by labelling, the "bottom-up" approach is applied to quantitatively assess the oxidized sites at single residue resolution by utilizing state-of-art LC-MS/MS approaches.[137–140] Typically, time course labelling is performed at various exposure times and correlated with data from a reference. The control can be an individual protein or a protein complex. Dose-response plots are used to determine the rate of oxidation of individual peptides and the effective dose received is calibrated using an internal reference.[129,141] Detected differential labelling can be utilized to investigate protein–protein interactions, epitope mapping or protein–DNA complexes.[142–144] More importantly, HRF has been successfully implemented for challenging membrane proteins like GPCRs,[145–147] ion-channels[148] and ion transporters.[149] Recent advances in labelling strategies and MS instrumentation together with MS-based data analysis tools[150,151] have matured HRF such that it is a fairly routine technique for complex biological systems where conventional structural elucidation approaches may have a low chance of success.

5.2.2.3 Photoaffinity Labelling

Photoaffinity labelling (PAL) combined with modern MS techniques is also one of the exciting technologies being applied to studying ligand–protein interactions. The ligands used in this technique comprise photoreactive groups that can be irradiated to generate highly reactive intermediates (typically a nitrene, carbene or diradical) upon activation by a specific wavelength of light. The commonly used photoreactive groups in PAL experiments are aromatic azides, benzophenones and aliphatic and aromatic diazirines.[152–155] Due to the unique properties of each group, a thorough understanding is needed of these labels regarding their chemistry, reactivity, labelling efficiency and selectivity.[156] The general approach in PAL-MS is to irradiate ligand–protein complexes with UV-light to initiate photolysis, resulting in the formation of intermediates that bind irreversibly to the protein. Further purification is performed to remove excess ligand from the buffer using LC, followed by intact mass analysis of the ligand–protein complex to assess the stoichiometry and specificity of the binding event. In the next step, ligand-bound protein samples are analysed *via* enzymatic digestion using a specific protease (usually trypsin) to cleave the entire protein into smaller peptide fragments. Since the peptides are irreversibly labelled due to the formation of covalent bonds with the photoprobe, the identification of labelled sites is achieved by tandem MS analysis with instruments that can generate thousands of high mass accuracy readings per sample. In order to facilitate selective purification/identification of the labelled targets, several reporter tags are integrated into the structure of the photolabels. In particular, radiolabels, affinity tags or fluorophores have been widely used as reporter tags. Radiolabels (radioactive isotopes, *e.g.*, ^{125}I, ^{3}H) are the preferred choice for profiling known targets where the direct isolation of labelled targets is not required. The use of radiolabels is particularly advantageous because of their small size and ease of detection, however special handling is required for their use because typically they possess very short half-lives. On the other hand, affinity tags (*e.g.*, biotin and epitope tags) and fluorophores, commonly used in proteomics workflows, permit easier enrichment and detection of photolabelled peptides. Nonetheless, these tags can be very bulky and thus they may affect the biological activity and cell permeability of many ligands. This technique provides a particularly promising method for finding novel and alternative ligand binding sites, aiding in investigating structures, functions and conformational changes of proteins and more importantly identifying the targets of biologically active compounds.[156,157]

5.2.3 HDX-MS

HDX is one of the techniques in the NMR arsenal to study protein dynamics,[158] however in recent years the HDX field has largely shifted towards HDX-MS due to advances in HDX methodology and MS instrumentation.[159] The attractiveness of HDX-MS resides in its high sensitivity, versatility (no protein size limitation), low protein consumption (μg) and ability to run experiments under close to physiological conditions. Together with these benefits, HDX-MS is also an extremely reliable method for studying protein conformation and dynamics.[18,160] Due to the increasing popularity of this technique, there is a massive amount of literature available on the topic. Covering every aspect of this technology in detail is out of the scope of this chapter and excellent reviews can be found elsewhere.[17,159,161] However, some representative examples will be mentioned to illustrate the progress on this technique including its relevant applications.

5.2.3.1 Fundamentals

HDX-MS is based on the MS quantitation of the exchange rate of protein hydrogens with deuterium in solution. The analysis workflow for a differential HDX-MS experiment is illustrated in Figure 5.4.

The protein of interest is incubated in a D_2O-buffered solution to allow exchange of hydrogen atoms by deuterium (Figure 5.4a). After a preset time (exchange time), the reaction is quenched by addition of a cold, low pH solution, followed by proteolytic digestion of the exchanged protein. The resulting peptides are analysed by HPLC-MS, and the isotopic distribution profiled by MS allows determination of their molecular weight (Figure 5.4b). Deuterium uptakes are computed for different exchange times (Figure 5.4c) and the obtained results are plotted for differential conditions in the experiment. The deuterium uptake in the apo-protein is a direct measure of the exposure of the hydrogens to the solvent and their participation in hydrogen bonding. The protein heat map is the representation of the uptake rates at the peptide level or mapped on the protein structure (Figure 5.4d). In a perturbation study, the differences in deuterium uptakes are used for epitope mapping of the sites of interaction (Figure 5.4c).

Due to the fast back-exchange of side chains and N-terminal hydrogens in solution after the quenching step, only exchange of backbone amide hydrogens can be monitored using HDX-MS. Each of these atoms becomes an ideal sensor for protein structure and dynamics, as a typical exchange relaxation time at physiological conditions ranges from milliseconds (ms) to hours.[159] The amide hydrogens close to the surface of the protein and those involved only in weak hydrogen bonds exchange rapidly whereas those in the

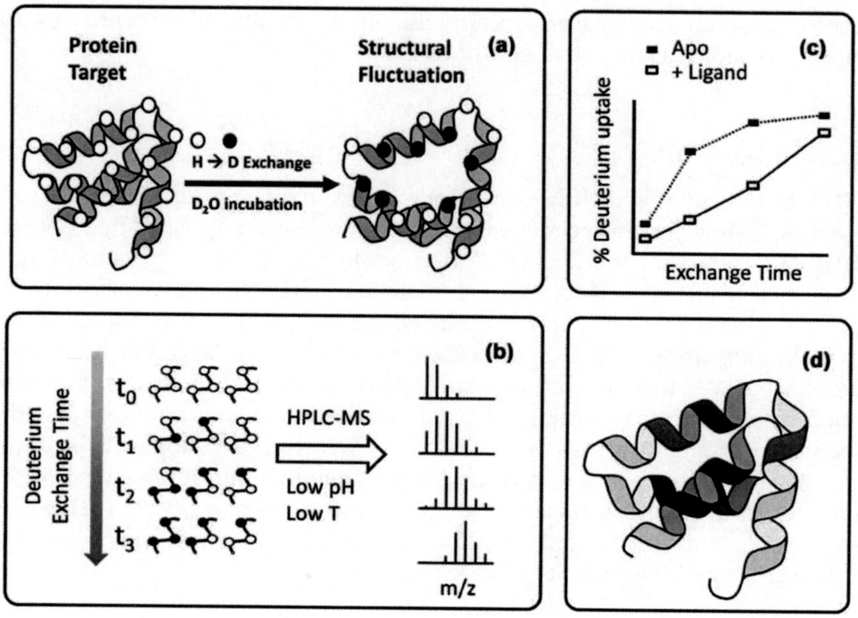

Figure 5.4 Workflow of a differential HDX experiment: (a) the protein is incubated with D_2O for a given period of time (exchange time) to allow exchange of hydrogen atoms (○) by deuterium atoms (●); (b) the reaction is quenched (usually by adding a low pH buffer) and the protein is digested to yield peptides that are analysed by HPLC-MS: (c) deuterium uptake is quantified and plotted under the two conditions used for comparison and (d) differences in deuterium uptake at the peptide level are deconvoluted and mapped on the protein structure for visualization.

inner regions of the protein or participating in strong hydrogen bonds exchange at a lower rate. Structural information on protein dynamics and conformation can be extracted at the peptide level. Spatial resolution from the peptide to near amino acid residue level has been achieved by the application of electron-capture dissociation (ECD)[162] and electron-transfer dissociation (ETD)[163] fragmentation methods. These methods prevent deuterium scrambling in the gas phase and are preferred over other fragmentation methods like collision-induced dissociation (CID). The resolution can be further improved by using computational methods to enable incorporation of enveloped-shape information and performing of close to residue-resolved back-exchange corrections.[164,165]

5.2.3.2 *HDX-MS in The Study of Protein Structure and Dynamics*

A thorough description of exchange mechanisms and dynamics has been reported elsewhere.[18,160,166–168] Two different kinetic exchange regimes (known as EX1 and EX2, which are monomolecular and bimolecular exchange, respectively) have been established as models to describe the dynamic behaviour of the protein. EX2 mechanism has a strong pH-dependent behaviour,[168] and under close to physiological conditions, it is the predominant one. EX2 is more informative than EX1; in the former, the exchange rate is related to the equilibrium between the open and closed form of the protein, thus providing a measure of the structural stability of the protein in the local environment of the exchangeable hydrogens.[159]

HDX-MS has been reported to investigate the conformational dynamics of molecules spanning from stapled peptides[169] to membrane proteins[170,171] and antibody–drug conjugates.[172] HDX-MS has also been applied to intrinsically disordered proteins. Combined with other techniques, it has provided insight into folding into β-hairpins and their relation with amyloid fibril formation,[173] and pulsed HDX-MS has proved successful in interrogating conformational changes in amyloid beta (Aβ) peptide aggregation.[174] ETD HDX-MS has proved successful in the study of higher order structures and the dynamics of whole protein molecules (top-down approach)[175] and top-down pulsed HDX-MS has been used to characterize short-lived protein folding intermediates.[176] Sub-ms pulsed HDX-MS is further pushing the limits of this technique by characterizing rapid conformational transitions[177] and revealing allosteric effects in enzyme inhibition.[178]

5.2.3.3 HDX-MS in The Study of Protein Interactions

A fundamental question in biophysics, especially within the context of drug discovery, is how a protein interacts with a given ligand. Although there is a plethora of biophysical tools to study protein–ligand interactions, only a few of them can provide localized structural information. X-ray crystallography and NMR play a central role in this arena, but both have limitations. Protein crystallization remains a challenging task in certain classes of proteins (*i.e.*, membrane proteins and intrinsically disordered proteins) and many proteins are above the size limit for analysis by NMR. Furthermore, only a limited set of conditions can be used in these techniques (solid state for crystallography, high protein concentration for NMR) hindering their general applicability. As described previously, HDX-MS has been proven extremely useful in the study of protein structure and dynamics, and the same principles can be applied to study ligand binding in the pocket and conformational changes induced by a binding event (allosteric effects).[17,179] This technique can be used under virtually any set of conditions or protein concentration, thus keeping the protein in solution during the experiment and generally under close to physiological conditions. The low throughput of this technique had been one of its caveats; however, advances in automation and data analysis have reduced this bottleneck.[180,181] Another concern in the application of HDX-MS was its reproducibility, but this has been recently assessed and HDX-MS has been demonstrated to yield highly reproducible results.[182–184]

Ligands under study can include chemical fragments, small organic molecules, nucleic acids, peptides and proteins.[17,180,185–187] Konermann *et al.*

proposed different scenarios that can result in a decrease or increase in deuterium uptake rate upon ligand binding, showing the potential of HDX-MS to reveal binding-induced changes in the protein tertiary structure.[188] Kinases have been extensively studied by HDX-MS.[189–191] Burke and Williams used HDX to evaluate activation of class IA phosphoinositide 3-kinases (PI3Ks) p85 subunit mediated by interaction with inhibitory contacts,[190] and Lorenzen and Pawson showed numerous applications in unravelling kinase dynamics following binding to ATP and/or phosphorylation.[191] The interaction of lipophilic small molecules with proteins has been recently investigated to understand mechanisms of transmembrane signaling[192] and Griffin *et al.* published several papers on the study of binding to protein receptors using HDX-MS, such as estrogen receptor (ERα), peroxisome proliferator-activated receptor (PPARγ) and retinoid X receptor (RXR).[193–195] Within the field of protein–protein interactions, epitopes are the interacting regions of binding partners. Epitope mapping is one of the most widely used applications of HDX in biopharmaceuticals,[159,196,197] with a strong presence in the development of antibodies.[197–199] In the small molecule area, binding of small molecules to pro-inflammatory cytokine IL-17A with the potential to disrupt interaction with its receptor has been recently demonstrated.[200] The creative application of HDX-MS to study protein interactions has been one of the great advancements in the field of biophysics and we expect it to continue providing solutions to challenging problems when other biophysical techniques and biochemical approaches fail.

5.2.4 Mass Spectrometry Techniques for Drug Screening

Multiple direct and indirect MS-based techniques have been applied to screen libraries of fragments and small molecules. Features of the most popular MS-based biophysical screening techniques are summarized in Table 5.1.

The MS-based biophysical screening approaches are illustrated in Figure 5.5. Direct MS (covalent and noncovalent protein–ligand complexes, (a) and (b), respectively) and AS-MS (c) utilize the interaction in solution to monitor changes in protein molecular weight after complexation (a) and (b), or differences in the physicochemical properties of the ligands between the free and bound state to yield measurable peaks by HPLC-MS (c). Frontal analysis (FA) and weak affinity chromatography (WAC) continuously infuse (d) or inject as a short plug (e) the sample through a column with the immobilized protein to detect interactions as a shift in the retention time (RT) or column volumes needed to elute the ligand peak.

5.2.4.1 ESI-Mass Spectrometry of Noncovalent Protein Complexes

ESI is a very gentle ionization technique that, under carefully selected ionization conditions, preserves the complex integrity of the molecule under

Table 5.1 Mass Spectrometry techniques used for drug screening.

Technique	Method	Description	Advantages	Disadvantages
ESI-MS of noncovalent complexes	Screening of libraries of fragments	Gentle ionization conditions preserve complex integrity in the gas phase	Detects weak binders with as low as mM binding affinities Interrogates large chemical diversity Small sample volume	Small molecule library size limited Challenging data interpretation of gas phase MS observations to understand biological interactions
ESI-MS of covalent complexes	Fragment-based lead assembly or screening of covalent binders	Fragments or small molecules bind covalently to the target. The covalent complexes are identified by MS	Selective chemical functionality combined with target engineering allows identification of site-specific modifications	Reactive site close to binding site required Low ionization efficiencies of modified proteolytic peptides
Affinity Selection Mass Spectrometry (AS-MS)	Size exclusion chromatography (SEC)	Target–ligand complexes elute faster than free ligands in SEC. Complexes are denatured to release small molecules followed by identified by LC–MS	Very high throughput Amenable to automation	Sequential processing of samples Needs optimization of protein transfer through SEC column
	Gel permeation chromatography (GPC)	SEC-based methodology in 96-well plate format followed by LC-MS	Very high throughput Plate format	Needs optimization of protein transfer through GPC column
	Ultrafiltration	Target–ligand complexes concentrated by ultrafiltration, followed by ligand dissociation & detection by LC-MS	Very high throughput Applicable to targets with poor chromatographic performance	Relatively high sample volume requirements Difficult automation
Solid-phase immobilized target	Frontal analysis (FA) and weak affinity chromatography (WAC)	Compounds are continuously infused (FA) or injected as short plugs (WAC) in a column with an immobilized target. The plateaus (FA) or peaks (WAC) of weak binders elute early, strong binders elute later	Generates affinity ranking Detects weak binders with as low as mM binding affinities	Need to immobilize the target High sample consumption in FA

study from its condensed state to the gas phase. Despite initial concerns over the rationalization of these ESI-MS observations to the interactions in the condensed phase,[201] a solid body of literature has emerged supporting the usefulness of these studies in addressing complex stoichiometry, shape, intersubunit connectivity and protein dynamics.[202] The application of ESI-MS to study biomolecular structure and dynamics has been discussed previously in this chapter, so in this section we will focus on this technique as a drug screening approach, particularly on its relevance in identifying weak binders.[203,204] In this approach, the protein is incubated with a pool of MS-encoded compounds and infused into MS under mild conditions. Assuming no change in charge state (z) takes place after complex formation, a shift is observed to higher m/z values for the complexes with respect to the unbound protein. The difference between the complex and the unbound protein m/z values multiplied by z yields directly the molecular weight (MW) of the bound ligand.[204] Protein consumption, a major limitation in other biophysical techniques, is relatively low when utilizing standard MS instrumentation; a further 10- to 20-fold decrease can be achieved with the use of a nano-ESI device.[203]

Due to its ability to detect interactions that are as weak as those in the mM range, native MS can assist fragment-based drug design (FBDD) and small molecule efforts. Native MS has been used within FBDD since its inception in 2002 by Swayze *et al.*, who coined the term "SAR by MS" when they screened compound libraries against a subdomain of 23S rRNA.[205] Since then, the studied targets have included stromelysin, bCA II, hCA II, XIAP, PNMT, Hsp90, Bcl-x, PfdUTPase and endothiapepsin.[203] The Triversa Nanomate platform (Advion, Ithaca, NY, USA) has been extensively reported in the literature within FBDD screening campaigns.[206,207] Native MS has consolidated itself as a well-established technique for fragment screening, but its application for screening in the small molecule area is difficult due to fundamental limitations (see Section 5.2.1). The resolving power of MS instruments for a given m/z value is impressive, but most proteins are highly charged and, accordingly, m/z shifts following complexation of small molecules are relatively small. Consistently, mass differences in the pooled small molecules need to be relatively high so that the complex can be unequivocally assigned, and this results in a low number of compounds per mixture (typically up to 8–12). This seems reasonable for FBDD screening as the chemical space sampled by a small collection of fragments is high, but as chemical complexity is increased, the number of molecules required to explore a similar chemical space rises exponentially. In this environment, ligand-based MS detection techniques allow multiplexing of a much higher number of compounds per pool and therefore they fit better in small molecule screening approaches.

5.2.4.2 *Affinity Selection Mass Spectrometry*

Affinity Selection Mass Spectrometry (AS-MS) comprises several techniques based on the physical separation of protein–ligand complexes from the

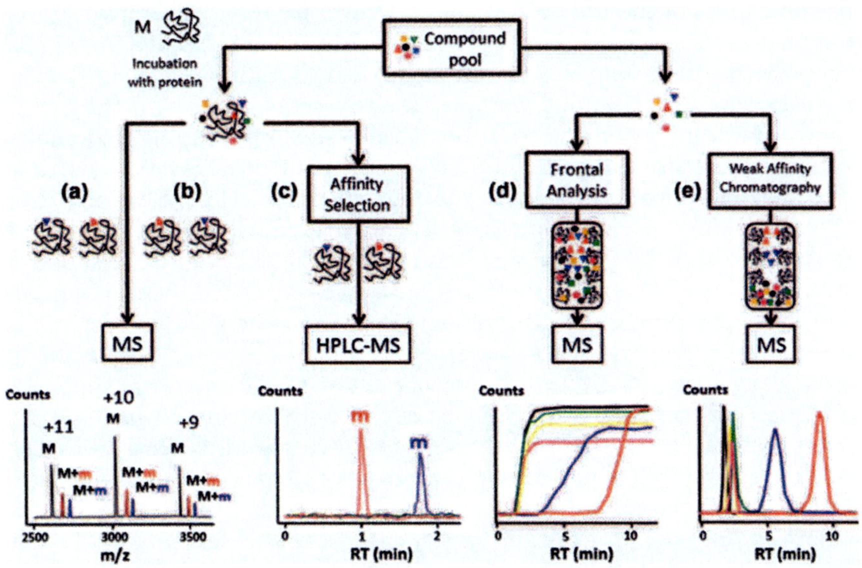

Figure 5.5 MS-based biophysical screening approaches. Direct approaches monitor changes in the observed *m*/*z* for the protein after non-covalent complexation (a) or covalent modification (b) to identify hits. AS-MS is a ligand-enabled technology that separates protein–ligand complexes from free small molecules. These complexes are denatured and the released ligands are identified by MS (c). In frontal analysis (FA), the sample is continuously infused through an affinity column with the immobilized target. Plateaus for the interacting compounds are shifted due to retention by the column (d). Weak affinity chromatography (WAC) shares the same principle but the compounds for testing are injected in pools as a small plug, so they are detected as peaks (e). RT: retention time.

unbound small molecule fraction and the subsequent MS detection of the released ligands following complex denaturation. In contrast with the direct study of ligand–protein interactions by MS under native conditions, here the solvents and modifiers used for the analysis can be selected without the need to preserve complex integrity, so the technique does not compromise the sensitivity of the MS detector. Furthermore, MS detection of fragments and small molecules is quite straightforward and therefore implementation of the experiment is very fast.[8]

The most popular AS-MS techniques couple liquid chromatography-MS with size-based separation methods to enable MS-based high throughput screening (HTS).[22] ALIS (Automated Ligand Identification System) is an online approach that utilizes a size exclusion chromatography (SEC) column to separate bound from unbound compounds.[208–214] The initial scope of this technique was the identification of ligands bound to the target of interest, however, several related methods have been reported that take advantage of the robustness of the fully automated platform for the quantitative study of protein–ligand interactions.[215–219] This platform can handle up to 2500 compounds per sample[209] and can be applied to multiple protein families, but it has some limitations. Samples are run sequentially and the need to couple multidimensional separations (size exclusion, trapping and analysis of released ligands) impedes the achievement of ultrafast analysis times. SpeedScreen is an alternative assay format where the gel permeation process takes place off-line, through a 96-well plate loaded with the active material for size exclusion.[220–223] Samples pass through the packing material and the eluting complexes are denatured by adding a strong solvent. These steps can be easily automated and run independently from the final analysis phase, thus allowing ultrafast HPLC-MS at the end of the process. Disposable gel permeation plates facilitate routine operation of the platform supporting multiple HTS campaigns that can be successfully integrated into the drug discovery process.[221] These methodologies have proved successful across a wide variety of targets including receptors, G-proteins, kinases and enzymes among others,[22] but protein transmission through the packing material remains a challenge for certain targets and assay enablement may be cumbersome. A method with a broader application range is ultrafiltration-based AS-MS, where the samples are centrifuged through a filter membrane, which does not allow molecules above a set molecular weight (molecular weight cutoff) to pass through.[224–227] As a consequence, the free protein and protein–ligand complexes are retained and concentrated. Ultrafiltration is not based on protein transmission through a packing material so the implementation time is very short and it can be applied to virtually any target with ease. This technique is particularly suitable for screening very complex samples, and has been successfully applied to identify active compounds in herbal and marine extracts.[228–232] Another advantage is that the sample over the membrane gets concentrated and consequently there is an increase in the protein concentration during the ultrafiltration process. This prevents dissociation of weakly-bound complexes and opens up the possibility to pick very weak binders by AS-MS, allowing the integration of this technique into FBDD lead discovery pipelines.[233,234] As discussed, AS-MS is now a well consolidated screening technique that allows HTS of large chemical collections as primary assay or follow-up and validation of compounds resulting from conventional HTS screening. Its application within FBDD has been relatively limited, but recently there has been promising progress in this field too.

5.2.4.3 *Drug Screening with Immobilized Target*

MS has also utilized immobilization of the target in the packing material of a column to screen new chemical entities. The two common techniques using this strategy are frontal analysis (FA) and weak affinity chromatography (WAC). These techniques exploit the differential retention of interacting ligands compared to non-interacting partners that typically elute from the column with the void volume. They can be applied in a broad range of biological systems, and are considered to be label-free methods. The ability of these techniques to rank ligands and determine K_d positions them as universal tools to enable convenient and efficient screening.[21,235–237] FA coupled to MS was initially introduced by Schiermer,[238] who made possible the simultaneous analysis of multiple compounds due to the extra dimension provided by the MS detector. In FA, the mixture containing potential ligands is continuously infused through the column. Unbound compounds are detected in the void volume, whereas active compounds are detected only when the binding capacity of the column is exceeded. The volume needed to reach this point, known as the break-through volume, is characterized as a sigmoidal front and can be used to indicate the relative affinities of the components of the mixture.[235] Calculation of K_d from breakthrough volumes is straightforward and soon after technique inception, numerous applications arose for the characterization of ligands to antibodies,[239,240] lectins[241] and receptors.[242–244] One example of the successful combination of virtual screening with FA screened a set of 500 compounds from which inhibitors of erythropoietin-producing hepatocellular B2 receptor tyrosine kinase domain were identified.[245] Another research group demonstrated FA suitability for medium throughput biophysical screening with a dedicated dual column system by screening a set of 100 compounds with a run time of 30 minutes.[246] An excellent review covered the state-of-the-art FA utilization before the irruption of WAC.[247] FA is a powerful and mature technique but it has some limitations. It cannot overcome the solubility issues of chemicals and the concomitant need to use relatively high DMSO percentages. Moreover, the rate of false negatives can increase because of buffer incompatibilities and ionization suppression events when assayed pools contain a large number of compounds. Sample consumption is relatively high in FA and extensive washing of the column between runs is needed if very potent compounds are present in the tested mixture. WAC (also known as zonal affinity chromatography) can diminish these drawbacks as, in this technique, the sample is not continuously infused but injected as a small plug. Concurrence of compounds in the MS spectrometer is reduced as, in general, the compounds elute in peaks or narrow bands. Furthermore, the column is exposed to smaller volumes of sample, thus reducing the risk of impacting protein folding, extending column lifetime and minimizing the wash time between injections. Since it was first described as a new approach for fragment screening with model trypsin/thrombin columns,[248] WAC has been combined with virtual screening to screen up to 1700 fragments with cyclin G-associated kinase[249] testifying to its usefulness in FBDD. Despite being highly suitable for the automated fast screening of fragments, WAC cannot provide the same level of information as FA and these two complementary techniques coexist to support drug discovery efforts.[237,250–255]

5.3 Perspectives

Recent developments in MS have paved the way for its increased application as a biophysical technique. MS roles currently span from the characterization of macromolecule structure and dynamics to the identification and confirmation of its interacting partners. Despite the huge progress achieved in this century, the unveiling of deeper levels of complexity within biological systems and the subsequent increasing demand for higher quality information have required the utilization of more powerful techniques and imaginative procedures to satisfy the field. The repertoire of applications developed in recent years and the innovations available today in related fields point to future directions. Future instrumentation will benefit from advances in sample introduction into the mass spectrometer and hyphenation to existing separation techniques. Direct acoustic droplet ejection allows MS data acquisition with a very high frequency (exceeding 10 000 data points per hour), extremely low volume consumption and minimum cross contamination between samples.[256–258] Progress on the acoustic nebulization device, such as incorporation of dual interdigitated transducers to generate standing waves, can improve its performance and boost sensitivity (10^2- to 10^3-fold gain in ion intensity) compared to travelling wave devices.[259] It has been shown that this device can easily produce mass spectra of small proteins (<20 kDa) and the higher abundance of homodimers observed as compared to ESI suggests its potential utility in native MS applications. Although there remain some doubts yet on the applicability of acoustic dispensing to samples suffering strong matrix effects, the potential of this sampling technique will definitely revolutionize current paradigms in drug discovery. Lab-on-a-Chip hyphenation with MS can overcome the limited resolving power of acoustic dispensing while sharing some of its advantages at the expense of a lower throughput. Microfluidic chips can be prepared with very diverse layouts to integrate sample treatment while maintaining extremely low sample consumption and reasonable speed, especially compared to traditional approaches.[178,260] Furthermore, recent application in bioanalysis has demonstrated the applicability of this technique within this field.[261,262] IMS coupled with MS permits analysis of very complex samples, opening new paths to interrogate biological systems. Consequently, it will certainly remain a hot topic in the area,[263] expanding its use as a detector from well-established separation techniques such as HPLC to others where it is currently less popular, such as capillary electrophoresis.[264,265] We envision a gradual yet very relevant shift towards more sophisticated, high resolution MS instrumentation (orbitrap and time-of-flight MS instruments) for multiple applications, including among others HDX-MS, AS-MS and WAC. Methodological and procedural improvements not directly related to technical advances will also play a critical role in coming years. As an example, the use of covalent inhibitors in core drug-discovery processes has not been well established because of the possibility of greater off target activity, which may lead to toxicity issues.[266] However, these off-target interactions can be monitored and minimized with the aid of modern MS and biophysical techniques. Thus, the covalent approach can address challenges of difficult-to-drug protein families either by increasing the specificity to target one member of the conserved protein family or by preventing interactions in PPI targets with sizable binding surfaces.[266–269] MS has also emerged in this field as a solid technology to screen covalent modifiers[270] and monitor covalent modifications to a protein at single amino acid resolution. Cellular thermal shift assay (CETSA) is a new biophysical technique that empowers researchers to directly assess target engagement in cells and tissues at various stages of drug development.[271,272] When combined with MS detection after proteolytic digestion, it allowed quantitation of more than 5000 proteins from the human K562 chronic myeloid leukemia cell line, showing the first description of the melting proteome of a human cell.[273] Differential studies in the presence of a drug of interest are used to confirm target engagement through ligand-induced stabilization or destabilization.[273] In summary, the tremendous momentum observed in the advancement of MS instrumentation and MS-based biophysical analysis in recent years forecasts a revolution in the utilization of these cutting edge techniques in modern pharmaceutical drug-discovery.

Acknowledgements

We thank Devon Thompson for her insight and comments that greatly improved the manuscript.

References

1. J. J. Thomson, *Rays of positive electricity and their application to chemical analyses*, Longmans, Green and Company, New York, 1913.
2. E. P. Diamandis, *Mol. Cell. Proteomics*, 2004, **3**, 367.
3. http://www.ddw-online.com/enabling-technologies/p146734-mass-spectrometry-in-drug-discovery-and-development-fall-10.html (last accessed August 2016).
4. D. I. Papac and Z. Shahrokh, *Pharm. Res.*, 2001, **18**, 131.
5. J. S. Janiszewski, T. E. Liston and M. J. Cole, *Curr. Drug Metab.*, 2008, **9**, 986.
6. *Mass Spectrometry in Drug Discovery*, ed. D. T. Rossi and M. W. Sinz, Marcel Dekker Inc., New York, 2002.
7. W. A. Korfmacher, *Using Mass Spectrometry for Drug Metabolism Studies*, CRC Press/Taylor Francis Group, Boca Raton, 2010.
8. *Mass Spectrometry in Medicinal Chemistry*, ed. K. Wanner and G. Höfner, Wiley-VCH, Weinheim, 2007, p. 36.
9. I. A. Kaltashov, S. J. Eyles, D. M. Desiderio and N. M. Nibbering, *Mass Spectrometry in Structural Biology and Biophysics: Architecture, Dynamics, and Interaction of Biomolecules*, 2 edn, Wiley, Hoboken, 2012.
10. M. Karas and F. Hillenkamp, *Anal. Chem.*, 1988, **60**, 2299.

11. C. M. Whitehouse, M. Mann, C. K. Meng, S. F. Wong and J. B. Fenn, *Science*, 1989, **246**, 64.
12. A. J. Heck, *Nat. Methods*, 2008, **5**, 927.
13. J. L. Benesch and B. T. Ruotolo, *Curr. Opin. Struct. Biol.*, 2011, **21**, 641.
14. F. Stengel, R. Aebersold and C. V. Robinson, *Mol. Cell. Proteomics*, 2012, **11**, R111. 014027.
15. A. Leitner, T. Walzthoeni, A. Kahraman, F. Herzog, O. Rinner, M. Beck and R. Aebersold, *Mol. Cell. Proteomics*, 2010, **9**, 1634.
16. B. Deng, C. Lento and D. J. Wilson, *Anal. Chim. Acta*, 2016, **940**, 8.
17. M. J. Chalmers, S. A. Busby, B. D. Pascal, G. M. West and P. R. Griffin, *Expert Rev. Proteomics*, 2011, **8**, 43.
18. L. Konermann, J. Pan and Y. H. Liu, *Chem. Soc. Rev.*, 2011, **40**, 1224.
19. J. Hardouin and C. Lange, *Curr. Org. Chem.*, 2005, **9**, 317.
20. M. M. Siegel, *Curr. Top. Med. Chem.*, 2002, **2**, 13.
21. N. Jonker, J. Kool, H. Irth and W. M. Niessen, *Anal. Bioanal. Chem.*, 2011, **399**, 2669.
22. D. A. Annis, E. Nickbarg, X. Yang, M. R. Ziebell and C. E. Whitehurst, *Curr. Opin. Chem. Biol.*, 2007, **11**, 518.
23. C. E. Whitehurst and D. A. Annis, *Comb. Chem. High Throughput Screening*, 2008, **11**, 427.
24. R. Apweiler, A. Bairoch and C. H. Wu, *Curr. Opin. Chem. Biol.*, 2004, **8**, 76.
25. A. Leitner, M. Faini, F. Stengel and R. Aebersold, *Trends Biochem. Sci.*, 2016, **41**, 20.
26. M. R. Chance, A. Fiser, A. Sali, U. Pieper, N. Eswar, G. Xu, J. E. Fajardo, T. Radhakannan and N. Marinkovic, *Genome Res.*, 2004, **14**, 2145.
27. R. B. Russell, F. Alber, P. Aloy, F. P. Davis, D. Korkin, M. Pichaud, M. Topf and A. Sali, *Curr. Opin. Struct. Biol.*, 2004, **14**, 313.
28. A. Sali, R. Glaeser, T. Earnest and W. Baumeister, *Nature*, 2003, **422**, 216.
29. T. Gheyi, L. Rodgers, R. Romero, J. M. Sauder and S. K. Burley, *J. Am. Soc. Mass Spectrom.*, 2010, **21**, 1795.
30. A. Bierzynski, *Acta Biochim. Pol.*, 2001, **48**, 1091.
31. A. Ilari, A. Bonamore, S. Franceschini, A. Fiorillo, A. Boffi and G. Colotti, *Proteins*, 2008, **71**, 2065.
32. V. Tugarinov, P. M. Hwang and L. E. Kay, *Annu. Rev. Biochem.*, 2004, **73**, 107.
33. M. L. Pusey, Z. J. Liu, W. Tempel, J. Praissman, D. Lin, B. C. Wang, J. A. Gavira and J. D. Ng, *Prog. Biophys. Mol. Biol.*, 2005, **88**, 359.
34. M. Grabowski, A. Joachimiak, Z. Otwinowski and W. Minor, *Curr. Opin. Struct. Biol.*, 2007, **17**, 347.
35. N. O'Toole, M. Grabowski, Z. Otwinowski, W. Minor and M. Cygler, *Proteins*, 2004, **56**, 201.
36. A. Wernimont and A. Edwards, *PLoS One*, 2009, **4**, e5094.
37. A. Sinz, C. Arlt, D. Chorev and M. Sharon, *Protein Sci.*, 2015, **24**, 1193.
38. P. Walian, T. A. Cross and B. K. Jap, *Genome Biol.*, 2004, **5**, 215.

39. F. Mancia and J. Love, *Curr. Opin. Struct. Biol.*, 2011, **21**, 517.
40. F. Chen, B. Gulbakan, S. Weidmann, S. R. Fagerer, A. J. Ibanez and R. Zenobi, *Mass Spectrom. Rev.*, 2016, **35**, 48.
41. P. C. Havugimana, G. T. Hart, T. Nepusz, H. Yang, A. L. Turinsky, Z. Li, P. I. Wang, D. R. Boutz, V. Fong, S. Phanse, M. Babu, S. A. Craig, P. Hu, C. Wan, J. Vlasblom, V. U. Dar, A. Bezginov, G. W. Clark, G. C. Wu, S. J. Wodak, E. R. Tillier, A. Paccanaro, E. M. Marcotte and A. Emili, *Cell*, 2012, **150**, 1068.
42. A. J. Heck and R. H. Van Den Heuvel, *Mass Spectrom. Rev.*, 2004, **23**, 368.
43. C. S. Kaddis and J. A. Loo, *Anal. Chem.*, 2007, **79**, 1778.
44. M. Sharon and C. V. Robinson, *Annu. Rev. Biochem.*, 2007, **76**, 167.
45. C. Uetrecht, C. Versluis, N. R. Watts, W. H. Roos, G. J. Wuite, P. T. Wingfield, A. C. Steven and A. J. Heck, *Proc. Natl. Acad. Sci. U. S. A.*, 2008, **105**, 9216.
46. J. M. Elkins, I. J. Clifton, H. Hernandez, L. X. Doan, C. V. Robinson, C. J. Schofield and K. S. Hewitson, *Biochem. J.*, 2002, **366**, 423.
47. H. Hernandez, A. Dziembowski, T. Taverner, B. Seraphin and C. V. Robinson, *EMBO Rep.*, 2006, **7**, 605.
48. M. Sharon, T. Taverner, X. I. Ambroggio, R. J. Deshaies and C. V. Robinson, *PLoS Biol.*, 2006, **4**, e267.
49. C. A. Keetch, E. H. Bromley, M. G. McCammon, N. Wang, J. Christodoulou and C. V. Robinson, *J. Biol. Chem.*, 2005, **280**, 41667.
50. F. Sobott, J. L. Benesch, E. Vierling and C. V. Robinson, *J. Biol. Chem.*, 2002, **277**, 38921.
51. J. Marcoux and C. V. Robinson, *Structure*, 2013, **21**, 1541.
52. E. van Duijn, *J. Am. Soc. Mass Spectrom.*, 2010, **21**, 971.
53. S. L. Bernstein, N. F. Dupuis, N. D. Lazo, T. Wyttenbach, M. M. Condron, G. Bitan, D. B. Teplow, J. E. Shea, B. T. Ruotolo, C. V. Robinson and M. T. Bowers, *Nat. Chem.*, 2009, **1**, 326.
54. C. Bleiholder, N. F. Dupuis, T. Wyttenbach and M. T. Bowers, *Nat. Chem.*, 2011, **3**, 172.
55. N. F. Dupuis, C. Wu, J. E. Shea and M. T. Bowers, *J. Am. Chem. Soc.*, 2009, **131**, 18283.
56. M. Grabenauer, T. Wyttenbach, N. Sanghera, S. E. Slade, T. J. Pinheiro, J. H. Scrivens and M. T. Bowers, *J. Am. Chem. Soc.*, 2010, **132**, 8816.
57. J. I. Baumbach and G. A. Eiceman, *Appl. Spectrosc.*, 1999, **53**, 338A.
58. G. A. Eiceman and Z. Karpas, *Ion Mobility Spectrometry*, CRC Press, Boca Raton, FL, 2nd edn, 2004.
59. R. Guevremont and R. W. Purves, *J. Am. Soc. Mass Spectrom.*, 1999, **10**, 492.
60. A. B. Kanu, P. Dwivedi, M. Tam, L. Matz and H. H. Hill, Jr., *J. Mass Spectrom.*, 2008, **43**, 1.
61. G. A. Anderson, A. A. Shvartsburg and R. D. Smith, *J. Mass Spectrom. Soc. Jpn.*, 2013, **2**, 1.
62. A. A. Shvartsburg, *Differential Ion Mobility Spectrometry: Nonlinear Ion Transport and Fundamentals of FAIMS*, CRC Press, Boca Raton, FL, 2008.

63. A. A. Shvartsburg and R. D. Smith, *Anal. Chem.*, 2008, **80**, 9689.
64. A. A. Shvartsburg and R. D. Smith, *J. Am. Soc. Mass Spectrom.*, 2008, **19**, 1286.
65. A. A. J. Shvartsburg and M. F. Jarrold, *Chem. Phys. Lett.*, 1996, **261**, 86.
66. C. Uetrecht, R. J. Rose, E. van Duijn, K. Lorenzen and A. J. Heck, *Chem. Soc. Rev.*, 2010, **39**, 1633.
67. S. Mehmood, T. M. Allison and C. V. Robinson, *Annu. Rev. Phys. Chem.*, 2015, **66**, 453.
68. E. A. S. Mason and H. W. Schamp, *Ann. Phys.*, 1958, **4**, 233.
69. B. T. Ruotolo, J. L. Benesch, A. M. Sandercock, S. J. Hyung and C. V. Robinson, *Nat. Protoc.*, 2008, **3**, 1139.
70. I. Michaelevski, M. Eisenstein and M. Sharon, *Anal. Chem.*, 2010, **82**, 9484.
71. C. Atmanene, S. Petiot-Becard, D. Zeyer, A. Van Dorsselaer, V. Vivat Hannah and S. Sanglier-Cianferani, *Anal. Chem.*, 2012, **84**, 4703.
72. S. J. Hyung, C. V. Robinson and B. T. Ruotolo, *Chem. Biol.*, 2009, **16**, 382.
73. J. Stojko, S. Fieulaine, S. Petiot-Becard, A. Van Dorsselaer, T. Meinnel, C. Giglione and S. Cianferani, *Analyst*, 2015, **140**, 7234.
74. Y. Zhao, A. Singh, L. Li, R. J. Linhardt, Y. Xu, J. Liu, R. J. Woods and I. J. Amster, *Analyst*, 2015, **140**, 6980.
75. S. Niu, J. N. Rabuck and B. T. Ruotolo, *Curr. Opin. Chem. Biol.*, 2013, **17**, 809.
76. E. van Duijn, A. Barendregt, S. Synowsky, C. Versluis and A. J. Heck, *J. Am. Chem. Soc.*, 2009, **131**, 1452.
77. A. Politis, A. Y. Park, S. J. Hyung, D. Barsky, B. T. Ruotolo and C. V. Robinson, *PLoS One*, 2010, **5**, e12080.
78. T. L. Pukala, B. T. Ruotolo, M. Zhou, A. Politis, R. Stefanescu, J. A. Leary and C. V. Robinson, *Structure*, 2009, **17**, 1235.
79. S. C. Wang, A. Politis, N. Di Bartolo, V. N. Bavro, S. J. Tucker, P. J. Booth, N. P. Barrera and C. V. Robinson, *J. Am. Chem. Soc.*, 2010, **132**, 15468.
80. C. Arlt, C. H. Ihling and A. Sinz, *Proteomics*, 2015, **15**, 2746.
81. K. Bosse, S. Haneder, C. Arlt, C. H. Ihling, T. Seufferlein and A. Sinz, *Proteomics*, 2016, **16**, 2801.
82. L. D. Gallego, M. Ghodgaonkar Steger, A. A. Polyansky, T. Schubert, B. Zagrovic, N. Zheng, T. Clausen, F. Herzog and A. Kohler, *Proc. Natl. Acad. Sci. U. S. A.*, 2016, **113**, 10553.
83. G. Guaitoli, F. Raimondi, B. K. Gilsbach, Y. Gomez-Llorente, E. Deyaert, F. Renzi, X. Li, A. Schaffner, P. K. Jagtap, K. Boldt, F. von Zweydorf, K. Gotthardt, D. D. Lorimer, Z. Yue, A. Burgin, N. Janjic, M. Sattler, W. Versees, M. Ueffing, I. Ubarretxena-Belandia, A. Kortholt and C. J. Gloeckner, *Proc. Natl. Acad. Sci. U. S. A.*, 2016, **113**, E4357.
84. H. Liu, H. Zhang, D. A. Weisz, I. Vidavsky, M. L. Gross and H. B. Pakrasi, *Proc. Natl. Acad. Sci. U. S. A.*, 2014, **111**, 4638.
85. P. J. Robinson, M. J. Trnka, D. A. Bushnell, R. E. Davis, P. J. Mattei, A. L. Burlingame and R. D. Kornberg, *Cell*, 2016, **166**, 1411.

86. F. Wu and S. Minteer, *Angew. Chem., Int. Ed. Engl.*, 2015, **54**, 1851.
87. B. X. Huang, H. Y. Kim and C. Dass, *J. Am. Soc. Mass Spectrom.*, 2004, **15**, 1237.
88. T. Walzthoeni, L. A. Joachimiak, G. Rosenberger, H. L. Rost, L. Malmstrom, A. Leitner, J. Frydman and R. Aebersold, *Nat. Methods*, 2015, **12**, 1185.
89. A. Sinz, *Expert Rev. Proteomics*, 2014, **11**, 733.
90. A. Politis, F. Stengel, Z. Hall, H. Hernandez, A. Leitner, T. Walzthoeni, C. V. Robinson and R. Aebersold, *Nat. Methods*, 2014, **11**, 403.
91. M. M. Young, N. Tang, J. C. Hempel, C. M. Oshiro, E. W. Taylor, I. D. Kuntz, B. W. Gibson and G. Dollinger, *Proc. Natl. Acad. Sci. U. S. A.*, 2000, **97**, 5802.
92. B. F. Cravatt, G. M. Simon and J. R. Yates, 3rd, *Nature*, 2007, **450**, 991.
93. J. Rappsilber, S. Siniossoglou, E. C. Hurt and M. Mann, *Anal. Chem.*, 2000, **72**, 267.
94. K. L. Bennett, M. Kussmann, P. Bjork, M. Godzwon, M. Mikkelsen, P. Sorensen and P. Roepstorff, *Protein Sci.*, 2000, **9**, 1503.
95. T. J. Mendez, J. V. Johnson, L. S. Nichols, G. H. Lang, J. R. Eyler, D. H. Powell and D. E. Richardson, *Bioconjugate Chem.*, 2000, **11**, 182.
96. A. N. Holding, *Methods*, 2015, **89**, 54.
97. E. V. Petrotchenko and C. H. Borchers, *Mass Spectrom. Rev.*, 2010, **29**, 862.
98. F. Chu, P. R. Baker, A. L. Burlingame and R. J. Chalkley, *Mol. Cell. Proteomics*, 2010, **9**, 25.
99. C. W. Combe, L. Fischer and J. Rappsilber, *Mol. Cell. Proteomics*, 2015, **14**, 1137.
100. M. Gotze, J. Pettelkau, S. Schaks, K. Bosse, C. H. Ihling, F. Krauth, R. Fritzsche, U. Kuhn and A. Sinz, *J. Am. Soc. Mass Spectrom.*, 2012, **23**, 76.
101. M. Grimm, T. Zimniak, A. Kahraman and F. Herzog, *Nucleic Acids Res.*, 2015, **43**, W362.
102. J. Kosinski, A. von Appen, A. Ori, K. Karius, C. W. Muller and M. Beck, *J. Struct. Biol.*, 2015, **189**, 177.
103. H. Xu, P. H. Hsu, L. Zhang, M. D. Tsai and M. A. Freitas, *J. Proteome Res.*, 2010, **9**, 3384.
104. C. Zheng, C. R. Weisbrod, J. D. Chavez, J. K. Eng, V. Sharma, X. Wu and J. E. Bruce, *J. Proteome Res.*, 2013, **12**, 1989.
105. Y. J. Lee, L. L. Lackner, J. M. Nunnari and B. S. Phinney, *J. Proteome Res.*, 2007, **6**, 3908.
106. A. Maiolica, D. Cittaro, D. Borsotti, L. Sennels, C. Ciferri, C. Tarricone, A. Musacchio and J. Rappsilber, *Mol. Cell. Proteomics*, 2007, **6**, 2200.
107. O. Rinner, J. Seebacher, T. Walzthoeni, L. N. Mueller, M. Beck, A. Schmidt, M. Mueller and R. Aebersold, *Nat. Methods*, 2008, **5**, 315.
108. A. Zelter, M. R. Hoopmann, R. Vernon, D. Baker, M. J. MacCoss and T. N. Davis, *J. Proteome Res.*, 2010, **9**, 3583.

109. Z. A. Chen, A. Jawhari, L. Fischer, C. Buchen, S. Tahir, T. Kamenski, M. Rasmussen, L. Lariviere, J. C. Bukowski-Wills, M. Nilges, P. Cramer and J. Rappsilber, *EMBO J.*, 2010, **29**, 717.
110. I. Karadzic, J. Maupin-Furlow, M. Humbard, L. Prunetti, P. Singh and D. R. Goodlett, *Proteomics*, 2012, **12**, 1806.
111. H. Liu, H. Zhang, D. M. Niedzwiedzki, M. Prado, G. He, M. L. Gross and R. E. Blankenship, *Science*, 2013, **342**, 1104.
112. K. Kramer, T. Sachsenberg, B. M. Beckmann, S. Qamar, K. L. Boon, M. W. Hentze, O. Kohlbacher and H. Urlaub, *Nat. Methods*, 2014, **11**, 1064.
113. A. P. Frei, O. Y. Jeon, S. Kilcher, H. Moest, L. M. Henning, C. Jost, A. Pluckthun, J. Mercer, R. Aebersold, E. M. Carreira and B. Wollscheid, *Nat. Biotechnol.*, 2012, **30**, 997.
114. L. A. Joachimiak, T. Walzthoeni, C. W. Liu, R. Aebersold and J. Frydman, *Cell*, 2014, **159**, 1042.
115. G. Xu and M. R. Chance, *Chem. Rev.*, 2007, **107**, 3514.
116. G. Xu and M. R. Chance, *Anal. Chem.*, 2005, **77**, 4549.
117. M. R. Chance, *Biochem. Biophys. Res. Commun.*, 2001, **287**, 614.
118. S. Gupta, H. Cheng, A. K. Mollah, E. Jamison, S. Morris, M. R. Chance, S. Khrapunov and M. Brenowitz, *Biochemistry*, 2007, **46**, 9886.
119. J. G. Kiselar and M. R. Chance, *J. Mass Spectrom.*, 2010, **45**, 1373.
120. J. G. Kiselar, P. A. Janmey, S. C. Almo and M. R. Chance, *Proc. Natl. Acad. Sci. U. S. A.*, 2003, **100**, 3942.
121. J. G. Kiselar, R. Mahaffy, T. D. Pollard, S. C. Almo and M. R. Chance, *Proc. Natl. Acad. Sci. U. S. A.*, 2007, **104**, 1552.
122. S. D. Maleknia and K. M. Downard, *Chem. Soc. Rev.*, 2014, **43**, 3244.
123. L. Wang and M. R. Chance, *Anal. Chem.*, 2011, **83**, 7234.
124. K. Takamoto and M. R. Chance, *Annu. Rev. Biophys. Biomol. Struct.*, 2006, **35**, 251.
125. K. D. Asmus, *Methods Enzymol.*, 1984, **105**, 167.
126. M. G. Simic, *Methods Enzymol.*, 1990, **186**, 89.
127. C. von Sonntag and H. P. Schuchmann, *Methods Enzymol.*, 1994, **233**, 47.
128. P. Wardman, *Strahlentherapie, Sonderb.*, 1978, **75**, 20.
129. S. Gupta, M. Sullivan, J. Toomey, J. Kiselar and M. R. Chance, *J. Synchrotron Radiat.*, 2007, **14**, 233.
130. M. A. J. a. F. Rogers, *Radiation Chemistry: Principles and Applications*, VCH, 1987.
131. B. Sclavi, S. Woodson, M. Sullivan, M. Chance and M. Brenowitz, *Methods Enzymol.*, 1998, **295**, 379.
132. F. S. Dainton, *J. Am. Chem. Soc.*, 1956, **78**, 1278.
133. D. H. C. Volman, *J. Am. Chem. Soc.*, 1959, **81**, 4144.
134. J. L. M. Weeks and M. S. Matheson, *J. Am. Chem. Soc.*, 1956, **78**, 1273.
135. T. D. Tullius and B. A. Dombroski, *Proc. Natl. Acad. Sci. U. S. A.*, 1986, **83**, 5469.
136. C. S. McClintock and R. L. Hettich, *Anal. Chem.*, 2013, **85**, 213.

137. B. C. Gau, J. Chen and M. L. Gross, *Biochim. Biophys. Acta*, 2013, **1834**, 1230.
138. C. C. Jumper, R. Bomgarden, J. Rogers, C. Etienne and D. C. Schriemer, *Anal. Chem.*, 2012, **84**, 4411.
139. X. Li, Z. Li, B. Xie and J. S. Sharp, *J. Am. Soc. Mass Spectrom.*, 2013, **24**, 1767.
140. C. S. McClintock, J. M. Parks, M. Bern, P. K. Ghattyvenkatakrishna and R. L. Hettich, *J. Proteome Res.*, 2013, **12**, 3307.
141. A. Baud, F. Gonnet, I. Salard, M. Le Mignon, A. Giuliani, P. Mercere, B. Sclavi and R. Daniel, *Biochem. J.*, 2016, **473**, 1805.
142. L. M. Jones, J. B. Sperry, J. A. Carroll and M. L. Gross, *Anal. Chem.*, 2011, **83**, 7657.
143. A. N. Schorzman, L. Perera, J. M. Cutalo-Patterson, L. C. Pedersen, L. G. Pedersen, T. A. Kunkel and K. B. Tomer, *DNA Repair*, 2011, **10**, 454.
144. H. Zhang, B. C. Gau, L. M. Jones, I. Vidavsky and M. L. Gross, *Anal. Chem.*, 2011, **83**, 311.
145. T. E. Angel, S. Gupta, B. Jastrzebska, K. Palczewski and M. R. Chance, *Proc. Natl. Acad. Sci. U. S. A.*, 2009, **106**, 14367.
146. T. Orban, B. Jastrzebska, S. Gupta, B. Wang, M. Miyagi, M. R. Chance and K. Palczewski, *Structure*, 2012, **20**, 826.
147. P. S. Padayatti, L. Wang, S. Gupta, T. Orban, W. Sun, D. Salom, S. R. Jordan, K. Palczewski and M. R. Chance, *Mol. Cell. Proteomics*, 2013, **12**, 1259.
148. S. Gupta, V. N. Bavro, R. D'Mello, S. J. Tucker, C. Venien-Bryan and M. R. Chance, *Structure*, 2010, **18**, 839.
149. S. Gupta, J. Chai, J. Cheng, R. D'Mello, M. R. Chance and D. Fu, *Nature*, 2014, **512**, 101.
150. P. Kaur, J. Kiselar, S. Yang and M. R. Chance, *Mol. Cell. Proteomics*, 2015, **14**, 1159.
151. P. Kaur, J. G. Kiselar and M. R. Chance, *Anal. Chem.*, 2009, **81**, 8141.
152. J. Das, *Chem. Rev.*, 2011, **111**, 4405.
153. L. Dubinsky, B. P. Krom and M. M. Meijler, *Bioorg. Med. Chem.*, 2012, **20**, 554.
154. M. Hashimoto and Y. Hatanaka, *Bioorg. Med. Chem. Lett.*, 2008, **18**, 650.
155. Y. Sadakane and Y. Hatanaka, *Anal. Sci.*, 2006, **22**, 209.
156. D. J. Lapinsky, *Bioorg. Med. Chem.*, 2012, **20**, 6237.
157. M. Hashimoto and Y. Hatanaka, *Eur. J. Org. Chem.*, 2008, **15**, 2513.
158. I. R. Kleckner and M. P. Foster, *Biochim. Biophys. Acta, Proteins Proteomics*, 2011, **1814**, 942.
159. B. Deng, C. Lento and D. J. Wilson, *Anal. Chim. Acta*, 2016.
160. H. Wei, J. Mo, L. Tao, R. J. Russell, A. A. Tymiak, G. Chen, R. E. Iacob and J. R. Engen, *Drug Discovery Today*, 2014, **19**, 95.
161. G. F. Pirrone, R. E. Iacob and J. R. Engen, *Anal. Chem.*, 2015, **87**, 99.
162. R. R. Abzalimov, C. E. Bobst and I. A. Kaltashov, *Anal. Chem.*, 2013, **85**, 9173.

163. R. R. Landgraf, M. J. Chalmers and P. R. Griffin, *J. Am. Soc. Mass Spectrom.*, 2012, **23**, 301.
164. Z.-Y. Kan, B. T. Walters, L. Mayne and S. W. Englander, *Proc. Natl. Acad. Sci. U. S. A.*, 2013, **110**, 16438.
165. R. R. Abzalimov and I. A. Kaltashov, *J. Am. Soc. Mass Spectrom.*, 2006, **17**, 1543.
166. R. Y.-C. Huang and G. Chen, *Anal. Bioanal. Chem.*, 2014, **406**, 6541.
167. A. J. Percy, M. Rey, K. M. Burns and D. C. Schriemer, *Anal. Chim. Acta*, 2012, **721**, 7.
168. M. M. Krishna, L. Hoang, Y. Lin and S. W. Englander, *Methods*, 2004, **34**, 51.
169. X. E. Shi, T. E. Wales, C. Elkin, N. Kawahata, J. R. Engen and D. A. Annis, *Anal. Chem.*, 2013, **85**, 11185.
170. A. Khanal, Y. Pan, L. S. Brown and L. Konermann, *J. Mass Spectrom.*, 2012, **47**, 1620.
171. O. Vadas and J. E. Burke, *Biochem. Soc. Trans.*, 2015, **43**, 773.
172. L. Y. Pan, O. Salas-Solano and J. F. Valliere-Douglass, *Anal. Chem.*, 2014, **86**, 2657.
173. W. Hosia, N. Bark, E. Liepinsh, A. Tjernberg, B. Persson, D. Hallén, J. Thyberg, J. Johansson and L. Tjernberg, *Biochemistry*, 2004, **43**, 4655.
174. Y. Zhang, D. L. Rempel, J. Zhang, A. K. Sharma, L. M. Mirica and M. L. Gross, *Proc. Natl. Acad. Sci. U. S. A.*, 2013, **110**, 14604.
175. R. R. Abzalimov, D. A. Kaplan, M. L. Easterling and I. A. Kaltashov, *J. Am. Soc. Mass Spectrom.*, 2009, **20**, 1514.
176. J. Pan, J. Han, C. H. Borchers and L. Konermann, *Anal. Chem.*, 2010, **82**, 8591.
177. D. Resetca and D. J. Wilson, *FEBS J.*, 2013, **280**, 5616.
178. T. Rob, P. K. Gill, D. Golemi-Kotra and D. J. Wilson, *Lab Chip*, 2013, **13**, 2528.
179. S. M. Schermann, D. A. Simmons and L. Konermann, *Expert Rev. Proteomics*, 2005, **2**, 475.
180. M. J. Chalmers, S. A. Busby, B. D. Pascal, Y. He, C. L. Hendrickson, A. G. Marshall and P. R. Griffin, *Anal. Chem.*, 2006, **78**, 1005.
181. B. D. Pascal, S. Willis, J. L. Lauer, R. R. Landgraf, G. M. West, D. Marciano, S. Novick, D. Goswami, M. J. Chalmers and P. R. Griffin, *J. Am. Soc. Mass Spectrom.*, 2012, **23**, 1512.
182. D. J. Cummins, A. Espada, S. J. Novick, M. Molina-Martin, R. E. Stites, J. F. Espinosa, H. Broughton, D. Goswami, B. D. Pascal, J. A. Dodge, M. J. Chalmers and P. R. Griffin, *Anal. Chem.*, 2016, **88**, 6607.
183. R. E. Iacob and J. R. Engen, *J. Am. Soc. Mass Spectrom.*, 2012, **23**, 1003.
184. J. Ahn, M. J. Cao, Y. Q. Yu and J. R. Engen, *Biochim. Biophys. Acta*, 2013, **1834**, 1222.
185. E. S. Gallagher and J. W. Hudgens, in *Methods Enzymol.*, ed. K. Zvi, Academic Press, 2016, vol. 566, pp. 357.
186. K. J. Pacholarz, R. A. Garlish, R. J. Taylor and P. E. Barran, *Chem. Soc. Rev.*, 2012, **41**, 4335.

187. D. P. Marciano, V. Dharmarajan and P. R. Griffin, *Curr. Opin. Struct. Biol.*, 2014, **28**, 105.
188. L. Konermann, A. D. Rodriguez and M. A. Sowole, *Analyst*, 2014, **139**, 6078.
189. J. Zhang, F. J. Adrián, W. Jahnke, S. W. Cowan-Jacob, A. G. Li, R. E. Iacob, T. Sim, J. Powers, C. Dierks and F. Sun, *Nature*, 2010, **463**, 501.
190. J. E. Burke and R. L. Williams, *Adv. Biol. Regul.*, 2013, **53**, 97.
191. K. Lorenzen and T. Pawson, *Biochem. Soc. Trans.*, 2014, **42**, 145.
192. C. S. Fabilane, P. N. Nguyen, R. V. Hernandez, S. Nirudodhi, M. Duong, C. S. Maier and V. Narayanaswami, *Protein Pept. Lett.*, 2016, **23**, 404.
193. M. J. Chalmers, Y. Wang, S. Novick, M. Sato, H. U. Bryant, C. Montrose-Rafizdeh, P. R. Griffin and J. A. Dodge, *ACS Med. Chem. Lett.*, 2012, **3**, 207.
194. T. S. Hughes, P. K. Giri, I. M. S. de Vera, D. P. Marciano, D. S. Kuruvilla, Y. Shin, A.-L. Blayo, T. M. Kamenecka, T. P. Burris and P. R. Griffin, *Nat. Commun.*, 2014, **5**, 3571.
195. L. J. Boerma, G. Xia, C. Qui, B. D. Cox, M. J. Chalmers, C. D. Smith, S. Lobo-Ruppert, P. R. Griffin, D. D. Muccio and M. B. Renfrow, *J. Biol. Chem.*, 2014, **289**, 814.
196. A. Dailing, A. Luchini and L. Liotta, *Expert Rev. Proteomics*, 2015, **12**, 457.
197. D. Pandit, S. J. Tuske, S. J. Coales, A. Liu, J. E. Lee, J. A. Morrow, J. F. Nemeth and Y. Hamuro, *J. Mol. Recognit.*, 2012, **25**, 114.
198. N. Clementi, N. Mancini, M. Castelli, M. Clementi and R. Burioni, *Drug Discovery Today*, 2013, **18**, 464.
199. A. M. Sevy, J. F. Healey, W. Deng, P. C. Spiegel, S. L. Meeks and R. Li, *J. Thromb. Haemostasis*, 2013, **11**, 2128.
200. A. Espada, H. Broughton, S. Jones, M. J. Chalmers and J. A. Dodge, *J. Med. Chem.*, 2016, **59**, 2255.
201. J. A. Loo, *Mass Spectrom. Rev.*, 1997, **16**, 1.
202. G. R. Hilton and J. L. Benesch, *J. R. Soc., Interface*, 2012, **9**, 801.
203. L. Pedro and R. J. Quinn, *Molecules*, 2016, **21**, 984.
204. S.-A. Poulsen, *Aust. J. Chem.*, 2014, **66**, 1495.
205. E. E. Swayze, E. A. Jefferson, K. A. Sannes-Lowery, L. B. Blyn, L. M. Risen, S. Arakawa, S. A. Osgood, S. A. Hofstadler and R. H. Griffey, *J. Med. Chem.*, 2002, **45**, 3816.
206. V. Vivat Hannah, C. Atmanene, D. Zeyer, A. Van Dorsselaer and S. Sanglier-Cianférani, *Future Med. Chem.*, 2010, **2**, 35.
207. H. J. Maple, R. A. Garlish, L. Rigau-Roca, J. Porter, I. Whitcombe, C. E. Prosser, J. Kennedy, A. J. Henry, R. J. Taylor and M. P. Crump, *J. Med. Chem.*, 2012, **55**, 837.
208. D. A. Annis, J. Athanasopoulos, P. J. Curran, J. S. Felsch, K. Kalghatgi, W. H. Lee, H. M. Nash, J.-P. A. Orminati, K. E. Rosner, G. W. Shipps, G. R. A. Thaddupathy, A. N. Tyler, L. Vilenchik, C. R. Wagner and E. A. Wintner, *Int. J. Mass Spectrom.*, 2004, **238**, 77.

209. C. E. Whitehurst, N. Nazef, D. A. Annis, Y. Hou, D. M. Murphy, P. Spacciapoli, Z. Yao, M. R. Ziebell, C. C. Cheng, G. W. Shipps, Jr., J. S. Felsch, D. Lau and H. M. Nash, *J. Biomol. Screening*, 2006, **11**, 194.
210. C. E. Whitehurst, Z. Yao, D. Murphy, M. Zhang, S. Taremi, L. Wojcik, J. M. Strizki, J. D. Bracken, C. C. Cheng and X. Yang, *Comb. Chem. High Throughput Screening*, 2012, **15**, 473.
211. X. Huang, C. C. Cheng, T. O. Fischmann, J. S. Duca, X. Yang, M. Richards and G. W. Shipps Jr., *ACS Med. Chem. Lett.*, 2012, **3**, 123.
212. V. D. Kutilek, C. L. Andrews, M. P. Richards, Z. Xu, T. Sun, Y. Chen, A. Hashke, N. Smotrov, R. Fernandez and E. B. Nickbarg, *J. Biomol. Screening*, 2016, **21**, 608.
213. Y. Deng, G. W. Shipps, Jr., A. Cooper, J. M. English, D. A. Annis, D. Carr, Y. Nan, T. Wang, H. Y. Zhu, C. C. Chuang, P. Dayananth, A. W. Hruza, L. Xiao, W. Jin, P. Kirschmeier, W. T. Windsor and A. A. Samatar, *J. Med. Chem.*, 2014, **57**, 8817.
214. D. A. Annis, C. C. Cheng, C.-C. Chuang, J. D. McCarter, H. M. Nash, N. Nazef, T. Rowe, R. J. Kurzeja, J. Shipps and W. Gerald, *Comb. Chem. High Throughput Screening*, 2009, **12**, 760.
215. T. N. O'Connell, J. Ramsay, S. F. Rieth, M. J. Shapiro and J. G. Stroh, *Anal. Chem.*, 2014, **86**, 7413.
216. T. Arai, M. Uehata, H. Akatsuka and T. Kamiyama, *Anal. Biochem.*, 2013, **436**, 109.
217. L. Z. Vilenchik, P. R. Sheth, C. C. Chuang and H. V. Le, *Anal. Biochem.*, 2011, **418**, 10.
218. G. C. Adam, J. Meng, J. Athanasopoulos, X. Zhang and K. T. Chapman, *Bioorg. Med. Chem. Lett.*, 2007, **17**, 2404.
219. J. Flarakos, K. L. Morand and P. Vouros, *Anal. Chem.*, 2005, **77**, 1345.
220. I. Muckenschnabel, R. Falchetto, L. M. Mayr and I. Filipuzzi, *Anal. Biochem.*, 2004, **324**, 241.
221. N. Brown, H. Zehender, K. Azzaoui, A. Schuffenhauer, L. M. Mayr and E. Jacoby, *J. Biomol. Screening*, 2006, **11**, 123.
222. M. A. Ashwell, J. M. Lapierre, C. Brassard, K. Bresciano, C. Bull, S. Cornell-Kennon, S. Eathiraj, D. S. France, T. Hall, J. Hill, E. Kelleher, S. Khanapurkar, D. Kizer, S. Koerner, J. Link, Y. Liu, S. Makhija, M. Moussa, N. Namdev, K. Nguyen, R. Nicewonger, R. Palma, J. Szwaya, M. Tandon, U. Uppalapati, D. Vensel, L. P. Volak, E. Volckova, N. Westlund, H. Wu, R. Y. Yang and T. C. Chan, *J. Med. Chem.*, 2012, **55**, 5291.
223. M. M. Siegel, K. Tabei, G. A. Bebernitz and E. Z. Baum, *J. Mass Spectrom.*, 1998, **33**, 264.
224. K. M. Comess, M. E. Schurdak, M. J. Voorbach, M. Coen, J. D. Trumbull, H. Yang, L. Gao, H. Tang, X. Cheng, C. G. Lerner, J. O. McCall, D. J. Burns and B. A. Beutel, *J. Biomol. Screening*, 2006, **11**, 743.
225. K. M. Comess, J. D. Trumbull, C. Park, Z. Chen, R. A. Judge, M. J. Voorbach, M. Coen, L. Gao, H. Tang, P. Kovar, X. Cheng,

M. E. Schurdak, H. Zhang, T. Sowin and D. J. Burns, *J. Biomol. Screening*, 2006, **11**, 755.

226. K. M. Comess, C. Sun, C. Abad-Zapatero, E. R. Goedken, R. J. Gum, D. W. Borhani, M. Argiriadi, D. R. Groebe, Y. Jia, J. E. Clampit, D. L. Haasch, H. T. Smith, S. Wang, D. Song, M. L. Coen, T. E. Cloutier, H. Tang, X. Cheng, C. Quinn, B. Liu, Z. Xin, G. Liu, E. H. Fry, V. Stoll, T. I. Ng, D. Banach, D. Marcotte, D. J. Burns, D. J. Calderwood and P. J. Hajduk, *ACS Chem. Biol.*, 2011, **6**, 234.
227. V. Mulabagal and A. I. Calderón, *Anal. Chem.*, 2010, **82**, 3616.
228. H. Li, F. Song, J. Xing, R. Tsao, Z. Liu and S. Liu, *J. Am. Soc. Mass Spectrom.*, 2009, **20**, 1496.
229. H. Cao, R. Yu, Y. Choi, Z.-Z. Ma, H. Zhang, W. Xiang, D. Y.-W. Lee, B. M. Berman, K. D. Moudgil and H. H. Fong, *Pharmacol. Res.*, 2010, **61**, 519.
230. Y. Choi, K. Jermihov, S.-J. Nam, M. Sturdy, K. Maloney, X. Qiu, L. R. Chadwick, M. Main, S.-N. Chen and A. D. Mesecar, *Anal. Chem.*, 2010, **83**, 1048.
231. X. Fu, Z. Wang, L. Li, S. Dong, Z. Li, Z. Jiang, Y. Wang and W. Shui, *Sci. Rep.*, 2016, **6**, 29680.
232. Y. Choi, K. Jermihov, S. J. Nam, M. Sturdy, K. Maloney, X. Qiu, L. R. Chadwick, M. Main, S. N. Chen, A. D. Mesecar, N. R. Farnsworth, G. F. Pauli, W. Fenical, J. M. Pezzuto and R. B. van Breemen, *Anal. Chem.*, 2011, **83**, 1048.
233. X. Chen, S. Qin, S. Chen, J. Li, L. Li, Z. Wang, Q. Wang, J. Lin, C. Yang and W. Shui, *Sci. Rep.*, 2015, **5**, 8361.
234. S. Qin, Y. Ren, X. Fu, J. Shen, X. Chen, Q. Wang, X. Bi, W. Liu, L. Li, G. Liang, C. Yang and W. Shui, *Anal. Chim. Acta*, 2015, **886**, 98.
235. J. J. Slon-Usakiewicz, W. Ng, J.-R. Dai, A. Pasternak and P. R. Redden, *Drug Discovery Today*, 2005, **10**, 409.
236. Z. Tong, K. S. Joseph and D. S. Hage, *J. Chromatogr. A*, 2011, **1218**, 8915.
237. E. Meiby, H. Simmonite, L. le Strat, B. Davis, N. Matassova, J. D. Moore, M. Mrosek, J. Murray, R. E. Hubbard and S. Ohlson, *Anal. Chem.*, 2013, **85**, 6756.
238. D. C. Schriemer, D. R. Bundle, L. Li and O. Hindsgaul, *Angew. Chem., Int. Ed.*, 1998, **37**, 3383.
239. L. Zhu, L. Chen, H. Luo and X. Xu, *Anal. Chem.*, 2003, **75**, 6388.
240. J. Wang, B. Zhang, J. Fang, K. Sujino, H. Li, A. Otter, O. Hindsgaul, M. M. Palcic and P. G. Wang, *J. Carbohydr. Chem.*, 2003, **22**, 347.
241. B. Zhang, M. M. Palcic, H. Mo, I. J. Goldstein and O. Hindsgaul, *Glycobiology*, 2001, **11**, 141.
242. J. J. Slon-Usakiewicz, W. Ng, J. E. Foster, J.-R. Dai, E. Deretey, L. Toledo-Sherman, P. R. Redden, A. Pasternak and N. Reid, *J. Med. Chem.*, 2004, **47**, 5094.
243. C. Temporini, S. Ceruti, E. Calleri, S. Ferrario, R. Moaddel, M. P. Abbracchio and G. Massolini, *Anal. Biochem.*, 2009, **384**, 123.

244. E. Calleri, S. Ceruti, G. Cristalli, C. Martini, C. Temporini, C. Parravicini, R. Volpini, S. Daniele, G. Caccialanza, D. Lecca, C. Lambertucci, M. L. Trincavelli, G. Marucci, I. W. Wainer, G. Ranghino, P. Fantucci, M. P. Abbracchio and G. Massolini, *J. Med. Chem.*, 2010, **53**, 3489.
245. L. Toledo-Sherman, E. Deretey, J. J. Slon-Usakiewicz, W. Ng, J.-R. Dai, J. E. Foster, P. R. Redden, M. D. Uger, L. C. Liao and A. Pasternak, *J. Med. Chem.*, 2005, **48**, 3221.
246. W. Ng, J.-R. Dai, J. J. Slon-Usakiewicz, P. R. Redden, A. Pasternak and N. Reid, *J. Biomol. Screening*, 2007, **12**, 167.
247. E. Calleri, C. Temporini, G. Caccialanza and G. Massolini, *ChemMedChem*, 2009, **4**, 905.
248. M. D. Duong-Thi, E. Meiby, M. Bergstrom, T. Fex, R. Isaksson and S. Ohlson, *Anal. Biochem.*, 2011, **414**, 138.
249. E. Meiby, S. Knapp, J. M. Elkins and S. Ohlson, *Anal. Bioanal. Chem.*, 2012, **404**, 2417.
250. R. Marti-Arbona, M. Teshima, P. S. Anderson, K. L. Nowak-Lovato, E. Hong-Geller, C. J. Unkefer and P. J. Unkefer, *J. Mol. Microbiol. Biotechnol.*, 2012, **22**, 205.
251. X. Zhao, H. Lu, J. Huang, J. Zheng, X. Zheng and Y. Zhang, *Chromatographia*, 2012, **75**, 411.
252. M. D. Duong-Thi, M. Bergstrom, T. Fex, R. Isaksson and S. Ohlson, *J. Biomol. Screening*, 2013, **18**, 160.
253. K. Vuignier, D. Guillarme, J. L. Veuthey, P. A. Carrupt and J. Schappler, *J. Pharm. Biomed. Anal.*, 2013, **74**, 205.
254. M.-D. Duong-Thi, G. Bergström, C.-F. Mandenius, M. Bergström, T. Fex and S. Ohlson, *Anal. Biochem.*, 2014, **461**, 57.
255. M.-D. Duong-Thi, M. Bergstrom, K. Edwards, J. Eriksson, S. Ohlson, J. ToYiu Ying, J. Torres and V. Agmo Hernandez, *Analyst*, 2016, **141**, 981.
256. I. Sinclair, R. Stearns, S. Pringle, J. Wingfield, S. Datwani, E. Hall, L. Ghislain, L. Majlof and M. Bachman, *J. Lab. Autom.*, 2016, **21**, 19.
257. T. P. Forbes, *Rapid Commun. Mass Spectrom.*, 2015, **29**, 19.
258. D. A. Thomas, L. Wang, B. Goh, E. S. Kim and J. Beauchamp, *Anal. Chem.*, 2015, **87**, 3336.
259. Y. Huang, S. R. Heron, A. M. Clark, J. S. Edgar, S. H. Yoon, D. Kilgour, F. Turecek, A. Aliseda and D. R. Goodlett, *J. Mass Spectrom.*, 2016, **51**, 424.
260. X. Feng, B. F. Liu, J. Li and X. Liu, *Mass Spectrom. Rev.*, 2015, **34**, 535.
261. H. Vu, R. J Quinn and N.-T. Nguyen, *Micro Nanosyst.*, 2015, **7**, 74.
262. A. Oedit, P. Vulto, R. Ramautar, P. W. Lindenburg and T. Hankemeier, *Curr. Opin. Biotechnol.*, 2015, **31**, 79.
263. J. C. May, C. R. Goodwin and J. A. McLean, *Curr. Opin. Biotechnol.*, 2015, **31**, 117.
264. G. G. Mironov, C. M. Clouthier, A. Akbar, J. W. Keillor and M. V. Berezovski, *Nat. Chem. Biol.*, 2016, **12**, 918.

265. J. D. Barr, L. Shi, D. H. Russell, D. E. Clemmer and A. E. Holliday, *Anal. Chem.*, 2016, **88**, 10933.
266. J. Singh, R. C. Petter, T. A. Baillie and A. Whitty, *Nat. Rev. Drug Discovery*, 2011, **10**, 307.
267. L. Bonanno, A. Jirillo and A. Favaretto, *Curr. Drug Targets*, 2011, **12**, 922.
268. D. C. Swinney, *Curr. Opin. Drug Discovery Dev.*, 2009, **12**, 31.
269. C. H. Yun, K. E. Mengwasser, A. V. Toms, M. S. Woo, H. Greulich, K. K. Wong, M. Meyerson and M. J. Eck, *Proc. Natl. Acad. Sci. U. S. A.*, 2008, **105**, 2070.
270. I. D. Campuzano, T. San Miguel, T. Rowe, D. Onea, V. J. Cee, T. Arvedson and J. D. McCarter, *J. Biomol. Screening*, 2016, **21**, 136.
271. D. Martinez Molina and P. Nordlund, *Annu. Rev. Pharmacol. Toxicol.*, 2016, **56**, 141.
272. A. J. Jensen, D. M. Molina and T. Lundbäck, *Future Med. Chem.*, 2015, **7**, 975.
273. M. M. Savitski, F. B. M. Reinhard, H. Franken, T. Werner, M. F. Savitski, D. Eberhard, D. M. Molina, R. Jafari, R. B. Dovega, S. Klaeger, B. Kuster, P. Nordlund, M. Bantscheff and G. Drewes, *Science*, 2014, **346**, 1255784.

CHAPTER 6

Characterization of Pharmaceutical Solids Combining NMR, X-ray diffraction and Computer Modelling

M. SARDO,[a] I. C. B. MARTINS,[a,b] C. VLADISKOVIC,[c,d] M. TERESA DUARTE,[b] J. ROCHA,[a] N. MASCIOCCHI[c] AND L. MAFRA*[a]

[a] University of Aveiro, CICECO – Aveiro Institute of Materials, Department of Chemistry, 3810-193 Aveiro, Portugal; [b] University of Lisbon, CQE – Centro de Química Estrutural, Instituto Superior Técnico, 1049-001 Lisboa, Portugal; [c] University of Insubria, Department of Science and High Technology & To. Sca. Lab, Via Valleggio, 11, 22100 Como, Italy; [d] Dipharma Francis Srl, Via Bissone 5, 20021 Baranzate (MI), Italy
*Email: lmafra@ua.pt

6.1 General Introduction

Solid pharmaceuticals exhibit a wide range of specific mechanical, thermodynamic and pharmacokinetic properties that impact their therapeutic performance. Unveiling the structural and molecular dynamic features of these solids over a range of length- and time scales is strategically important for rationalizing, engineering and improving these properties, either in bulk or formulated drugs. Traditional solid-state characterization techniques encompass vibrational spectroscopy, thermal analysis and diffraction. Although these techniques possess many advantages, none of them can be easily applied to characterize the multiple components found in a formulated drug. *Solid-state nuclear magnetic resonance* (SSNMR) spectroscopy is a well-established method for studying inter- and intramolecular interactions based on the magnetic and electric properties of the molecule with extreme sensitivity and nuclear specificity, and over recent years it has successfully been used in tandem with *single-crystal* (SCXRD) and/or *X-ray powder diffraction* (XRPD) to obtain unambiguous structural information. SSNMR spectroscopy has been very effective in characterizing amorphous dispersions, disorder and localizing light atoms (*e.g.*, ^{1}H, $^{6/7}Li$), and in helping to discriminate among isoelectronic species, which often aids in XRPD structure determination. *Crystal structure prediction* (CSP) algorithms may also be combined with diffraction and NMR crystallography methods. This chapter aims to provide an overview of selected applications where SSNMR can be used alone or combined with diffraction and computational methods. For this purpose, the analytical methods (SSNMR and diffraction) are described (Section 6.2), followed by a selection of case studies (Section 6.3).

6.2 Methods

6.2.1 SSNMR Spectroscopy in Pharmaceutical Research

SSNMR spectroscopy is a well-established technique for studying interactions based on the magnetic and electric properties of molecules, and is currently an important tool in the pharmaceutical industry, largely owing to its unique capabilities in distinguishing between polymorphic forms, anhydrates, hydrates, and other solvate forms of *active pharmaceutical ingredients* (APIs). Although SSNMR spectroscopy is a technique requiring strong user expertise and involving high maintenance costs, it has several strengths over other solid-state techniques, some of which are worth mentioning. (i) SSNMR is able to probe the structure of amorphous as well as crystalline systems; (ii) it is able to selectively "observe" a target "nuclear probe" within a mixture of other nuclides contained in the same or distinct phases; making the technique independent of matrix effects once the spectrometer is tuned to a specific target nucleus. This feature is particularly advantageous in the study of complex solid mixtures without further sample preparation; for instance, when studying an API or an excipient in a commercial formulation. (iii) SSNMR is not particle size-dependent, which is an issue in diffraction techniques. These unique capabilities allow us to use SSNMR spectroscopy in a number of topics of interest for the pharma community, as highlighted below:

- Qualitative and quantitative analysis of drug polymorphs and/or impurities;

- Acquisition of the number of molecules in the crystallographic asymmetric unit by direct counting of the resonances (provides space group restraints);
- Identification of packing interactions and supramolecular synthons (*e.g.*, *hydrogen bonds* (H-bonds), $\pi\cdots\pi$ and CH$\cdots\pi$ contacts), and extraction of interatomic distances, torsion angles and coordination numbers;
- Identification of drug$\cdots$host interactions in drug delivery materials;
- Probe molecular dynamics through line shape analysis and relaxation studies.

Various reviews are available in the literature, describing and highlighting the relevance of SSNMR methods in pharmaceutical research.[1–4] However, not many of them focus on the complementarity between NMR and diffraction techniques.

6.2.1.1 SSNMR Basics and Main Techniques

SSNMR produces much broader peaks compared to liquid-state NMR, owing to strong anisotropic interactions, such as dipolar and quadrupolar couplings, *chemical shift anisotropy* (CSA) and quadrupolar coupling, that are eliminated or reduced by the fast random molecular motions in liquid samples.

The NMR technique measures spectral information resulting from the total interactions perturbing the spin system. These interactions can be external or internal.[5] The former includes the interaction of the nuclear spin system with (i) the strong externally applied static magnetic field (B_0), designated the Zeeman interaction and (ii) *radio frequency* (RF) irradiation pulses.

The internal interactions (*e.g., chemical shift* (CS), through-space and through-bond dipole–dipole couplings, quadrupolar coupling, paramagnetic coupling and knight shifts) encompass the sources of fluctuating magnetic and electric fields internally generated in the sample by the nuclear spin, which strongly depend on the local chemical environment. Such interactions can be interpreted as perturbations of the Zeeman interactions and may cause extensive line broadening in the solid-state due to the effect of their anisotropic contributions. Atomic-scale information such as nuclear proximities, torsion angles, coordination numbers, molecular packings, and so on may also be obtained *via* these local interactions.[6]

Several line-narrowing strategies are available to remove the anisotropic broadening. *Magic angle spinning* (MAS) was the first line-narrowing experiment in solids.[7] MAS induces a time-modulation caused by sample mechanical rotation (around an axis tilted by 54.7°), which averages out the first-order broadening effects caused by the anisotropic interactions. For quadrupolar nuclei, MAS alone is not able to eliminate the second-order quadrupolar anisotropic broadening. In this case, other approaches have been developed to obtain isotropic lineshapes.[8–10]

Beside the time modulation induced by MAS, RF pulses are a second source of time modulation generated by employing special RF pulse sequence blocks, which may either cause line narrowing, also known as decoupling, or help in reintroducing anisotropies under MAS (also known as "recoupling" sequences), through interference between both the RF and MAS modulation frequencies. Over the last three decades, an extraordinary number of RF pulse sequences combined with MAS have emerged to produce line narrowing,[11,12] sensitivity enhancement (*cross-polarization* (CP), *Carr–Purcell–Meiboom–Gill* (CPMG)) and recoupling of dipolar or CSA interactions.[13–24]

1D NMR Techniques

The isotropic CS of nuclei in different chemical environments can be measured through one-dimensional (1D) high-resolution spectroscopy. For the case of diluted spins (low natural abundance), 1D spectra are obtained either by CP or single-pulse direct excitation, combined with MAS and heteronuclear decoupling. In the field of pharmaceutics, the ^{13}C CPMAS experiment is one of the most frequent experiments. The sensitivity gain obtained by CP experiments is due to the transfer of polarization from ^{1}H (abundant) to dilute spins (*e.g.*, ^{13}C, ^{15}N, *etc.*), involving the application of two simultaneous RF pulses fulfilling the Hartmann–Hahn matching condition during the contact time.[25] Heteronuclear decoupling is applied during signal acquisition to remove the dipolar interaction with protons, with multi-pulse sequences, such as TPPM[26] or SPINAL-64[27] or others.[28,29]

Variations of these 1D methods may be explored when spectral editing is necessary to facilitate assignment of resonances. For instance, the non-quaternary suppression[30–32] sequence is particularly effective in selectively polarizing quaternary carbons and CPPI-based sequences help to distinguish CH, CH_2, CH_3 and quaternary carbons in ^{13}C spectra.[33,34] On the other hand, spectral editing techniques making use of heteronuclear scalar couplings yield unambiguous identification of carbon types in natural abundance samples[35] and the $^{1}J_{^{15}N-^{1}H}$ spectral editing method can readily differentiate the number of protons directly bonded to a nitrogen site and is not susceptible to motional effects.[36]

Two methods have been presented recently to overcome the inherently non-quantitative nature of CPMAS: a combination of CP with a broadband homonuclear recouping technique (DARR),[37,38] and a simple method based on multiple CP blocks for transferring magnetization, proposed by Schmidt-Rohr.[39]

Unlike in liquids, the ^{1}H MAS NMR spectra of solids suffer from strong broadening due to strong ^{1}H–^{1}H dipolar couplings (typically 50–100 kHz). Over recent decades, several line narrowing techniques have made ^{1}H NMR amenable to the characterization of solid pharmaceuticals. The high natural abundance and occurrence of ^{1}H in organic solids is a fundamental feature, leading to several resolution enhancement approaches that have been developed for ^{1}H homonuclear decoupling, either by *combining rotation and multiple-pulse sequences* (CRAMPS)[40] (*e.g.*, FSLG, PMLG[41,42] and DUMBO[43,44]), or by applying ultra-fast spinning speed rates (up to 120 kHz using 0.75 mm rotors). Recently, very fast MAS in combination with proton detection-based methods have also been explored to overcome sensitivity/resolution limitations in small molecule pharmaceuticals.[45] These developments have made ^{1}H SSNMR very attractive for the direct observation of intermolecular interactions such as strong and weak H-bonding as well as other interactions, *e.g.*, π–π and CH···π contacts, which usually determine the conformation and functionality of molecules in crystal packing.

Multidimensional NMR Techniques

Two-dimensional (2D) SSNMR experiments provide additional structural information, directly probing intra- or intermolecular interactions through correlations between specific nuclear sites. Spectral assignment, internuclear distances and information on intermolecular interactions may be retrieved through different homo- and *heteronuclear correlation* (HETCOR) 2D SSNMR experiments.

Dipolar-based 2D ^{1}H–X {X = ^{13}C, ^{15}N, ^{19}F} CP HETCOR experiments[46–49] are among the most commonly used techniques to analyse pharmaceuticals as they allow the study of the spatial proximities of unlike spin pairs, regardless of their chemical connectivity.

Two major variants of the HETCOR experiment exist: (i) dipolar-based 2D HETCOR using CP or LG-CP HETCOR or *D*-HMQC[50] for magnetization transfer by means of through-space dipolar couplings; (ii) J-coupling-based 2D HETCOR NMR experiments using through-bond connectivities; successful examples of such techniques are MAS-J-HMQC[51] and REPT-HSQC.[52] In both cases, CRAMPS can be added to one or both evolution times of the 2D experiment, further enhancing the ^{1}H resolution.

Double Quantum (DQ) spectroscopy is another widely used method in detecting ^{1}H–^{1}H proximities.[53] The most commonly used pulse sequences are symmetry-based DQ recoupling methods and the through-bond INADEQUATE[54,55] sequence. The most commonly used 1D and 2D SSNMR techniques applied to pharmaceutical systems are summarized in Table 6.1.

The major drawback of SSNMR spectroscopy is its intrinsic low sensitivity. *Dynamic nuclear polarization* (DNP) NMR spectroscopy is a technique that can increase the sensitivity of SSNMR by up to *ca.* 2 orders of magnitude at high fields.[56,57] In a DNP experiment, microwave irradiation is used, under strong magnetic fields and at temperatures typically below 4 K, to drive the transfer of polarization from unpaired electrons to nuclei.[56] DNP has recently been shown to be efficient in polarizing powdered organic solids.[58]

This technique opens up exciting possibilities, particularly for the study of ^{17}O-containing compounds at natural isotopic abundance[59] and ^{13}C and ^{15}N studies on diluted pharmaceuticals encapsulated in *drug delivery systems* (DDSs).

6.2.1.2 Molecular Dynamics

Molecular motions are intimately connected with fundamental properties, such as the stability of the solid phases, solid–solid phase transitions,

Table 6.1 Common SSNMR experiments (adapted from ref. 2 and 149).

Pulse sequence	Description
1D techniques	
^{1}H MAS	Fast structural fingerprint of the compound.
CPMAS Cross polarization/magic-angle spinning	High-resolution NMR spectra from dilute, spin-1/2 nuclei.
DPMAS Direct polarization	Used to detect liquid species. Consists of a single pulse and acquisition with ^{1}H decoupling.
SPMAS Single pulse/MAS	High-resolution NMR spectra from abundant spin systems.
Dipolar dephasing, CPMAS	Detection of non-protonated resonances.
CPPI Cross polarization combined with polarization inversion	Distinguish different types of carbon resonances.
TOSS Total suppression of spinning sidebands	A sideband suppression experiment used to simplify the spectra.
Relaxation methods	
T_1 measurement	Determines the spin lattice relaxation time.
$T_{1\rho}$ measurement	Measures the spin lattice time in the rotating frame.
2D techniques	
PASS, FIREMAT 2D phase adjustments of the spinning sidebands, Five π replicated magic angle turning	Correlates anisotropic and isotropic CS, which provides information on the molecular conformation.
^{1}H–^{13}C HETCOR	Study the spatial proximity of rare-abundant pairs of nuclei, independently of their chemical connectivity.
^{1}H–^{19}F HETCOR	Detects dipolar connectivity between ^{1}H and ^{19}F. LGCP variant minimizes spin diffusion.
^{1}H–^{1}H DQ-CRAMPS	Dipolar correlation between H-bonding protons and neighbouring protons.

intra- and intermolecular interactions and chemical reactivity. SSNMR spectroscopy is one of the most powerful techniques for the study of molecular motions providing unrivalled information on the degree of mobility, geometry of motion, correlation times and activation energies. This wealth of information arises from the possibility of measuring different nuclear properties, which are sensitive to motions occurring over a wide range of frequencies, and from exploiting different nuclei.[60] Relaxation times (spin–lattice relaxation – T_1, spin–spin relaxation – T_2, and spin–lattice relaxation in the rotating frame – $T_{1\rho}$), CSA, residual dipolar couplings, and the lineshape of motionally-averaged dipolar and quadrupolar patterns may all be explored in dynamic studies. Lineshape analysis of ^{2}H NMR signals from 1D static spin–echo experiments has traditionally been the most popular approach to probe local order, structure, and dynamics.[61] Dynamics may also have a deep impact on well-established SSNMR techniques used to recouple spin pairs. Molecular motions may, for example, affect the CP dynamics in various ways[62] and internuclear distances may be estimated from this perturbation.

Proton relaxation times are often averaged in solids by the process of spin diffusion, which leads to a loss of local dynamics information. Spin diffusion may, however, be used to extract other parameters, mainly structural phase dimensions in heterogeneous materials.[63] In pharmaceutical systems, T_1 of the water protons may be used as an indicator of water mobility.[64] Many examples employing NMR to study dynamics in small molecules, with a special focus on pharmaceutical molecules, have been reported in the literature.[1]

A detailed characterization of the internal rotations and interconformational jumps of ibuprofen in the solid-state was performed by Carignani *et al.*[60] through the analysis of different NMR observables (^{13}C CS, ^{13}C lineshape, ^{1}H and ^{13}C spin-lattice relaxation times at different temperatures).

The effects of pharmaceutical processing (namely, compaction, lyophillization, spray drying, cryogrinding) have been assessed through ^{13}C CPMAS and ^{1}H T_1 experiments for the case of lactose.[65] The authors found that pressing a tablet from the bulk crystalline form reduced the ^{1}H T_1 threefold, indicating that tablet compression could have a significant negative impact on formulation stability.[65]

On a different topic, Facey and co-workers[66] showed how SSNMR (namely dipolar-dephasing ^{13}C CPMAS and ^{2}H static experiments) can provide complementary information in the full characterization of a clathrate, indicating that the guest molecule exhibits considerable motional freedom in the confined space of the cage.

Relaxation measurements are often relevant when performing dynamic studies of water molecules in pharmaceutical hydrates.[64,67] Several SSNMR techniques (^{1}H–^{13}C HETCOR, ^{2}H MAS, variable-temperature ^{13}C/^{1}H) coupled with first-principles computation have provided a detailed picture of the interaction of water molecules with the host sildenafil citrate structure in a non-stoichiometric hydrate.[67] The water molecules were found to be undergoing C_2 flips and exchanging between water sites (information obtained from the ^{13}C spectrum), even at the lowest temperatures studied (−58 °C).

During pharmaceutical production, often, desolvation of solvate forms is necessary and almost inevitably produces fine powders, turning their characterization into a challenge. In a recent report on the solvates and desolvates of formoterol fumarate, Apperley *et al.*[68] used ^{13}C magnetisation transfer experiments to measure the dynamics of the phenylene group ring flips, while ^{13}C T_1 allowed methyl group rotation to be characterised. Remarkably important for structural characterisation, ^{13}C and ^{2}H relaxation times from a sample labelled in deuterium at the CH site in the fumarate ion were carefully analysed using exact calculations of spin–lattice relaxation times to obtain a motional model of the fumarate ions in the desolvate, even though the structure of the material was unknown.[68]

6.2.1.3 *Computational Approaches*

The structural characterization of pharmaceuticals plays a significant role in drug development. Even though reference structures from SCXRD studies are commonly available, when the sample is in powder form, characterization becomes difficult.

SSNMR spectroscopy is sensitive to the local structural environment and very weak molecular packing interactions, but interpreting the obtained high-resolution spectra is very often not straightforward, presenting a considerable challenge in organic systems with a variety of similar functional groups. In this regard, the use of *ab initio* calculations of NMR parameters plays a key role in a number of important tasks: (i) resonance assignment, (ii) *in silico* study of crystal packing interactions (*e.g.*, *nucleus-independent chemical shifts* – NICS),[69] (iii) crystal structure validation comparing experimental and theoretical isotropic CSs, quadrupolar parameters[66,70–73] and principal components of the CSA tensor[74,75] and (iv) structure refinement.[76–78]

Much progress has been made in the recognition of the power of a combined approach linking experimental SSNMR with first principle calculations of NMR parameters. For example, *the gauge-including projector augmented wave* (GIPAW)[79] using the plane wave pseudopotential formalism within *density functional theory* (DFT) has been widely applied for calculating the NMR parameters, energies and properties of periodic solid systems incorporating crystallographic repetition.[80]

6.2.2 Single-crystal X-ray Diffraction

Structure and function are intimately related, so an accurate knowledge of the molecular structure is a prerequisite for rational drug design and structure-based functional studies. SCXRD is the technique that ultimately provides unambiguous, accurate and reliable three dimensional structural parameters.[81]

The need for single crystals of good quality, obtained from a bulk crystallization of the compound, is a pre-requisite in the use of SCXRD. The information acquired from this technique includes unit cell dimensions, interatomic distances and angles, molecular conformations, packing and intermolecular interactions, such as H-bonds and π–π stacking. SCXRD is widely used in structure determination, from so-called small molecules to extremely larger proteins and viruses. The technique is of particular relevance in pharmaceutical research and development.

Pharmaceutical solids can be either crystalline or amorphous, crystalline being the preferred form due to physicochemical stability considerations. Also, APIs are known to crystallize in different arrangements, promoting polymorphism, and often crystallize as hydrates, solvates, salts or cocrystals. The information retrieved from SCXRD analysis is of particular importance, understanding supramolecular interactions and crystal packing provides important insights into physicochemical properties.

Nowadays, the development of more sophisticated facilities equipped with high energy X-ray sources is proving very useful, especially for collecting data on very small crystals. Many factors, including the use of recombinant technologies for sample production, the cryogenic cooling of crystals, the availability of more sensitive detectors, and progress in computer hard- and software, have paved the way for progress in SCXRD.[82]

More recently, a new generation of laser-driven plasma accelerators, which accelerate high-charge electron beams to high energy in short distances to produce directionally, spatially coherent, intrinsically ultrafast beams of hard X-rays, has been developed. These accelerators drastically reduce the size of the source, increasing its brightness by 1000 times, thus opening up spectral application to both the biological and chemical fields.[83,84]

6.2.2.1 *Structure Determination: The Phase Problem and Methods for Structure Solution*

The theory of X-ray diffraction is well known and presented in different text books.[85] Diffraction is based on Bragg's Law ($2d\sin\theta = n\lambda$), and transforms the crystal into a diffraction pattern, that is, the electron density to the structure factor, from the real space to the reciprocal space, using a Fourier Transform (FT).[86]

The main problem in solving a structure by SCXRD is deriving the phases φ_{hkl} (or directly the positions of the atoms) from the $|F(hkl)|$.

$$\rho(xyz) = \frac{1}{V}\sum_{hkl-\infty}^{+\infty} |F(hkl)| \cdot \mathrm{e}^{-2\pi i[hx+ky+lz-\varphi(hkl)]} \tag{6.1}$$

Electron density function

Throughout the years, different mathematical methods have been developed and used to solve the phase problem and determine the crystal structure: heavy atom methods, with a special emphasis on the Patterson method, were the first to be developed, when heavier atoms were present in the molecule; direct methods and, more recently, dual-space methods gained high importance in structure solution through the implementation of new algorithms that use both reciprocal and direct spaces.[85]

For macromolecules, methods such as multiple isomorphous replacement, multi-wavelength anomalous dispersion, and shake-and-bake (*ab initio* multi-trial method) are being used. Nowadays, due to the huge size of the *Protein Data Bank* (PDB), molecular replacement is the most popular.

From these techniques, both the Paterson, working predominately in the direct space, and direct methods, working predominately in the reciprocal space, have been widely used and so we will briefly introduce the most recent development in the area: the Charge Flipping method.

Dual space method: charge flipping

In the case of the dual-space method, none of the direct and reciprocal spaces is dominant in structure determination. It is the combination of the results obtained from the analysis in each space that makes it possible to solve the structure.[87,88]

The operations of dual space algorithms can be described as constraints, projections and interaction schemes. A constraint is an electron density property that is defined by amplitude (Fourier space) and positivity (direct space). In the amplitude constraint, the Fourier coefficients of the electron density need to be equal to the experimental ones. The positivity constraint requires that the density is positive everywhere. Regarding projection, this is an operation that transforms a random electron density into a density that satisfies one of the constraints, requiring the application of both domains (amplitude and positivity). A positivity projection results in a simple operation in which all pixels with negative values are set to zero. However, in the case of the amplitude projection, several steps are required:

(i) Calculation of the FT of the starting electron density;
(ii) Replacement of the structure factor amplitudes by the experimentally determined amplitudes;
(iii) Calculation of the new electron density by inverse FT (FT^{-1}) of the modified structure factors.

These projections can be combined into more or less sophisticated iteration schemes. Different dual-space algorithms use different types of projections and iteration schemes.[87,89]

Charge flipping is a well-known representative algorithm used in the dual-space. In this method, the iterative process starts by assigning random phases to the experimental structure-factor amplitudes; then, an electron density is calculated by FT^{-1}; once the electron density has been defined, all of the grid points g, under a certain positive threshold δ, are multiplied by -1 (flipped), causing a modification in the electron density and allowing the calculation of new temporary structure factors (using FT). In the last step of the first cycle, the generated phases are combined with the experimental amplitudes and a set of structure factors enters the next cycle of the iteration (Figure 6.1).[89]

The *charge flipping* algorithm is, in the language of projections, a repeated application of the amplitude projection and positivity. In this context, a reflection is an operation defined as $R = 2P - I$, where P is the projection operation and I is an intensity.

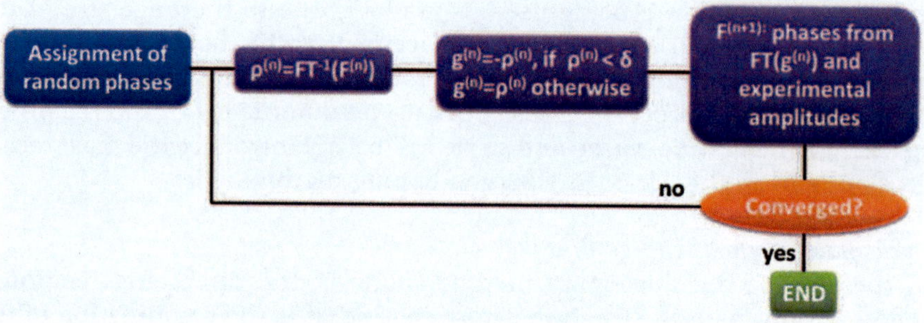

Figure 6.1 Schematic representation of the charge-flipping algorithm operation. FT represents the Fourier transform, ρ is the density and g the grid points. Data from ref. 87 and 89.

Other methods using the dual-space approach are low density elimination, iterated overprojections and the recently published algorithm used in the SHELXT program.[90]

6.2.3 Powder X-ray Diffraction

While SCXRD is generally the method of choice for structural determination, yielding structures of better quality and reliability, in pharmaceutical solids this is usually not a viable option for several reasons. First of all, compounds with pharmacological activity are generally relatively large organic molecules which tend to crystallize with difficulty, thus yielding small crystals with poor diffraction power. Secondly, pharmaceutical compounds are often polymorphic, and the crystal form of interest might be obtained by desolvation, grinding, heating or other methods which make it impossible to grow a single crystal of suitable size and quality. In these cases, powder diffraction methods can be successfully employed to obtain structural information on moderately complex pharmaceutical solids.[91] Moreover, powder diffraction has become a necessary technique for identification purposes, as each crystalline form has a unique powder diffraction profile. The polymorphic nature of most pharmaceutical compounds, coupled with quality and intellectual property issues, makes it mandatory to have a quick, cheap and widely available technique to identify the crystal form on hand and to ascertain its purity (the absence of peaks attributable to different crystalline forms). Obviously, while this is not as fundamental in phase identification, high quality powder diffraction data of a monophasic sample need to be available before structure solution can be attempted.

6.2.3.1 *Bragg/Rietveld Approach to Powder Structures*

Many different methods for solving a structure from powder data have been successfully employed in recent years, both in reciprocal space and real space.[92] The methods best suited to pharmaceutical solids are generally considered to be real-space methods, because extensive data on the atom connectivity, chemical nature and content of the phase examined are usually available before structure solution from powder data is attempted. This means that the variables to be explored are the relative positions of the molecular fragments in the unit cell and the torsion angles that best describe their conformation. This is feasible and efficient for pharmaceutical solids for several reasons. Firstly, the chemical composition is known and the initial (gas phase) geometries are easily obtained with widely available molecular mechanics programs, database searches or even more sophisticated quantum-mechanical methods. Secondly, the useful structural information in these cases is not so much the precise bond lengths or angles but the relative positions of molecular fragments in the structure and their intermolecular bonding, for example electrostatic and π–π interactions, or H-bond motifs.

Structure solution from powders, once the initial information regarding chemical content and atom connectivity is available, can be divided into three main steps:

(1) Indexing and space group determination;
(2) The obtaining of an approximate structure solution;
(3) Refinement of the obtained solution (by the Rietveld method).

Indexing and Space Group Determination

The aim of the indexing step is to determine six parameters which describe the unit cell and to attribute each reflection observed in the reciprocal space (each peak present in the powder diffractogram) to a unique *hkl* triplet. This can be tricky using powder data because of the paucity of clearly defined single peaks and of the severe overlap of many reflections. Firstly, peak positions need to be very precisely determined and expressed as Q_{obs}, as defined in eqn (6.2).

$$Q_{obs}(hkl) = \frac{1}{d_{hkl}^2} \tag{6.2}$$

(note that this Q value has nothing to do with the momentum transfer Q used in the Total Scattering section 6.2.3.3). Secondly, a set of values of Q_{obs} (usually, about 20) must be used to satisfy the set of conditions described in eqn (6.3), where a_{ij} (with $i,j=1,3$) are six parameters related to the lattice metrics, h, k and l are integers and Δ is a predetermined tolerance value.

$$Q_{obs} - \Delta < h^2a_{11} + k^2a_{22} + l^2a_{33} + hka_{12} + hla_{13} + kla_{23} < Q_{obs} + \Delta \tag{6.3}$$

This implies that, for a number of experimentally derived Q_{obs} values (N), the number of variables to be determined is $6+3N$. The problem is therefore *always* underdetermined and yields an infinite number of possible solutions. The correct solution must therefore be chosen on the basis of several criteria. Knowledge of the chemical composition is crucial to estimate the approximate unit cell volume, and can be calculated using tables of average atomic volumes in the solid state or, more roughly, by considering 18 Å^3 for each non-hydrogen atom.

The correct unit cell is also chosen on the basis of numerical criteria which are specifically devised. In particular, two figures of merit are often used: M_{20} or F_N (De Wolff and Smith/Snyder figures of merit, respectively). Both values give an indication of the difference between the observed and calculated values of the peak positions. Finally, the space group must be assigned, taking into consideration unit cell volume and systematic absences, using the correlation tables present in the International Tables for Crystallography, Vol. A, which is very similar to the process for SCXRD.

Structure Solution

Once the unit cell and the space group have been determined, an approximate structure solution can be found by using a trial molecular moiety, based on known atom connectivity and free floating in real space. This moiety and other necessary chemical species, like counter ions or solvent molecules, are moved as rigid bodies, and, where necessary, dihedral angles are also changed in the search for the correct molecular conformation. For each iteration, the intensity for each point of the diffractogram is then calculated, and the difference between the experimental and calculated intensity is evaluated using a cost function which needs to be minimized. The parameters to be explored are then changed with a well defined protocol (typically, Monte Carlo, Simulated Annealing or Genetic Algorithms) and the intensities and the cost function are calculated again. This process can be described as moving (randomly or otherwise) in an *N*-dimensional space (N equals the number of variables explored), searching for that unique single point (corresponding to a well defined set of parameters) which minimizes the cost function. When found, these parameters represent an approximate structure solution. To avoid the needle in a haystack problem, the number of parameters explored should be kept as low as possible, and the choice of the parameters to be tested in the new iteration(s) depends on the method used.

Rietveld Refinement

Once an approximate solution is found, it needs to be refined using the Rietveld method. This is based on evaluating the difference between the calculated and experimental diffraction profile and minimizing a cost function (analogous to the one used in the structure solution step) using a non-linear least squares approach. While during structure solution the number of variables explored should be kept as low as possible, during Rietveld refinement many more parameters can be refined at once (or sequentially), including molecular fragment position, torsion angles, unit cell parameters, instrument, background and peak shape description.

As an example, Figure 6.2 shows a typical Rietveld refinement plot for a fully refined crystal structure. Specifically defined *profile* agreement factors (R_p and R_{wp}) are used in assessing the quality of the fit, jointly with the R_{Bragg}

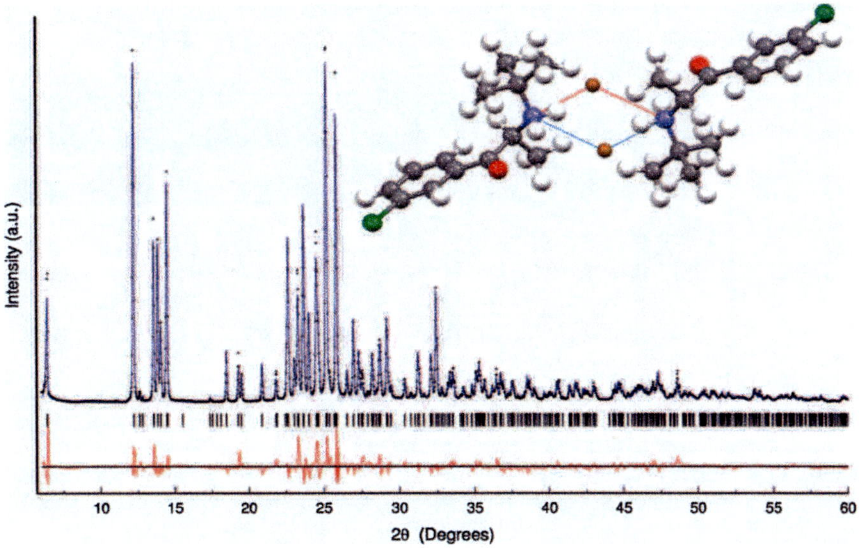

Figure 6.2 The calculated and experimental curves for *Bupropion hydrobromide form I* (overlaid), with the difference plot at the bottom. Peak markers are shown as black vertical ticks. Horizontal axis: 2θ, °. The inset shows the pairing of two Bupropion molecules through the mutual embrace promoted by four evident H-bond interactions of the NH···Br type.

value, which becomes more comparable to the *R*-factor ubiquitously used in single-crystal X-ray analysis.

Variable Temperature XRPD

As powder diffraction is a quick and easy technique which requires little sample preparation, it is also well-suited to measurements at non-ambient conditions, particularly at variable temperatures and pressures. Particularly relevant to pharmaceutical solids is variable temperature XRPD, as it is often essential to understand the behaviour of a specific crystalline form upon heating or cooling. The sample may be heated to specific temperature directly on the sample holder, thus allowing the XRPD patterns to be collected at different temperatures. This technique allows not only the observation of thermal expansion and phase transitions, as shown in Figure 6.3, but also, by using the methods described in the previous paragraphs, determination of the structures of otherwise elusive crystal forms.[93]

6.2.3.2 Structure Solution from XRPD Data Coupled to SSNMR

As explained in the previous paragraphs, in solving crystal structures from powder diffraction data, acquiring information from complementary sources regarding the molecular species, their ionization, protonation, conformation, *etc.*, can be necessary. SSNMR is often an ideal technique to

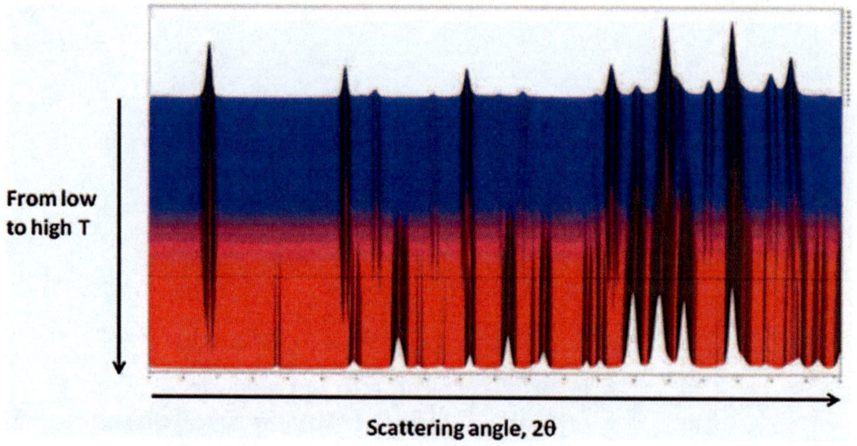

Figure 6.3 A typical VT-XRPD diagram (here, for *azelastine 2-butanone* solvate) showing the evolution of the XRPD trace upon heating, desolvation and molecular rearrangement in the solid becoming evident by the sudden change of the powder pattern.

complement XRPD due to the variety of information that can be obtained from different experiments.

Examples of information that may be obtained from SSNMR which is used during structure solution from powder data include the number of crystallographically independent molecules in the unit cell, information regarding the position of protons on the molecular species present (this can include information regarding proton transfer and its degree, protonation sites, tautomerism or the presence of zwitterionic forms) and information regarding the presence and position of solvent molecules such as water in hydrates. Examples of the synergy between structure solution from XRPD and SSNMR can be found in section 6.3.5 of the present chapter.

6.2.3.3 *The Total Scattering Approach to Powder Diffraction*

Much less known, particularly in the field of covalent or molecular solids, than the Bragg/Rietveld approach, some innovative methods for analysing the entire powder diffraction trace have been recently introduced thanks to the availability of a firm crystallographic theory and the simultaneous development of highly performing computational tools for data analysis. These methods, collectively referred to as Total Scattering methods, employ (as in the classical Rietveld approach) the entire measured digitized powder diffraction pattern, but avoid any kind of intensity partitioning in "peak contribution" over a smoothly changing, continuous background (the origin of which can be instrument- or sample-dependent). Two complementary methods, based on the use of the so-called *Pair Distribution Function* (PDF[94]) and the *Debye Scattering Equation* (DSE[95]), have been developed, operating in real (r) and reciprocal ($Q = 4\pi \sin\theta/\lambda$) space, respectively. Both methods require a thorough data-reduction analysis, in which all collected data, $I_{raw}(2\theta)$, must be numerically corrected for extra-sample contributions (sample-holder and air-scattering contributions) and inelastic (Compton) scattering, fluorescence and absorption effects. All of these features are easily accessible at dedicated synchrotron beamlines, optimized either for the PDF or DSE approaches. Once these corrections have been properly applied, a net dataset becomes available, normally proposed in the Q scale as $I_{net}(Q)$, which includes normalization to the number of atoms in the sample. The most interesting feature shared by both methods is the possibility of dealing with non-periodic systems, and, though periodicity facilitates the data analysis and interpretation of the results, they can be equally used on amorphous, disordered or nanosized materials, and also on liquids, solutions and gases, thus broadening the class of materials which can be investigated well beyond the single- or poly-crystalline methods inherent in the Bragg and Rietveld approaches.

The Pair Distribution Function Approach

The net data, $I_{net}(Q)$, which are normalized to the average scattering factors (for X-rays, Q-dependent), can be transformed into the total scattering function $S(Q)$, as shown in eqn (6.4):

$$S(Q) = \frac{I_{net}(Q) - \sum_i c_i \left| f_i(Q) \right|^2}{\left| \sum_i c_i f_i(Q) \right|^2} + 1 \tag{6.4}$$

Here, c_i are the molar fractions of elements possessing f_i scattering power. $S(Q)$ represents the intensity deviation (which can be either positive or negative) from that of an ideal monoatomic gas of the same composition and reflects the relative changes of $I_{net}(Q)$ due to the overall positive/negative interference of the waves scattered by each atom, with an amplitude scaled to (absolute) electron units. The continuous $S(Q)$ function, thus, replaces the collection of all measurable $|F_{hkl}|$, the ingredients of the Bragg and Rietveld approaches. As $S(Q)$ (a dimensionless quantity) at large Q approaches unity (Q values above 30 Å^{-1} are normally accessible with high-energy beams, with photon energies well above 25 keV), an ancillary function (*the normalized reduced total scattering function*) $F(Q) = Q[S(Q) - 1]$, oscillating around zero at large Q, is often employed.

It can be shown that, by sine-Fourier transforming $F(Q)$ as in eqn (6.5), the $G(r)$ function (defined in real space) can be obtained and used for subsequent data analysis and structure modelling.

$$G(r) = \frac{2}{\pi} \int_{Q_{min}}^{Q_{max}} Q[S(Q) - 1] \sin(Qr) \mathrm{d}Q \tag{6.5}$$

In simple words, $G(r)$ represents the histogram of interatomic distances in the sample, $r_{ij} = |\mathbf{r_i} - \mathbf{r_j}|$, between pairs of atoms located in $\mathbf{r_i}$ and $\mathbf{r_j}$ (weighed by the product of their scattering amplitudes, $f_i f_j$), blurred by thermal motion, static or dynamic disorder, limited instrumental resolution, *etc.*, and lying on a baseline with negative slope, as given by the constant $-4\pi\rho_0$ factor shown in eqn (6.6).

$$G(r) = 4\pi r\rho(r) - 4\pi r\rho_0, \tag{6.6}$$

where $\rho(r)$ is the pair density function. Figure 6.4 shows a representative example of this type of analysis, with the interpretation and labelling of the shortest interatomic contacts.

At this stage, modelling of $G(r)$ can be performed by an atomistic description, with an ensemble of atoms satisfying eqn (6.7)

$$G(r) + 4\pi r\rho_0 = \frac{1}{r}\sum_i \sum_j \frac{f(0)_i f(0)_j}{\langle f(0)\rangle^2}\delta(r - r_{ij}) \tag{6.7}$$

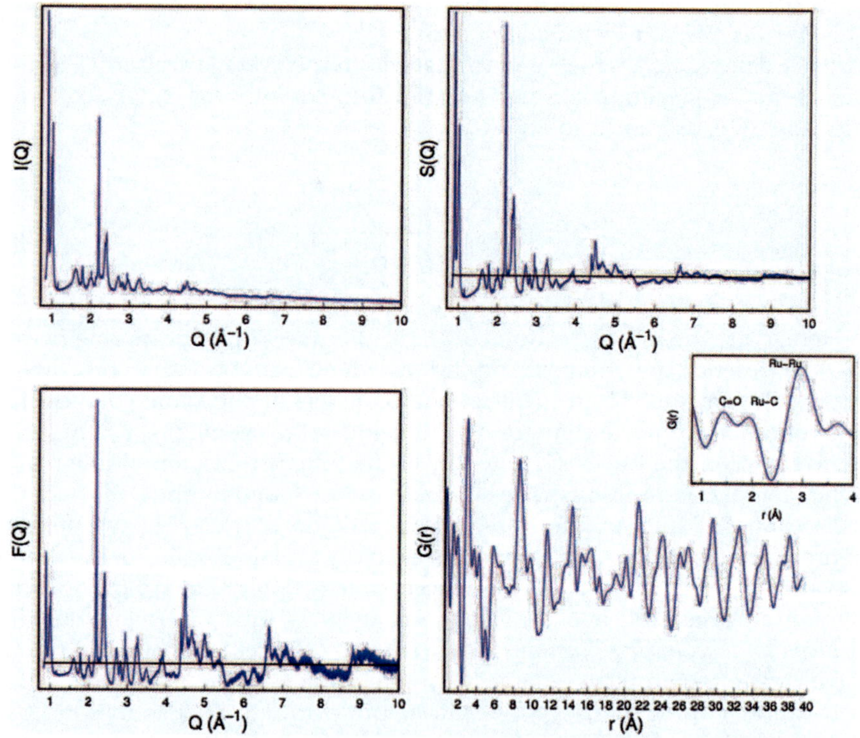

Figure 6.4 Top: Raw data (left) and $S(Q)$ (right) for polymeric ruthenium tetracarbonyl, collected with 1 Å radiation at the MS-X04SA beamline of the Swiss Light Source; bottom: the $F(Q)$ and $G(r)$ functions derived therein. Peaks in the $G(r)$ curve indicate highly probable interatomic distances (as in the inset). Horizontal black lines refer to the theoretical asymptotes for $S(Q) = 1$ and $F(Q) = 0$.

where $f(0)$ is the atomic scattering factor of the rest atoms at $Q=0$ (for X-rays, essentially this is Z, the atomic number).

Thus, numerical matching of the experimental and modelled $G(r)$ functions (eqn (6.5) and (6.7), respectively), upon varying the model parameters (mostly atomic configurations in the geometrical sense, also independently from periodic conditions), may allow the "local structure" to be effectively retrieved. Needless to say, a model is never proved, just substantiated by (or in agreement with) the experimental data.

However, the $G(r)$ function can also be easily used for fingerprinting purposes, very much similar to how the powder diffraction trace distinguishes different phases. The power of the PDF approach, however, resides in the fact that amorphous or liquid materials can also be characterized, and fruitfully compared in a qualitative manner with local structural features approximating those of crystalline phases. For the sake of completeness, the reader is directed to the series of papers and software programs collectively gathered in Billinge's webpage (*thebillingegroup.com*).

The Debye Scattering Equation Approach

As the sine-Fourier transform proposed in eqn (6.8) ideally requires integration over the whole $[0,\infty)$ range, limitations in the $Q_{\min}$ and $Q_{\max}$ values due to finite photon energy and inaccessible angular values (in the presence of a beamstop and a limited size of the X-ray detector) may induce severe truncation errors, and the occurrence, in the $G(r)$ curve, of noisy (high-frequency) ripples and spurious peaks. Therefore, it would be extremely desirable to devise a Total Scattering method in reciprocal space, not requiring the dangerous numerical transformation, and a direct pattern matching the unnormalized $I_{\text{net}}(Q)$ curve (of experimental origin) with a model curve, $I_{\text{Model}}(Q)$, which best describes the calculated X-ray Total Scattering trace, based on an atomistic model, with refinable parameters.

The Debye scattering equation, in its pristine form (eqn (6.8))

$$I_{\text{Model}}(Q) = 2\sum_i \sum_j f_i f_j \frac{\sin(Qr_{ij})}{Qr_{ij}} \tag{6.8}$$

or in the recently proposed modification[96] (eqn (6.9)),

$$I_{\text{Model}}(Q) = 2\sum_i o_i^2 f_i(Q)^2 + 2\sum_i \sum_{j>i} o_i o_j f_i(Q) f_j(Q) T_i(Q) T_j(Q) \frac{\sin(Qr_{ij})}{Qr_{ij}} \tag{6.9}$$

provides the tool for such an analysis. Here, i and j run over the N atoms in the portion of matter (typically a crystal or a nanocrystal) which has randomly oriented replicas in the sample, o_i and T_i are atomic site occupancy factors (for disordered or vacant sites within an average periodic system) and (temperature dependent) "crystallographic" Debye–Waller factors, respectively.

The direct calculation of the scattered intensity in reciprocal space bears another significant advantage, in that several traces (for example, of a multiphasic sample, or of a distribution of sizes, or, occasionally, of nanocrystal shapes) can be easily added, and matched, by pertinent scaling factors, with the experimental curve. Also, as the chemical composition may not be known *a priori*, the normalization step in the PDF data reduction, and the inelastic Compton scattering subtraction procedure can be easily avoided (and reinstated, during the complete Debye function analysis, as an additional term, properly scaled to the elastic scattering contribution).

The possibility of defining, and refining, parameters for analytical size distribution curves (typically, lognormal functions of the monovariate or bivariate type) makes the DSE approach more powerful than PDF in deriving microstructural features, mostly of the morphological type. As they can be simultaneously determined with structural parameters (lattice constants, site occupancy and Debye–Waller factors), this method adds other relevant dimensions to the full characterization of (real, defective and polydispersed) nanomaterials (typically addressing their strain, size, shape and core-shell structure). Examples of very recent applications of the DSE, in its rejuvenated form, are proposed in a following section, where the DebUsSY suite of programs[97] was extensively used.

6.3 Application Case-Studies

6.3.1 Polymorphism

SSNMR spectroscopy is currently an important tool in the pharmaceutical industry, largely owing to its unique capabilities in distinguishing between different polymorphic forms, anhydrates, hydrates, and other solvate forms of APIs.[1,2,98] Eighty percent of solid pharmaceuticals possess more than one polymorphic form and different polymorphs may have distinct physicochemical properties, such as solubility, melting point, dissolution rate, density, hardness, and crystal morphology, all of which may affect their bioavailability, handling, packing, shelf life, and patentability.[71] Characterization of polymorphs is often difficult, especially when single crystals suitable for XRD cannot be isolated. Solid-state ^{13}C NMR spectroscopy is sensitive to minor changes in molecular conformation and therefore is able to distinguish between different polymorphic forms, even when XRPD patterns are identical (see the case of neotame[99] shown in Figure 6.5). In this case, NMR detects changes in conformation while the unit cell parameters do not vary significantly.

Along with *ab initio* CSP, infrared (IR) spectroscopy and XRPD, SSNMR spectroscopy is an extremely powerful method for characterizing polymorphs and the combination of these techniques has been very successful both for validation of polymorph trial structures and for complete structure determination.[75,100–104] When ^{13}C NMR spectra are not able to solve

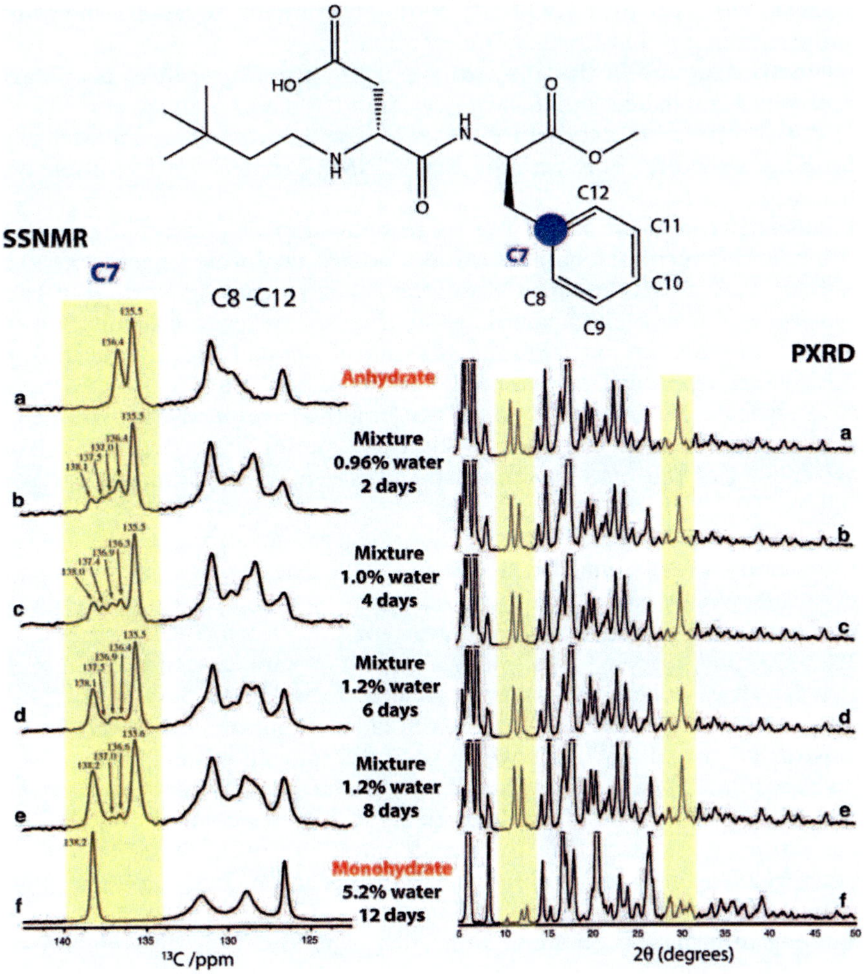

Figure 6.5 (Left) Phenyl region of the ^{13}C CPMAS spectra and (Right) XRPD patterns of (a) neotame anhydrate and (b–e) sample after being sealed in a jar for 2, 4, 6, and 8 days, respectively; (f) sample after being placed in a relative humidity environment of 84% for 12 days.
(Right) Reprinted with permission from ref. 99. Copyright 1999, American Chemical Society.

polymorphism problems, other nuclides have proved useful, including ^{1}H, ^{2}H, ^{15}N, ^{31}P, ^{19}F and even ^{23}Na, ^{17}O and ^{35}Cl.[1,71,73,105,106]

For instance, in the absence of diffraction techniques, a combination of isotope labelling, NMR and computational methods was implemented to investigate the structures of different crystal forms of cimetidine.[107] Distance measurements, C–H bond orientation analysis, and torsion angle determination were performed from ^{13}C CPMAS and *double-quantum heteronuclear local field* (DQ-HLF) NMR experiments to assess molecular conformations in polymorphs of cimetidine.[107]

The 3D structure of the anhydrous polymorph of anticancer drug paclitaxel was established by SSNMR ^{13}C and ^{15}N CST analysis and ^{1}H–^{13}C HETCOR correlation data by Heider *et al.*[75] without previous knowledge of the X-ray structure. This was the first SSNMR-based characterization of a drug with $Z' > 1$.[75]

Hamaed *et al.*[71] presented the first application of ^{35}Cl for the identification and structural characterization of hydrochloride anesthetics. The authors showed that the sensitivity of the ^{35}Cl *electric field gradient* (EFG) and *chemical shift tensor* (CST) parameters to the chlorine chemical environment allows for prediction in the number of short H-bonds involving the chlorine ion.[71] For this purpose, *ab initio* calculations of the ^{35}Cl EFG and CST (using the Restricted Hartree–Fock and B3LYP method, respectively) parameters were conducted and compared with the experimental ones taken from simulations of the MAS spectra of the ^{35}Cl central transition recorded at 21.1 T.

Much of the SSNMR literature has focused on strong intermolecular interactions, such as ionic and H-bonding. The true impact of weak H-bonds or van der Waals contacts may be obscured by other packing constraints. Although these interactions have lower energies, −1 to −5 kcal mol^{-1} for CH···π contacts *versus* −5 to −40 kcal mol^{-1} for moderate/strong H-bonds, they can drive polymorph transformations. In work by Uldry *et al.*[108] and Mafra *et al.*,[109] GIPAW CST and NICS calculations on the piece-wise deconstructed crystalline lattice were used to deconvolute these effects.

The 2D PASS (*phase-adjusted spinning sideband*) experiment together with DFT-based calculations of the local fields was used by Smith and co-workers[74] to analyse the structural information available in the SSNMR spectra of 5-methyl-2-[(2-nitrophenyl)amino]-3-thiophenecarbonitrile, producing red, orange and yellow (ROY) polymorphs. It was shown that CS and dipolar coupling information obtained with such experiments is different for the distinct polymorphic forms and was thus used to quantitatively determine aspects of the molecular structure, including the coplanar angle between the phenyl and thiophene rings. ^{13}C–^{14}N residual dipolar couplings were exploited to measure bond angles in the three ROY compounds and a good agreement was obtained with respect to the X-ray angle values. Results from DFT chemical shielding calculations for each carbon site were used to solve assignment ambiguities. In addition, the CST values were used as molecular constraints in determining the model structure that provides the best correspondence between theory and experiment using a least-squares minimization function for all three ROY polymorphs:

$$\chi^2 \equiv \sum_n \left[\sum_{i,j} \left(\frac{A^{(n)}_{\exp_{i,j}} - A^{(n)}_{\mathrm{model}_{i,j}}}{\varepsilon^{(n)}_{\exp_{i,j}}} \right) \right] \to 0$$

where $A_{\text{exp}}^{(n)}$ and $A_{\text{model}}^{(n)}$ are the experimental and calculated tensors for the n sites in the ROY molecules. $\varepsilon_{\text{exp}}^{(n)}$ is the error associated with the experimental tensor values. A plot of the χ^2 values derived from the different models *versus* the coplanar angle gave their minima at angles very close to the ones obtained by XRD (*i.e.*, 41.0° (modelling) *vs.* 43.6° (X-ray) for ROY-R; 58.0° *vs.* 50.1° for ROY-O and 91.9° *vs.* 91.8° for ROY-Y).

^{13}C and ^{15}N NMR CSs have proven to be useful in discriminating between salts and cocrystals of a given mixture, as in the case of two polymorphic forms of a sulfathiazole : oxalic acid 1 : 1 complex.[110] According to IR spectra and ^{13}C and ^{15}N NMR CSs, both polymorphs were salts where the proton of a –COOH group in oxalic acid was transferred to a –NH_2 group in sulfathiazole. Advanced SSNMR measurements (^{1}H CRAMPS and ^{1}H–^{13}C LG-CP HETCOR) were useful in elucidating the structure of the polymorphs, for which SCXRD data was not available.

A detailed structural investigation of the polymorphism of xaliproden hydrochloride, using SCXRD, synchrotron XRPD, and ^{13}C CPMAS SSNMR was presented by Chandrappa *et al.*[111] In particular, various CP transfer conditions (from ^{1}H to ^{13}C and from ^{19}F to ^{13}C) as well as different decoupling schemes (^{1}H single decoupling and ^{1}H–^{19}F double decoupling) were used to characterize the phases, from both a structural and a dynamical point of view. The results showed that the polymorphism between xaliproden Forms I and III mostly concerns the degree of freedom of the CF_3 group, higher in Form III than in Form I, and they clarified the status of Form III, which could then be assigned to a well-defined polymorph of xaliproden.

In a recent contribution, Wang *et al.*[112] performed structural modelling of a hydrate salt verified by SSNMR and ^{13}C CST calculations. Atorvastatin calcium Form I (ATC-I) is the most stable crystalline form known for this drug and no SCXRD structure had been solved. A complete ^{13}C, ^{19}F and ^{15}N NMR spectral assignment of ATC-I, containing two molecules in the asymmetric unit, was achieved. In addition, a possible local structure for ATC-I based on the simultaneous interpretation of the experimental SSNMR data and DFT calculations of ^{13}C CSs was proposed. ^{13}C CST principal values of all resolvable carbon resonances were measured through a 2D ^{13}C SUPER (*separation of undistorted powder patterns by effortless recoupling*) experiment, another variant that recovers the CSA profile under MAS rotation.

Although DNP-NMR techniques present great potential for the rapid characterization of polymorphs and formulated pharmaceuticals, open questions remain relating to sample formulation and how this may impact species that display polymorphism. A recent contribution showed how protocols can be developed to prepare samples for impregnation DNP while maintaining the desired polymorphic form of the sample, using theophylline as a model compound.[113] Grinding and impregnation procedures may cause polymorphic transitions or dehydration in certain situations. For each form, the authors were able to determine a procedure for grinding and an appropriate solvent for impregnation which prevented the conversion of theophylline to Form II. DNP-NMR was then applied to these polymorphs.

Natural abundance 2D correlation experiments such as ^{1}H–^{15}N, ^{1}H–^{13}C HETCOR or ^{13}C–^{13}C INADEQUATE were acquired in reasonable times, *i.e.*, in some cases, a gain factor over 100× in experimental time. This protocol may have a direct impact on the efficiency of NMR crystallography protocols for polymorphs.

6.3.2 NMR and X-ray Approaches to Study Amorphous Systems

6.3.2.1 Solid-State NMR

SSNMR is helpful for the characterization of amorphous dispersions because it allows access to detailed information on molecular structure, dynamics, and domain morphology in small molecule and macromolecular systems often inaccessible to XRD techniques. Paudel *et al.*[114] reviewed the most recent studies on a number of 1D and 2D SSNMR techniques for the analysis of molecular mobility, miscibility, drug-carrier interactions, crystallinity and crystallization in amorphous dispersions.

2D ^{14}N–^{1}H HMQC experiments recorded at a 60 kHz MAS rate were recently performed, in a drug (acetaminophen), a polymer (PVP) and polymer–drug dispersions, to assess the proximities between nitrogen and proton nuclei involved in H-bonds.[72] This NMR experiment considerably shortened the acquisition time required compared to a ^{15}N–^{1}H HETCOR spectrum (6 h *vs.* 48 h).[36] The GIPAW-DFT approach was employed to calculate ^{15}N isotropic CSs and ^{14}N EFG parameters (C_Q and η_Q) in order to assist in the resonance assignment process. The observed changes in the isotropic shifts for both ^{14}N and ^{15}N were evaluated because while ^{15}N shifts are only affected by the isotropic CS, ^{14}N shifts combine the effect of the isotropic CS and the additional second-order *quadrupole induced shift* (QIS) contribution. While ^{15}N CSs do not show significant changes between acetaminophen and the corresponding dispersion, The ^{14}N CS changes downfield by 65 ppm when incorporated in the solid amorphous dispersion.

More extensive use of 2D SSNMR (^{1}H–^{1}H, ^{1}H–^{13}C, ^{1}H–^{19}F) dipolar correlation methods can directly probe interactions between amorphous materials and small molecules by measuring direct dipolar interactions, spin diffusion, and ^{1}H T_1 relaxation times.[115] This approach was used to examine the local molecular environment of drug molecules (acetaminophen, indomethacin, diflusinal, telithromycin), the association and interactions between the drug and polymers, and longer range effects including the appearance of nano- and microsuspensions within the polymer.

^{1}H–^{13}C HETCOR NMR experiments were valuable in confirming the absence or presence of contact between both components of the dispersion, as PVP contains no aromatic protons. For the case of the PVP–indomethacin dispersions,[115] correlations between the aliphatic carbons of PVP and the aromatic protons were observed, unambiguously confirming molecular contact between the polymer and drug. The use of ^{1}H–^{19}F HETCOR experiments applied to amorphous dispersions of fluorinated APIs was also reported as a fast and efficient method to show the formation of a glass solution in the case of diflusinal–PVP dispersions.[115]

Other studies focused on the molecular mobility of amorphous dispersions, describing the use of ^{13}C CPMAS and the measurement of spin-lattice relaxation times (T_1 and $T_{1\rho}$) to assess the miscibility and physical instability of drug–polymer solid dispersions.[63,116]

In a recent contribution, SSNMR methods were able to detect interactions between ezetimibe and mesoporous silica.[117] XRPD was used to show the absence of crystalline ezetimibe in the mesoporous materials and ^{19}F SSNMR was used to assess the drug's physical state and study molecular mobility. ^{19}F–^{29}Si CPMAS methods were used to directly detect adsorbed ezetimibe. 2D ^{1}H–^{13}C, ^{1}H–^{19}F, ^{1}H–^{29}Si heteronuclear and ^{1}H–^{1}H homonuclear correlation experiments were used to investigate interactions between the drug and silica through ^{1}H environments.[117]

Policianova *et al.*[118] presented methodological approaches in the structure evaluation of solid dispersions. Due to the complexity of the solid dispersions, structural characterization involved the combination of DSC, WAXS, ^{13}C CPMAS NMR spectroscopy and the measurement of ^{13}C-detected $T_1(^{1}\mathrm{H})$ relaxation times. In this contribution, the existence of nanometer-sized motional heterogeneities in PEO$_x$–acetylsalicylic acid was revealed by ^{1}H–^{13}C WISE experiment; H-bonding was probed by ^{13}C and ^{15}N NMR experiments; ^{1}H–^{1}H and ^{1}H–^{13}C correlation experiments allowed the measurement of interatomic distances and the size of the molecular clusters. The experimental findings confirmed that structural properties of the amorphous dispersions are determined by a combination of two factors: (i) the accessibility and affinity of suitable molecular sites for forming H-bond interactions between the API and the polymer matrix, and (ii) the segmental dynamics and glass transition temperature (T_g) of the polymeric coformers.[118]

The mobility of amorphous simvastatin in different frequency scales was studied using conventional ^{1}H MAS and ^{13}C CPMAS NMR techniques over temperatures ranging from 293 to 333 K (below and above the T_g region) and line shape simulations.[119] The ^{13}C $T_{1\rho}$ was measured as a function of temperature, and the correlation time and activation energy data obtained for local motions in different frequency scales revealed strong dynamic heterogeneity, which appears to be essential for the stability of the amorphous form of simvastatin. Overall, the results confirm that simvastatin in glass form has potential for use in solid phase formulation in the pharmaceutical industry.[119]

For many years, it has been thought that X-ray characterization of amorphous materials was impracticable, as the very limited information residing in a smoothly changing diffraction trace appeared devoid of any relevant structural information (with some notable exceptions: see, for example, ref. 120). Moreover, no firm theory of crystal packing in amorphous organics (in contrast to vitreous metallic and oxidic materials) was available, making it almost impossible to quantitatively extract useful descriptors. Altogether, the proverb "*all cats are grey in the dark*" could be easily applied to the many amorphous patterns collected through the years, mostly for poorly water-soluble APIs requiring the individuation of new pathways to increase their bioavailability.[121] In this context, amorphization has been the method of choice, well before nanostructured drugs or cocrystallization with suitable conformers came on the scene. However, when *polyamorphism* became a (more) common occurrence in applied environments, devising new (diffraction-based) fingerprinting methods for nano-sized and glassy organics became necessary.

6.3.2.2 X-ray Scattering: The Real Space Approach

While the theoretical basis for PDF analysis was set a few decades ago, in the absence of suitable computational aids, total scattering methods went partially disregarded, particularly in the field of organics, where molecules are moderately complex and bond distances and angles show limited variability. Much more recently, using user-friendly tools tailored for PDF analysis, Billinge *et al.* (2010),[122] showed that the PDF curves of amorphous carbamazepine and indomethacine (derived by synchrotron X-ray total scattering experiments) were highly informative in the local packing of these common drugs, opening up a way to the meaningful study of glassy, disordered, molecular materials. Given that the crystal packing of different reference crystalline polymorphs was indeed known, the local environment of each molecule in the amorphous phase could be traced back by visual inspection; a numerical analysis also provided the extent of the X-ray effective coherence length (estimated, here, in the 4–5 nm range). Further examples include lactose, raffinose, dextran/trehalose/indomethacine – polymer hybrids, *etc.* Beyond fingerprinting, this work stimulated the development of new theoretical/computational advances, such as: (i) the separate modelling of intra- and intermolecular distances; (ii) structure solution and refinement of organic molecules from PDF data; (iii) phase quantification in binary mixtures of nanosized drugs, and (iv) the development of fast and reliable algorithms and programs for rapid PDF analysis.

As anticipated, the phenomenon of *polyamorphism* (the ability of a substance to exist in several different amorphous modifications, with a clear, discontinuous, first-order phase transition between them), originally evidenced in metallic and ionic compounds, has become an important scientific, as well as industrial issue, with significant consequences in the defence of intellectual property. An interesting application of PDF methods to the fingerprinting of differently organized amorphous materials was recently reported for the widely-used drug paracetamol,[123] which nicely demonstrated, by total scattering methods, the different nature and chemical behaviour of two distinct amorphous systems, eventually affording, upon homogeneous crystallization, two different crystalline (truly polymorphic) phases.

6.3.2.3 X-ray Scattering: The Reciprocal Space Approach

Characterization of amorphous materials by the reciprocal space approach, which needs the definition of atomistic clusters, or statistically generated atomic configurations on the basis of molecular dynamics or other competing methods, is a rather difficult task. While it has been adapted to inorganic and hybrid materials in Reverse Monte Carlo studies targeted towards PDF analysis[124] (mostly from neutron scattering data), we are not aware of similar studies for organic materials, neither in real nor in reciprocal space.

Nonetheless, a thorough DSE study of hybrids containing nanosized drug nanocrystals (Vinpocetine) and crosslinked polyvinylpyrrolidone (PVP-Cl) has been very recently presented.[125] Using a combination of real and reciprocal space analyses (on the same set of data, partitioned for the different components), it was shown that these composites possess size and size distributions of the nanocrystalline fraction, which depend on both the milling times and drug-to-polymer weight ratios. Interestingly, the occurrence of an additional amorphous component was interpreted, through real space data analysis, as a slight deformation (induced by mechanical grinding) on PVP-Cl, rather than as the originally purported formation of amorphized Vinpocetine (this having a major effect on the interpretation of *in vitro* and *in vivo* solubility curves). In addition, very much as proposed for the detection, separation and quantification of the different (nanocrystalline and amorphous) phases in titania/silica hybrids and polyphasic biomimetic calcium phosphates, estimation of the weight percentages of the different components (W_j, $j=1, \ldots, N_{\text{phases}}$) was based on the entire (independent) scattering contributions (which could be isolated after DSE analysis), taken as:

$$W_j \propto \int_{Q_{\min}}^{Q_{\max}} \frac{I_j(Q)}{\langle f_i(Q)\rangle^2} \mathrm{d}Q \tag{6.10}$$

Here, $\langle f_i \rangle^2$ is the weighted square average of the atomic scattering factors (f_i, $i=1, \ldots, N_{\text{elements}}$) nominally present in phase j. This method appears to be far more accurate than the ubiquitous usage of signal area ratios (forgetting about *chemistry* – the phase composition, and *physics* – the fading away of the X-ray atomic scattering factors), although (still) not completely rigorous.

Not strictly related to the field of organic molecules, but rather close to it in that molecular or polymeric crystals have been studied, are other contributions by Guagliardi's group, which involved embedding highly defective coordination polymers (*e.g.*, highly flexible Co/Fe 1,4-benzenedipyrazolates).[126] The analysis of their odd peak broadening effects, not following simple (though anisotropic) *hkl* dependence, suggested the presence of correlated displacement and/or stacking faults, which could be simulated by the DSE, once paracrystallinity effects were introduced. Needless to say, the standard Rietveld approach could have equally well described the experimental total scattering trace, at the expenses of an unphysical, phenomenological, description of awkward peak shapes and widths and of other annoying effects. Additional significant examples of the use of the DSE in the broader Debye function analysis frame have been reported for nanosized transition metals, iron oxides, rare earth borides, hybrid perovskites, lead chalcogenides and biomimetic nanoapatites.

Still related to the field of organics, the active use of the DSE in comparing differently organized nanoclusters of a few molecules, or even extended structures, both calculated from computational approaches and specifically tailored interatomic potentials, has been also proposed.[127] The qualitative, or even quantitative, match between XRD traces calculated by the different models (ranked in energy during crystal structure prediction studies), can then be used to verify structural similarities and "solution" clustering.

6.3.3 Drug Delivery Systems

One of the biggest challenges in the pharmaceutical sciences is finding biocompatible and biodegradable materials for the transportation and controlled release of APIs to target locations in the human body. In this field, SSNMR can answer several important structural questions regarding pores filling, topology of interactions, local molecular dynamics of API and/or DDS, and the nature of inter- or intramolecular interactions.[128] A recent review by Skorupska and co-workers[128] reported the applicability of SSNMR to study non-steroidal anti-inflammatory drugs, antibiotics and anticancer drugs confined in zeolites and mesoporous silica nanoparticles.

Azaïs *et al.*[129] reported a study of ibuprofen encapsulated in MCM-41 matrices with different pore diameters, involving ^{1}H, ^{13}C and ^{29}Si SSNMR experiments at ambient and low temperatures.[129] ^{1}H–^{13}C and ^{1}H–^{29}Si HETCOR NMR experiments were carried out to gain insight into drug···MCM-41 interactions. At ambient temperature, ibuprofen is extremely mobile inside the silica pores, as shown by measuring the apparent transverse relaxation time (T_2*); high T_2* values compared to those of a rigid sample were obtained, which is consistent with the high mobility of ibuprofen in MCM-41. Subsequently, the interaction between ibuprofen and the silica surface is weak, which favours fast drug release in simulated intestinal or gastric fluid. Further studies with ibuprofen were carried out by Tang *et al.*[130] involving the correlation between molecular dynamics and melting transition for bulk ibuprofen and confined ibuprofen in titania nanotubes. ^{13}C single-pulse, CPMAS and T_1 relaxation measurements were performed on both bulk and confined ibuprofen, suggesting that confinement substantially changes the segmental dynamics which directly participate in the melting transition.

Microporous (zeolites FAU, LTA and LTL) and mesoporous (SBA-15 and MCM-41) materials were investigated by Amorim[131] and Vilaça[132,133] regarding their suitability as DDS for *in vitro* models of solid tumors.

Anticancer drugs α-cyano-4-hydroxycinnamic acid,[131] 5-fluorouracil[133] and salicylic acid[132] were encapsulated in the various materials, characterized by FTIR, $^{1}H/^{13}C/^{27}Al$ SSNMR, UV-vis, chemical analysis, XRPD, and SEM, and the effect of the new DDS was tested on human cancer cell lines for viability assessment. SSNMR was used in the structural characterization of the new systems, evaluating the successful loading of the drug within the porous material as well as probing drug–host interactions.

Horcajada and co-workers proposed an alternative route for the controlled storage and release of Ibuprofen in metal–organic frameworks (MIL-53(Cr) and MIL-53(Fe)). ^{1}H SSNMR was employed to prove that Ibuprofen molecules hosted in MIL-53(Cr) exist in their neutral form. A strong H-bond was observed between the carboxylic acid group of Ibuprofen and the hydroxyl group of MIL-53(Fe). In addition, the ^{13}C CPMAS spectra of MIL-53(Cr or Fe)–Ibuprofen were very close to those obtained for pure Ibuprofen, revealing that Ibuprofen@MIL-53 exhibits a similar conformation to that of its pure form. XRPD performed before and after Ibuprofen release, showed that the MOF structure was retained, proving the flexibility of MOFs to adapt their porous size to the dimensions of a drug, optimizing drug–matrix interactions.[134]

Cyclodextrins are also important host molecules for drug delivery. In fact, more than 35 drugs in the world are currently marketed as solid- or solution-based cyclodextrin complex formulations. Vogt and Strohmeier tested α and β cyclodextrins to evaluate the applicability of this class of compounds as DDSs.[135] 2D SSNMR correlation experiments between ^{1}H, ^{13}C, ^{31}P, and ^{19}F nuclei were used to assess the inclusion of several drugs (adefovir dipivoxil, voriconazole, dexamethasone, prednisolone and diflusinal) in cyclodextrins. ^{1}H–$^{19}F/^{13}C$ CP/LG-CP HETCOR experiments demonstrated the effectiveness of SSNMR for describing API-DDS connectivity. A partial complex of diflunisal with β-CD was produced by co-grinding to illustrate the ability of ^{19}F 1D and 2D SSNMR experiments in separating and quantifying included *versus* free drug in the solid.

Interactions between ^{1}H and ^{13}C nuclei were successfully demonstrated for an amorphous 1 : 1 drug–CD complex prepared by cogrinding using the 2D ^{1}H–^{13}C CP-HETCOR NMR experiments.[136] In this study, drug inclusion in β-CD was demonstrated by detection of correlations between ^{1}H sites assigned to the aromatic regions of the drug and ^{13}C sites assigned to β-CD.[136]

6.3.4 Formulated Drugs

During formulation and manufacturing processes, a drug is exposed to several procedures that can compromise the solid-state properties of a given API: direct compression, drying, milling, granulation and compaction. The possibility of inducing phase conversions and polymorphism is high.

In a study reported by Nonappa *et al.*, an accidental discovery of a mixture of polymorphic forms in the commercial deoxycholic acid led to the study of the polymorphic nature of six natural bile acids. These compounds were identified and characterized using ^{13}C CPMAS. Two different polymorphic forms were found by comparison of the spectral patterns. One of the polymorphs displayed a *doublet* resonance pattern, in its ^{13}C CPMAS spectrum, originating from the two non-equivalent molecules present in the asymmetric unit. This observation, along with complementary studies of SCXRD and XRPD diffraction, explained the difference in the side chain conformation and H-bonding modes found for each polymorph.[137]

In a more recent study, Miclaus *et al.*[138] demonstrated how the combination of SSNMR and XRPD offers a convenient tool for solid-form determination of APIs directly on the marketed product, without the need for prior sample preparation procedures. The methods were used to discriminate what particular form was present in the tablet/capsule of quercetin dietary supplements,[138] given the fact that this was a complex mixture that also contained a number of excipients alongside the active ingredients. The authors found both methods to be complementary, as XRPD proved useful for quick identification of the major crystalline components and ^{13}C SSNMR was able to distinguish between the quercetin solid forms in their commercial formulation, even in the presence of additional active ingredients and excipients.

The 2D ^{1}H–^{1}H DQ-SQ NMR technique was also employed to identify a specific pseudo-polymorph of a given API directly on the tablet formulation, with no sample preparation.[139]

Considering that *ca.* 20–25% of currently developed pharmaceuticals contain at least one fluorine atom in the molecule, a fast and reliable characterization method of amorphous forms of APIs in drug products even at low concentrations (1–5%) was presented by Brus and co-workers.[140] Atorvastatin was studied using ^{19}F MAS NMR spectroscopy to distinguish between different amorphous forms of both the pure API and atorvastatin as a low concentration tablet formulation.[140]

6.3.5 From Crystal Packing Interactions to Crystal Structure Determination

Supramolecular interactions in crystalline solid drugs have an impact on their physicochemical properties, such as solubility, stability, manufacturing ease and bioavailability. The use of X-ray techniques to determine the crystal structure and their validation through combination with experimental SSNMR spectroscopy and calculated CST were applied to prove molecular association, identify H-bond donors and acceptors, detect aromatic ring-currents and report the strength of these interactions.

6.3.5.1 *Probing ^{15}N Nuclei to Characterize H-bonds and Proton Transfer in Acid–Base Drug Systems*

The ^{15}N nuclide ($I = 1/2$, 0.36% natural abundance) represents an important nucleus for studying pharmaceuticals and has been used to study distinct amine-based functional groups as well as to aid in the discrimination of protonated *vs.* nonprotonated nitrogens in acid–base complexes.[36,141–144] One of the earliest studies exploiting CST in multicomponent pharmaceuticals provides guidance for using ^{15}N CST to classify a complex as a salt or cocrystal, and to rank bond strength. In Li *et al.*'s study, the degree of proton transfer in three different adducts of a cancer treatment candidate, with three different organic acids, was examined.[143] Adducts of the drug candidate with succinic, malonic and maleic acids, as well as the free base, were studied *via* SCXRD and SSNMR. The ^{15}N isotropic CS for these complexes differed by up to −120 ppm from the crystalline free base, demonstrating the high sensitivity of ^{15}N to intermolecular interactions. A combination of solution and solid-state NMR experiments as well as DFT calculations of ^{15}N CSs were carried out to assign the ^{15}N CSs. For complexes where proton transfer occurred from the carboxylic acid to the amine group (salts), the ^{15}N isotropic CSs moved upfield by 80 to 120 ppm from the crystalline free base. In contrast, the cocrystal complexes produced upfield shifts of 20 to 40 ppm due to strong H-bonding between the nitrogen acceptor and the carboxylic acid donor. The observed isotropic CS differences correlate well with the ΔpK_a between the drug's two basic nitrogens and the acids, and with the rank ordering of the bond strengths based on donor–acceptor (D···A) bond distances extracted from the single crystal structures. Indeed, a nitrogen atom within a system of saturated bonds (*e.g.*, aliphatic nitrogen) often exhibits much smaller shifts with protonation ($\Delta\delta_{15N}$ up to 15 pm) than its aromatic/heterocyclic counterparts containing unsaturated bond systems (*e.g.*, amines *vs.* pyridines), where the ^{15}N CS "decreases" by typically $\Delta\delta \sim$ 50–100 ppm.[142,143,145]

Similar methods were applied by Kahn *et al.* to the quinidine methyl paraben cocrystal and quinidine 4-hydroxybenzoate salt.[146,147] Again, DFT tensor calculations were performed to assign the ^{15}N resonances and confirm bonding interactions. The ^{15}N $\Delta\delta$ values for the cocrystal and salt from the free quinidine were much smaller ($\Delta\delta_{15N} \sim$ 2–18 ppm) than seen in the previous study ($\Delta\delta_{15N} \sim$ 20–40 ppm), and can be explained by the formation of moderate to weaker bonds in these complexes compared to the free quinidine.

Another interesting study was developed by Pindelska *et al.*,[148] where the conformational changes of arylpiperazine derivatives (free base and salt) and their intermolecular interactions were explored. Arylpiperazine derivatives are psychoactive agents and interesting targets in the search for antidepressant drugs. It is known that conformational differences in these molecular structures may have an impact on the activity of serotonin and dopamine receptors. SCXRD and SSNMR were used, respectively, to determine the space group and the packing interactions of each compound, and to discriminate between salts and free bases. The results showed that protonation (checked by ^{15}N NMR) only slightly affects the geometrical parameters such as bond angles and distances. In addition, the arylpiperazine derivative conformation changes upon protonation and with different lengths of the linker.[148]

Other reports show that SSNMR methods have been explored to prove cocrystal formation during solid form screening activities.[149] ^{15}N spectral-editing methods based on short contact-time (limited to ~120 μs) CPMAS or interrupted decoupling, whereby the proton decoupling field is turned-off for a short period of 130 μs before acquisition to dephase signals originating from protonated nitrogen, can also be employed to distinguish protonated from nonprotonated nitrogens. An ingenious one-bond ^{15}N–^{1}H *J*-coupling ($^{1}J_{^{15}N-^{1}H}$) based NMR experiment showed potential to distinguish between a different number of protons covalently bonded to a nitrogen, and therefore readily differentiate ionized from non-ionized nitrogen species in distinct amine residues.[36]

6.3.5.2 *Characterization of Weak H-bonding and CH···π Interactions in Hydrates*

The impact of packing interactions (CH···π, CH···O$^{(-)}$, π···π and $^{(+)}$NH···O$^{(-)}$) and hydration on the ^{1}H and ^{13}C isotropic CSs was examined for the anhydrate (Form I) and channel hydrate (Form II) of the antibiotic ciprofloxacin.[109] To evaluate hydration effects, the water molecules were removed *in silico* from the optimized structure and the CST was recalculated. The crystal packing and hydration effects on the calculated ^{1}H isotropic CS of Forms I and II are plotted as blue and red stems in Figure 6.6. For instance, the CS difference ($\Delta\delta^{\text{hydration}}$) values (red stems) were determined from the GIPAW calculated values for the 'dehydrated' Form II subtracted from the water-occupied crystal. The presence of water in Form II results in a maximum $\Delta\delta^{\text{hydration}} = 2$ ppm and 5 ppm for ^{1}H and ^{13}C, respectively. Hydration has the greatest impact on the atoms in the cyclopropane and piperazine rings that interface with the water channel, as the polarizing effect of the water oxygens considerably deshields these protons.

The GIPAW calculated ^{1}H CS differences ($\Delta\delta^{\text{packing}}$) between the complete crystal and 1D chains (blue stems in Figure 6.6) show the packing contributions of both forms. It is not surprising to see the largest $\Delta\delta^{\text{packing}}$ is for the HN proton involved in the zwitterionic $^{(+)}$NH···O$^{(-)}$ H-bond. Large packing effects were observed for CH···π interactions, shielding ^{1}H CS values by up to 3.6 ppm (H3′ in Figure 6.6). These and the π···π interactions are also impacted by intermolecular ring currents, which influence ^{1}H and ^{13}C shielding. NICS[150] were determined to estimate the contribution from aromatic ring currents alone. The $\Delta\delta^{\text{packing}}$ values were corrected by NICS for selected crystal packing interactions. These corrected values show that weak intermolecular interactions, such as the CH···π contacts in ciprofloxacin, can contribute up to −4 ppm to the ^{1}H CS. Smaller packing contributions, up to 2.3 ppm, were calculated for the weak CH···O$^{(-)}$ H-bonds. Similar values were obtained for the weak CH···X H-bonding in maltose,[151] uracil and 4-cyano-4′-ethynylbiphenyl[108] by comparable CST calculation methods. The full assignment of the 17 ^{13}C resonances of I (Figure 6.7) was possible with the help of calculated CSs and 2D ^{1}H–^{13}C HETCOR experiments using

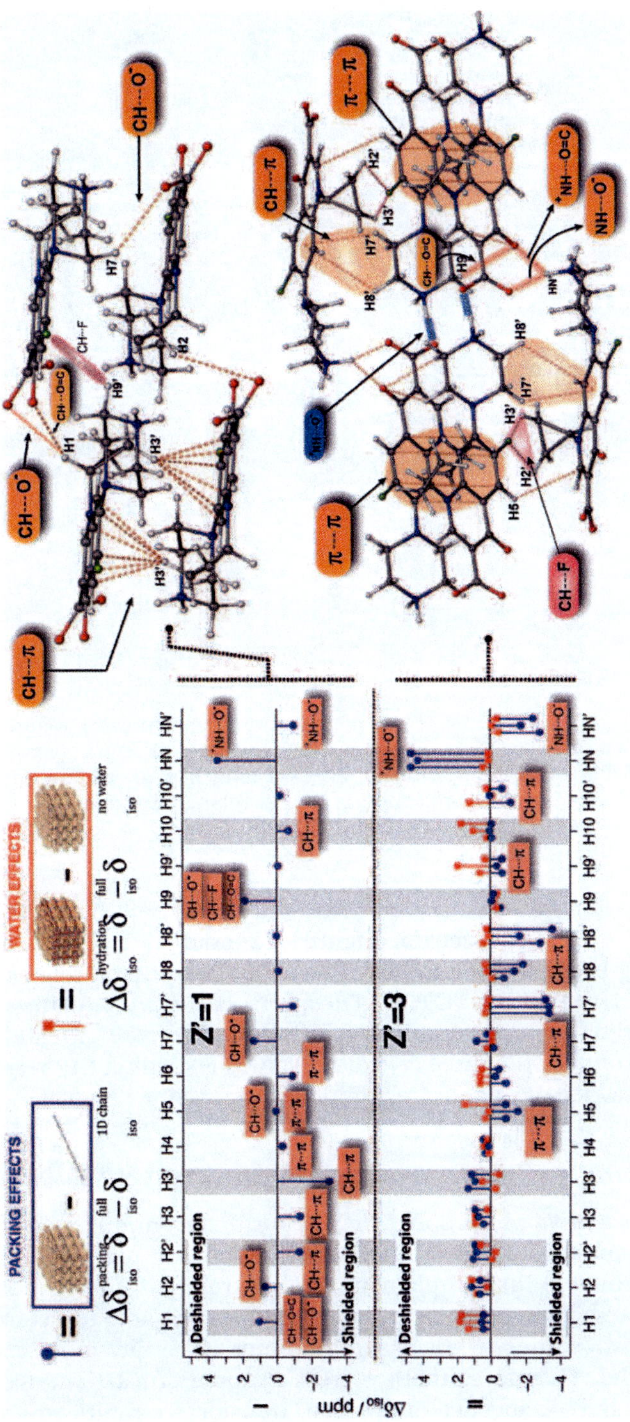

Figure 6.6 (left) Δδ (ppm) plot showing crystal packing contributions (blue stems) and hydration contributions (red stems) to the GIPAW calculated ^{1}H isotropic CS of ciprofloxacin Form I and Form II. In II, each of the three stems per proton corresponds to a crystallographic unique molecule ($Z' = 3$). Right: Detailed view of intermolecular interactions in the ciprofloxacin Forms I and II structures.

Reprinted with permission from ref. 109. Copyright 2012, American Chemical Society.

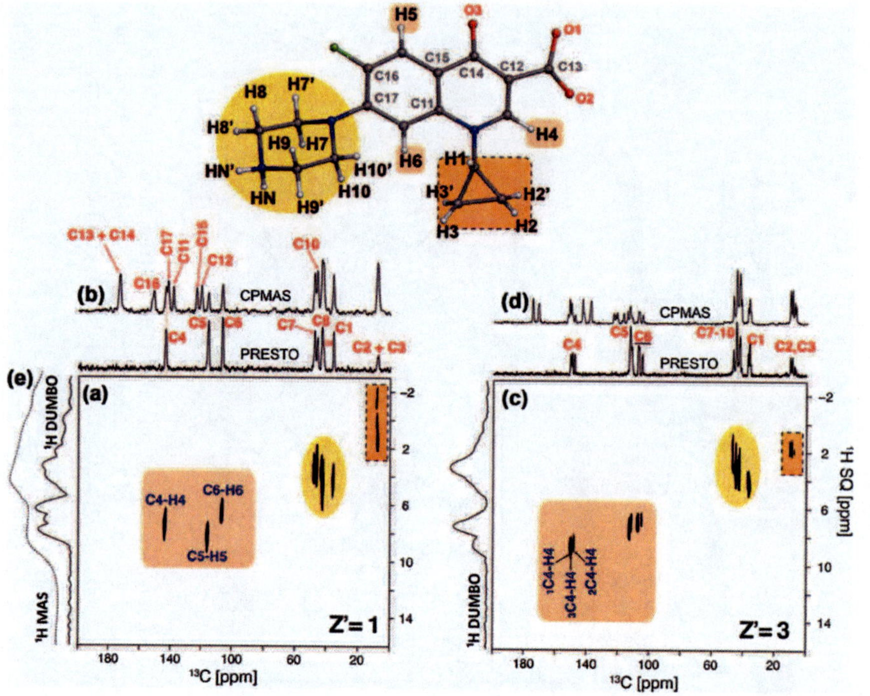

Figure 6.7 2D ^{1}H–^{13}C PRESTO-HETCOR of ciprofloxacin forms (a) I and (c) II recorded at 800 MHz. (b and d) ^{13}C CPMAS spectra recorded at 400 MHz. (e) ^{1}H MAS and wDUMBO spectra of I are shown for comparison along the F1 projection of (a and f) labelling scheme.
Adapted with permission from ref. 109. Copyright 2012, American Chemical Society.

the PRESTO transfer, which afforded one-bond (C–H) selectivity. The assignment of the ^{13}C CPMAS spectrum (Figure 6.7c and d) of II is substantially more challenging because the number of resonances is tripled relative to I. Similarly to form I, the 2D ^{1}H–^{13}C HETCOR spectrum (Figure 6.7c) identified the aromatic carbons and discriminated between all of the CH and CH_2 carbons and protons for the three crystallographically distinct CIP residues, except for the crowded 42–48ppm region of the piperazine residue.

6.3.5.3 *Combining XRD, DFT and NMR for Structure Validation*

Due to its high sensitivity in probing the chemical surroundings of nuclei, SSNMR can be used to obtain information regarding the structure of pharmaceutical solids, thus complementing the structural information obtained from XRD.[1,152] For example, although XRD is not able to determine precise hydrogen positions, NMR is notorious for its ability to detect the signal of ^{1}H nuclei. Therefore, another great example of XRD and SSNMR synergy regards the occurrence of proton transfer in multicomponent pharmaceutical salts and cocrystals (see Section 6.3.5.1) not only because properties like solubility and hygroscopicity can change drastically from neutral to ionic species, but also for regulatory considerations. In addition, an inability to form single crystals suitable for SCXRD analysis makes NMR one of the most complementary techniques to XRPD. In order to summarize the multitude of papers that show the combined (and fruitful) usage of diffraction and NMR experiments, a few case studies are hereafter reported.

A recent example of proton transfer analysis on pharmaceutical adducts is the work by Tatton and co-workers, where $^{1}J_{^{15}N-^{1}H}$-edited spectra were used instead of the simple ^{15}N CSs.[36] $^{1}J_{^{15}N-^{1}H}$ -edited spectra were used to distinguish protons directly bonded to a hydrogen in the anti-ulcer drug cimetidine and to confirm the formation of a zwitterion in the solid state of tenoxicam, a non-steroidal anti-inflammatory drug. Figure 6.8 shows the $^{1}J_{^{15}N-^{1}H}$-edited spectra of cimetidine recorded for different τ values. The initial ^{15}N CP-MAS experiment shown in Figure 6.8a confirms the presence of the expected six nitrogen sites and for $\tau = 1.6$ ms (Figure 6.8b), all signals remain positive. The absence of the fast dephasing typical of a nitrogen species with two directly attached protons indicates that none of the signals are from a primary amine, which is consistent with the structure shown above. However, dephasing at a significantly faster rate is observed for N10, N12 and N15, suggesting larger residual dipolar interactions due to proton(s) in close proximity, which are not completely averaged out by MAS. Longer τ values show a clear inversion of the N10, N12 and N3 resonances, confirming that these nitrogen species have a direct one-bond J coupling to a proton. For the remaining three resonances (N1, N12 and N14), the signal stays positive for all τ values.

Regarding cocrystal/salt formation, there are several interesting studies reporting the use of XRD combined with SSNMR and DFT calculations. Examples are the two cocrystals of theophylline with 4-aminosalicylic and 4-aminobenzoic acid, reported by Fernandes *et al.*[153] In this study, SSNMR was essential in confirming the absence, rather unexpectedly, of an intermolecular interaction engaging the amino groups. Moreover, calculated ^{1}H and ^{13}C NMR CSs from the resulting XRPD structure model, using first-principles periodic DFT techniques within the GIPAW approach, were used to assess the agreement with the corresponding experimental CSs. Another study reported by Martins *et al.*[141] focused on studying the cocrystal/salt paradigm of new multicomponent crystal forms of azelaic acid, an antibacterial used to treat skin disorders. In this study, five compounds were prepared with the aim of improving the physicochemical properties, such as solubility. The crystal structures of the four new multicomponent forms were characterized by SCXRD and SSNMR. For azelaic acid with morpholine (Figure 6.9), the use of DFT geometry optimization (Figure 6.9c), along with SSNMR was of paramount importance in order to define the exact position of the disordered hydrogen atoms (Figure 6.9b). High-resolution ^{1}H SSNMR at very-fast MAS rates (60 kHz, Figure 6.9a) showed an unusually strong H-bond interaction, which appeared as a disordered hydrogen in the XRD data. Two

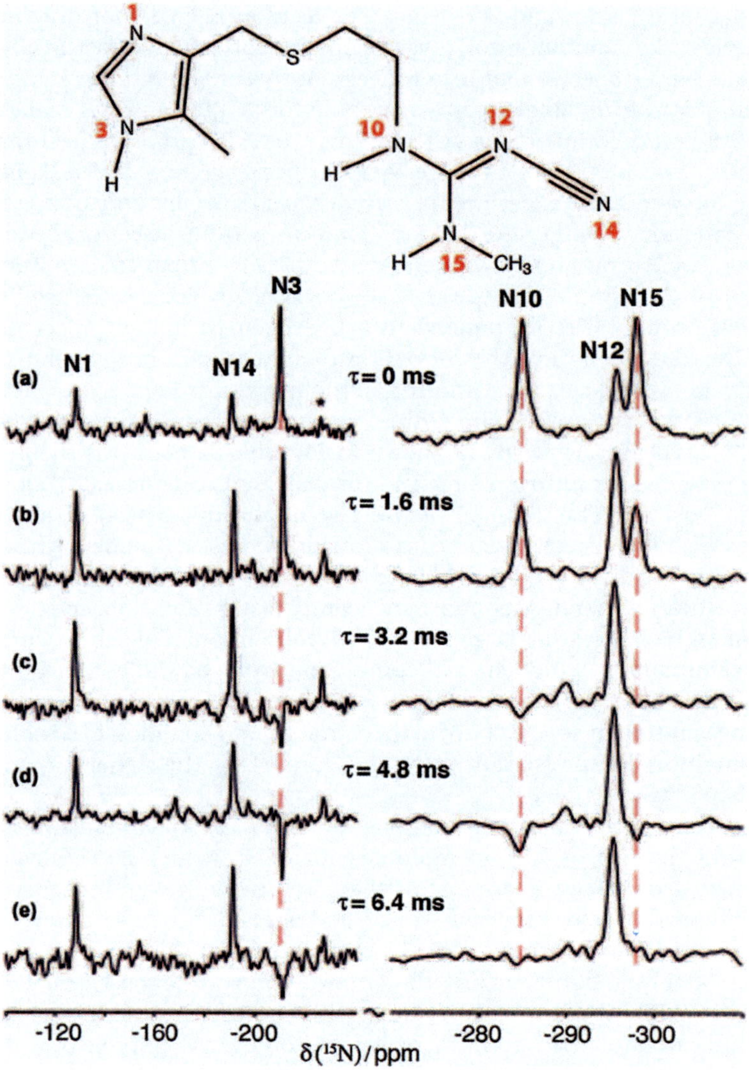

Figure 6.8 (a) ^{15}N CP-MAS and (b–e) $^{1}J_{^{15}N-^{1}H}$-edited spectra of cimetidine recorded for a range of rotor-synchronised τ values. Spinning sidebands are indicated by asterisks (*). The dashed vertical lines assist in the observation of a zero crossing for a peak due to modulation by a $^{1}J_{^{15}N-^{1}H}$ coupling. Spectra were obtained at 11.7 T and 295 K.
Adapted from ref. 36 with permission from The Royal Society of Chemistry.

particular hydrogens were equidistant from their H-bond donors and acceptors ($d_{Xacceptor\cdots H} \sim d_{Xdonor\text{-}H} = 1.21\text{–}1.22$ Å for $\delta_H = 20.1$ ppm; $d_{Xacceptor\cdots H} \sim d_{Xdonor\text{-}H} = 1.23$ Å for $\delta_H = 19.7$ ppm, Figure 6.9c) thus

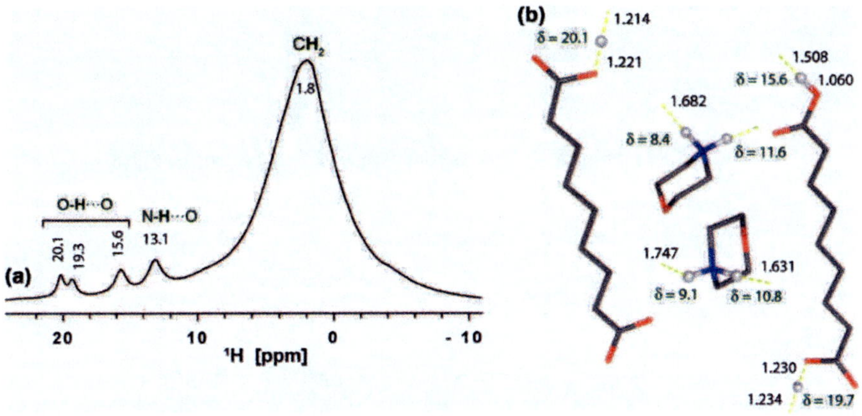

Figure 6.9 (a) 1D ^{1}H MAS NMR spectra of azelaic acid with morpholine recorded at $B_0 = 16.4$ T. (b) View along a of the packing structure of azelaic acid:-morpholine showing the supramolecular interactions. The atomic disorder in the hydrogen atoms is highlighted with green dashed lines. (c) ^{1}H GIPAW-DFT CSs shown on the asymmetric unit. Only CSs pertaining to H-bond involved protons are shown (white spheres); remaining protons have been omitted. X–H···O H-bonds (X = N, O) are shown as yellow dashes, along with the corresponding bond lengths in Å. Reprinted with permission from ref. 141. Copyright 2015, American Chemical Society.

inducing ^{1}H resonances shifted to extremely high frequencies (Figure 6.9a). ^{13}C CPMAS and ^{15}N CPMAS (Figure 6.10) analyses were used to confirm which type of multicomponent solid forms (cocrystal or molecular salt) was obtained with azelaic acid. Figure 6.10 depicts the ^{15}N CPMAS spectra of both a molecular salt (Figure 6.10a) and a cocrystal (Figure 6.10b) with a CS difference of around 250 ppm between both species.[141]

Very recently, Harris and co-workers studied cimetidine hydrochloride,[154] determining its structure by XRPD data using the direct-space genetic algorithm technique; the structure obtained was validated by calculating solid-state ^{13}C NMR data for the crystal structure using the GIPAW-DFT approach and assessing the agreement with the corresponding experimental data.

In another work, Filip and co-workers calculated ^{13}C/^{1}H CSs from the structure obtained from XRD not only to validate it, but especially to verify the correct conformation description for the hydroxyl groups of quercetin.[78] As exploring the conformational variables of molecules when solving structures form powder data can add too many variables to the problem to guarantee a solution, information on molecular conformation from SSNMR can, therefore, be a key contribution. In this case, the use of SSNMR was particularly crucial because not only was it impossible to grow a single crystal of quercetin suitable for XRD analysis, but powder diffraction alone did not allow determination of an unambiguous structure solution.

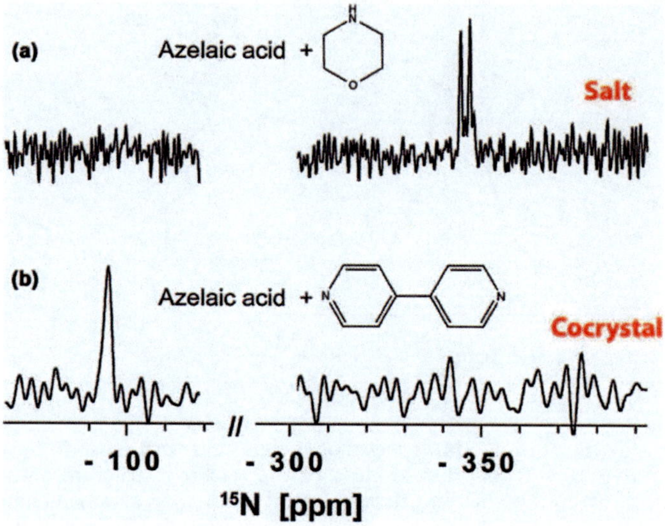

Figure 6.10 1D ^{15}N CPMAS NMR spectra of (a) molecular salt azelaic acid:morpholine and (b) cocrystal azelaic acid:bipiridine.
Reprinted with permission from ref. 141. Copyright 2015, American Chemical Society.

Dudenko *et al.* recently combined XRD, DFT and NMR for the structure validation of the 1:1 cocrystal of indomethacin and nicotinamide also comparing calculated GIPAW-DFT and measured ^{1}H and ^{13}C CSs to assess the validity and quality of the refined crystal structure.[155]

XRPD methods, lattice energy calculations, ^{13}C and ^{15}N SSNMR and a systematic search algorithm were applied to anhydrous theophylline.[156] The trial crystal structure generation unravelled two different H-bonding motifs involving N–H···N and N–H···O intermolecular interactions in a fixed unit cell. These structures were "indistinguishable" in terms of calculated lattice energy and XRD profile fitting to experimental data. SSNMR spectroscopy was applied successfully in this example to distinguish subtle conformational and packing differences by comparison of experimental and calculated isotropic ^{15}N and ^{13}C CSs. In fact, analysis of the NMR data showed that the N–H···N model accounts for the experimental ^{15}N CSs more adequately than the N–H···O structural model.

A high value approach for the structure determination of an enantiotropically-related dimorphic system having low solid–solid conversion temperatures was presented.[77] The crystal structure of the thermodynamically more stable form at room temperature was determined by SCXRD (polymorph 1, $Z'=4$, $Z=16$). The crystal structure of the other form (polymorph 2, $Z'=1$, $Z=4$) was determined using iterative XRPD structure solution methods, assisted by SSNMR experiments (dipolar connectivity and CS measurements). DFT geometry optimizations were used in tandem with Rietveld refinement and NMR CS calculations to improve and verify the structure of polymorph 2. Several experimental techniques were applied to the characterization of both polymorphs, namely, ^{1}H–^{19}F–^{13}C double-CP and ^{1}H–^{13}C CP-HETCOR with FSLG decoupling, ^{1}H–^{1}H DQ-SQ spectroscopy and relaxation measurements. Furthermore, the ^{13}C CST values were calculated and experimentally determined using a 2D ^{13}C CP-PASS NMR experiment.

6.3.5.4 NMR Crystallography Approaches

Crystallography is the field of science that studies the spatial organization of molecules/atoms in a solid. Usually, NMR is used in a passive way (structure validation); over recent years, some examples employing NMR crystallography have emerged, *i.e.*, where SSNMR plays an "active" role in the structure determination step. This section is divided into three representative sub-sections.

6.3.5.4.1 Structure Determination using Proton Spin-Diffusion 2D NMR Spectra. A complete NMR crystallography protocol for the structure determination of molecular crystals was presented by Salager and co-workers.[103] The protocol combines *proton spin-diffusion* (PSD) restraints with molecular modelling to determine the crystalline structure of thymol. The best structures then undergo structure refinement using DFT geometry optimization and are selected based on their agreement with experimental ^{1}H and ^{13}C CS. The build-up curves extracted from the 2D PSD spectra are used as constraints to evaluate the several trial structures generated by molecular modelling. This approach includes a standard force field to ensure reasonable geometries are generated combined with a PSD force field term (E_{PSD}), defined as the weighted goodness-of-fit (χ^2_{PSD}) between the calculated and experimental PSD build up curves. The weight ensures that the agreement with the PSD experimental data will be the dominant term in the modified initial force field. The build-up curves do not provide a direct measure of distance information, as each individual cross-peak (Figure 6.11a) contains contributions from tens of equivalent pairs of exchanging protons. To overcome this problem the authors proposed using a phenomenological rate matrix description of the build-up curves ($P_{ij}(\tau_{SD})$), inspired by previous contributions.[157] With this approach, the intra- and intermolecular distances r_{ij} between protons i and j in the generated candidate structures are computed to simulate $P_{ij}(\tau_{SD})$ defined as:

$$P_{ij}(\tau_{SD}) = \exp(-K\tau_{SD})_{ij} M_z^{j,0}$$

$M_z^{j,0}$ accounts for the volume of the diagonal peaks at $\tau_{SD} = 0$ s and the elements of the K matrix ($K \propto 1/r_{ij}$) may be found in ref. 103. Each cross-peak build-up curve shown in Figure 6.11b represents, in fact, a sum of different curves arising from dipolar interactions with many different equivalent molecules.[157] Each experimental and calculated cross-peak volume obtained from the PSD spectra is used to calculate χ^2_{PSD}. Globally, the reported molecular modelling approach discussed above encompasses three main steps,

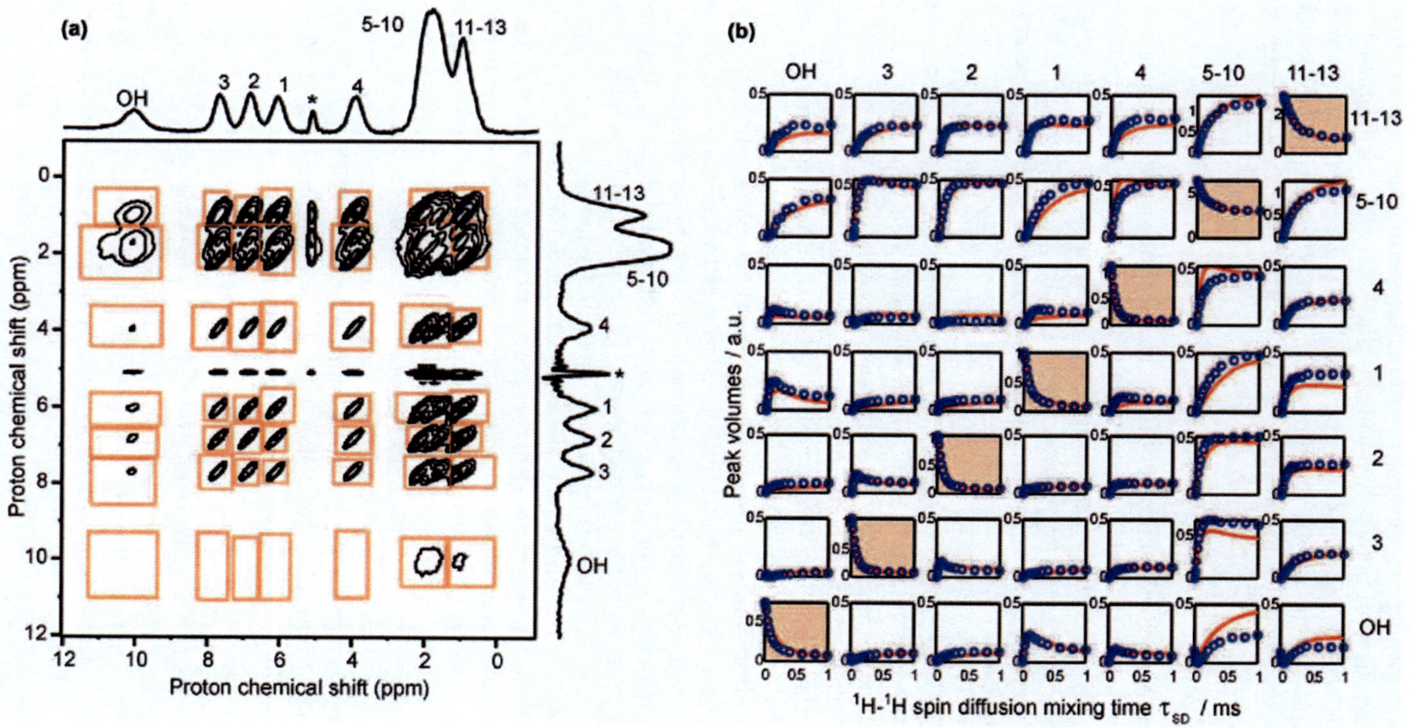

Figure 6.11 (a) 2D ^1H–^1H spin-diffusion spectrum of thymol. Boxes indicate the regions used for integration of each cross-peak. Asterisks indicate the carrier frequency artefacts; (b) ^1H–^1H spin-diffusion build-up curves. Experimental data points are represented by circles; the best fit from the rate matrix analysis using the X-ray structure is shown using solid lines.
Adapted from ref. 103 with permission from the PCCP Owner Societies.

starting from an initial population of 3000 trial structures and ending in the 42 lowest energy PSD-optimized structures (E_{PSD} < 2925 kJ mol^{-1}).[103]

6.3.5.4.2 Combining Crystal Structure Prediction Methods and SSNMR. A protocol for the *ab initio* crystal structure determination of naturally abundant powdered solids was introduced[158] combining CSP computational approaches and involving measured and calculated ^{1}H CSs. This method was also applied to thymol and the obtained structure was identical to the reference structure with an improved *root mean square deviation* (RMSD) compared to the previous PSD-optimization protocol. The same protocol has been applied more recently to four small drug molecules with higher flexibility: cocaine, flutamide, flufenamic acid and theophylline.[159] The correct structure of these four compounds was successfully determined based on the RMSD values resulting from the comparison of experimental and DFT calculated ^{1}H CSs. One of the main conclusions from this study is related to the ^{1}H CSs being much more sensitive to structural changes than ^{13}C CSs. For cocaine, flutamide, and flufenamic acid, the ^{1}H CSs lead to a clear structure determination by comparison with the ensemble of structure predictions, whereas in all of these cases, the ^{13}C CSs were not sufficiently sensitive to lead to structure determination.

6.3.5.4.3 Chemical Shift-Driven Crystal Structure Determination. Our group recently proposed a new strategy for the NMR crystallography of multiple-component molecular crystals in which ^{1}H NMR CSs enter directly in the structure generation step, governed by a genetic algorithm.[160] The methodology developed in this work avoids the use of time-consuming high-level DFT calculations and uses CSs in the structure-refinement step as pseudoforces acting on the models, leading to the lowest energy structure. The proposed protocol takes the unit cell parameters, space group information (both taken from XRPD experiments) and solid-state ^{1}H isotropic CSs as input information and processes the data in 3 steps: (i) generation of an ensemble of structures; (ii) refinement of the generated structures; (iii) high-level energy minimization of the refined structures. Step (i) employs a genetic algorithm where the fitness (E_{total}) of each structure is defined as the sum of the lattice energy ($E_{lattice}$) of the system with a term accounting for the deviation between the experimental (δ_{exp}) and the on-the-fly calculated (δ_{calc}) ^{1}H CSs (E_δ):

$$E_{total} = E_{lattice} + E_\delta$$

$$E_\delta = \sum_h \left\{ \begin{array}{cc} k\left(\delta^h_{calc} - \delta^h_{exp} + \varepsilon_\delta\right)^2, & \delta^h_{calc} \leq \delta^h_{exp} - \varepsilon_\delta \\ 0 & |\delta^h_{calc} - \delta^h_{exp}| < \varepsilon_\delta \\ k\left(\delta^h_{calc} - \delta^h_{exp} - \varepsilon_\delta\right)^2, & \delta^h_{calc} \geq \delta^h_{exp} - \varepsilon_\delta \end{array} \right\}$$

E_{lattice} is described by the classical biomolecular force-field (GAFF) which comprises intramolecular (bonds, angles and dihedrals) and intermolecular (Coulomb and van der Waals) components, the latter having contributions from all unit cells. In E_δ, the sum runs over all of the protons h for which experimental CSs (δ_{exp}) are available. The parameter ε_δ controls the width of the flat-well potential and defines a threshold for the accepted error in the δ_{calc} values. The on-the-fly calculation of δ_{calc} requires a fast methodology, amenable to integration within a classical molecular mechanics/dynamics code. The calculation of δ_{calc} is inspired by previous protein solution NMR work[161–163] and is parametrically calculated according to:

$$\delta_{\text{calc}} = \delta_{\text{local}} + \delta_{\text{ef}} + \delta_{\text{m}} + \delta_{\text{rc}}$$

where δ_{local} is a constant particular to each proton (assumed proportional to its atomic partial charge), and δ_{ef}, δ_{m}, and δ_{rc} are contributions due to electric field effects,[164] group magnetic anisotropy[165,166] and aromatic ring currents.[163,167] The genetic algorithm yields, in step (i), a set of generated structures, which moderately obey the input data. In step (ii), these are refined by means of molecular dynamics simulated annealing. Finally, in step (iii), the set of refined structures is energy minimized using tight-binding DFT. In the whole process, E_{lattice} is responsible for the physical and chemical validity of the generated structures, whereas E_δ is responsible for imposing the experimental constraints derived from SSNMR experiments.

6.4 Future Perspectives

The expected developments in this field involve improvements in the analytical techniques in use as well as the drug formulation process. Pharmaceuticals may be developed in many forms (salt, cocrystal, amorphous, crystallised from ionic liquids, *etc.*) that can induce changes in the existing intermolecular interactions, therefore impacting drug delivery. Amorphous systems, in particular, pose a challenge regarding structural characterization, which can be overcome by SSNMR, especially when combined with the diffraction methods also mentioned in this chapter.

The main limitation of SSNMR is its inherently low sensitivity and in the field of pharmaceutical research, its potential is often hampered by the still limited ^{1}H resolution attainable. Many physico-chemical properties of pharmaceutical systems are connected to hydrogens, which represent a challenge for SSNMR observation. On this topic, both high magnetic fields (1.2 GHz in development) and ultra-fast spinning NMR probes pushing the limits (MAS rates up to 120 kHz) to higher sensitivity and resolution, will certainly help to address the current limitations.

A better integration of the different solid-state characterization techniques is expected, through the interaction of both SSNMR and X-ray diffraction communities and the development of new software with improved crystal structure prediction algorithms, integrating NMR and diffraction parameters in the refinement process. In fact, a new Commission on "NMR Crystallography and Related Methods" was established in 2014 and is now part of the International Union of Crystallography (IUCr). This is a recognition of the importance of SSNMR spectroscopy as a relevant analytical tool for the elucidation of crystal structures.

Acronyms

SSNMR	Solid-state nuclear magnetic resonance
SCXRD	Single-crystal X-ray diffraction
XRPD	X-ray powder diffraction
CSP	Crystal structure prediction
API	Active pharmaceutical ingredient
H-bond/bonding	Hydrogen bond/bonding
CSA	Chemical shift anisotropy
RF	Radio-frequency
CS	Chemical shift
MAS	Magic angle spinning
CP	Cross-polarization
CPMAS	Cross-polarization magic angle spinning
CPMG	Carr–Purcell–Meiboom–Gill
CRAMPS	Combining rotation and multiple-pulse sequences
HETCOR	Heteronuclear correlation
DQ	Double-quantum
DNP	Dynamic nuclear polarization
DDS	Drug delivery system
NICS	Nucleus-independent chemical shifts
GIPAW	Gauge including projected augmented wave
DFT	Density functional theory
PDB	Protein data bank
FT	Fourier transform
PDF	Pair distribution function
DSE	Debye scattering equation
DQ-HLF	Double-quantum heteronuclear local field
EFG	Electric field gradient
CST	Chemical shift tensor
PASS	Phase-adjusted spinning sideband
IR	Infrared
SUPER	Separation of undistorted powder patterns by effortless recoupling
QIS	Quadrupole induced shift
PSD	Proton spin diffusion
RMSD	Root mean square deviation
MIL	Material Institute Lavoisier
MOF	Metal–organic framework.

Acknowledgements

The authors acknowledge Fundação para a Ciência e a Tecnologia (FCT) for funding through projects IF/00993/2012/CP0172/CT0013 and PTDC/QEQ-QAN/6373/2014. This work was developed in the scope of POCI-01-0145-FEDER-007679 | UID/CTM/50011/2013 (CICECO), financed by national funds through the FCT/MEC and co-financed by FEDER under the PT2020 Partnership Agreement. N. M. thanks all of the To.Sca.Lab. co-workers, who, through the years, developed the powerful DFA approach for nanosized and defective materials. Federica Bertolotti (University of Aarhus) is heartily thanked for fruitful discussions and technical support.

References

1. M. Geppi, G. Mollica, S. Borsacchi and C. A. Veracini, *Appl. Spectrosc. Rev.*, 2008, **43**, 202–302.
2. R. T. Berendt, D. M. Sperger, P. K. Isbester and E. J. Munson, *TrAC-Trends Anal. Chem.*, 2006, **25**, 977–984.
3. P. A. Tishmack, D. E. Bugay and S. R. Byrn, *J. Pharm. Sci.*, 2003, **92**, 441–474.
4. G. A. Monti, A. K. Chattah and Y. G. Linck, *Annu. Reports NMR Spectrosc.*, 2014, **83**, 221–269.
5. M. Mehring, in *eMagRes*, John Wiley & Sons, Ltd, Chichester, UK, 2007.
6. M. J. Duer, *Solid-state NMR spectroscopy: principles and applications*, Blackwell Science Ltd, Oxford, UK, 2002.
7. E. R. Andrew, A. Bradbury and R. G. Eades, *Nature*, 1958, **182**, 1659.
8. S. E. E. Ashbrook, *Prog. Nucl. Magn. Reson. Spectrosc.*, 2004, **45**, 53–108.
9. D. Freude, in *Encyclopedia of Analytical Chemistry*, John Wiley & Sons, Ltd, Chichester, UK, 2006.
10. A. Goldbourt and P. K. Madhu, *Annu. Reports NMR Spectrosc.*, 2004, **54**, 81–153.
11. P. Hodgkinson, *Prog. Nucl. Magn. Reson. Spectrosc.*, 2005, **46**, 197–222.
12. B. C. Gerstein, in *eMagRes*, John Wiley & Sons, Ltd, Chichester, UK, 2009.
13. G. De Paëpe, *Annu. Rev. Phys. Chem.*, 2012, **63**, 661–684.
14. J. M. Griffiths, A. E. Bennett and R. G. Griffin, in *eMagRes*, John Wiley & Sons, Ltd, Chichester, UK, 2007.
15. J. Schaefer, in *eMagRes*, John Wiley & Sons, Ltd, Chichester, UK, 2007.
16. D. M. Grant, in *eMagRes*, John Wiley & Sons, Ltd, Chichester, UK, 2007.
17. J. Z. Hu, W. Wang and R. J. Pugmire, in *eMagRes*, John Wiley & Sons, Ltd, Chichester, UK, 2007.
18. S.-F. Liu, J.-D. Mao and K. Schmidt-Rohr, *J. Magn. Reson.*, 2002, **155**, 15–28.
19. M. H. Levitt, *J. Chem. Phys.*, 2008, **128**, 52205-1–52205-25.
20. M. H. Levitt, in *eMagRes*, John Wiley & Sons, Ltd, Chichester, UK, 2007.
21. J. Courtieu, J. P. Bayle and B. M. Fung, *Prog. Nucl. Magn. Reson. Spectrosc.*, 1994, **26**, 141–169.

22. L. Frydman, G. C. Chingas, Y. K. Lee, P. J. Grandinetti, M. A. Eastman, G. A. Barrall and A. Pines, *J. Chem. Phys.*, 1992, **97**, 4800.
23. M. Tomaselli, B. H. Meier, M. Baldus, J. Eisenegger and R. R. Ernst, *Chem. Phys. Lett.*, 1994, **225**, 131–139.
24. R. Tycko, *J. Am. Chem. Soc.*, 1994, **116**, 2217–2218.
25. S. R. Hartmann and E. L. Hahn, *Phys. Rev.*, 1962, **128**, 2042–2053.
26. A. E. Bennett, C. M. Rienstra, M. Auger, K. V. Lakshmi and R. G. Griffin, *J. Chem. Phys.*, 1995, **103**, 6951.
27. B. M. Fung, A. K. Khitrin and K. Ermolaev, *J. Magn. Reson.*, 2000, **142**, 97–101.
28. Z. Gan and R. R. Ernst, *Solid State Nucl. Magn. Reson.*, 1997, **8**, 153–159.
29. A. Equbal, M. Bjerring, K. Sharma, P. K. Madhu and N. C. Nielsen, *Chem. Phys. Lett.*, 2016, **644**, 243–249.
30. R. K. Harris, *Nuclear magnetic resonance spectroscopy: a physicochemical view*, Longman Scientific & Technical, 1987.
31. K. Paradowska and I. Wawer, *J. Pharm. Biomed. Anal.*, 2014, **93**, 27–42.
32. S. J. Opella and M. H. Frey, *J. Am. Chem. Soc.*, 1979, **101**, 5854–5856.
33. X. L. Wu, S. T. Burns and K. W. Zilm, *J. Magn. Reson. Ser. A*, 1994, **111**, 29–36.
34. R. Sangill, N. Rastrupandersen, H. Bildsoe, H. J. Jakobsen and N. C. Nielsen, *J. Magn. Reson. Ser. A*, 1994, **107**, 67–78.
35. A. Lesage, S. Steuernagel and L. Emsley, *J. Am. Chem. Soc.*, 1998, **120**, 7095–7100.
36. A. S. Tatton, T. N. Pham, F. G. Vogt, D. Iuga, A. J. Edwards and S. P. Brown, *CrystEngComm*, 2012, **14**, 2654–2659.
37. K. Takeda, Y. Noda, K. Takegoshi, O. Lafon, J. Trébosc and J.-P. Amoureux, *J. Magn. Reson.*, 2012, **214**, 340–345.
38. G. Hou, F. Deng, S. Ding, R. Fu, J. Yang and C. Ye, *Chem. Phys. Lett.*, 2006, **421**, 356–360.
39. R. L. Johnson and K. Schmidt-Rohr, *J. Magn. Reson.*, 2014, **239**, 44–49.
40. B. C. Gerstein, R. G. Pembleton, R. C. Wilson and L. M. Ryan, *J. Chem. Phys.*, 1977, **66**, 361–362.
41. E. Vinogradov, P. Madhu and S. Vega, *Chem. Phys. Lett.*, 2002, **354**, 193–202.
42. E. Vinogradov, P. K. Madhu and S. Vega, *Chem. Phys. Lett.*, 1999, **314**, 443–450.
43. D. Sakellariou, A. Lesage, P. Hodgkinson and L. Emsley, *Chem. Phys. Lett.*, 2000, **319**, 253–260.
44. E. Salager, R. S. Stein, S. Steuernagel, A. Lesage, B. Elena and L. Emsley, *Chem. Phys. Lett.*, 2009, **469**, 336–341.
45. M. K. Pandey, H. Kato, Y. Ishii and Y. Nishiyama, *Phys. Chem. Chem. Phys.*, 2016, **18**, 6209–6216.
46. D. Burum and A. Bielecki, *An improved experiment for heteronuclear-correlation 2D NMR in Solids*, Academic Press, 1991, vol. 94.
47. P. Caravatti, L. Braunschweiler and R. R. Ernst, *Chem. Phys. Lett.*, 1983, **100**, 305–310.

48. P. Caravatti, G. Bodenhausen and R. R. Ernst, *Chem. Phys. Lett.*, 1982, **89**, 363–367.
49. B.-J. van Rossum, H. Förster and H. J. M. de Groot, *J. Magn. Reson.*, 1997, **124**, 516–519.
50. G. Tricot, J. Trébosc, F. Pourpoint, R. Gauvin and L. Delevoye, *Annu. Rep. NMR Spectrosc.*, 2014, **81**, 145–184.
51. A. Lesage, D. Sakellariou, S. Steuernagel and L. Emsley, *J. Am. Chem. Soc.*, 1998, **120**, 13194–13201.
52. K. Saalwächter, R. Graf and H. W. Spiess, *J. Magn. Reson.*, 1999, **140**, 471–476.
53. S. P. Brown, A. Lesage and B. N. D. Elena, *J. Am. Chem. Soc.*, 2004, **126**, 13230–13231.
54. A. Lesage, C. Auger, S. Caldarelli and L. Emsley, *J. Am. Chem. Soc.*, 1997, **119**, 7867–7868.
55. A. Lesage, M. Bardet and L. Emsley, *J. Am. Chem. Soc.*, 1999, **121**, 10987–10993.
56. Q. Z. Ni, E. Daviso, T. V. Can, E. Markhasin, S. K. Jawla, T. M. Swager, R. J. Temkin, J. Herzfeld and R. G. Griffin, *Acc. Chem. Res.*, 2013, **46**, 1933–1941.
57. V. S. Bajaj, C. T. Farrar, M. K. Hornstein, I. Mastovsky, J. Vieregg, J. Bryant, B. Eléna, K. E. Kreischer, R. J. Temkin and R. G. Griffin, *J. Magn. Reson.*, 2003, **160**, 85–90.
58. A. J. J. Rossini, A. Zagdoun, F. S. Hegner, M. Schwarzwälder, D. Gajan, C. Coperet, A. Lesage and L. Emsley, *J. Am. Chem. Soc.*, 2012, **134**, 16899–16908.
59. F. A. Perras, T. Kobayashi and M. Pruski, *J. Am. Chem. Soc.*, 2015, **137**, 8336–8339.
60. E. Carignani, S. Borsacchi and M. Geppi, *Chem. Phys. Chem.*, 2011, **12**, 974–981.
61. J. Hirschinger and A. D. English, *J. Magn. Reson.*, 1989, **85**, 542–553.
62. C. Fülber, D. E. Demco and B. Blümich, *Solid State Nucl. Magn. Reson.*, 1996, **6**, 213–223.
63. S. Schantz, P. Hoppu and A. M. Juppo, *J. Pharm. Sci.*, 2009, **98**, 1862–1870.
64. S. Yoshioka, Y. Aso, T. Osako and T. Kawanishi, *J. Pharm. Sci.*, 2008, **97**, 4258–4268.
65. J. W. Lubach, D. Xu, B. E. Segmuller and E. J. Munson, *J. Pharm. Sci.*, 2007, **96**, 777–787.
66. G. A. Facey, C. I. Ratcliffe and J. A. Ripmeester, *J. Phys. Chem.*, 1995, **99**, 12249–12256.
67. A. Abraham, D. C. Apperley, S. J. Byard, A. J. Ilott, A. J. Robbins, V. Zorin, R. K. Harris and P. Hodgkinson, *CrystEngComm*, 2016, **18**, 1054–1063.
68. D. C. C. Apperley, A. F. Markwell, I. Frantsuzov, A. J. Ilott, R. K. Harris and P. Hodgkinson, *Phys. Chem. Chem. Phys.*, 2013, **15**, 6422–6430.
69. P. von Ragué Schleyer, C. Maerker, A. Dransfeld, H. Jaio and N. J. R. van Eikema Hommes, *J. Am. Chem. Soc.*, 1996, **118**, 6317–6318.

70. D. L. L. Bryce, G. D. Sward and S. Adiga, *J. Am. Chem. Soc.*, 2006, **128**, 2121–2134.
71. H. Hamaed, J. M. Pawlowski, B. F. T. Cooper, R. Fu, S. H. Eichhorn and R. W. W. Schurko, *J. Am. Chem. Soc.*, 2008, **130**, 11056–11065.
72. A. S. Tatton, T. N. Pham, F. G. Vogt, D. Iuga, A. J. Edwards and S. P. Brown, *Mol. Pharm.*, 2013, **10**, 999–1007.
73. X. Kong, M. Shan, V. Terskikh, I. Hung, Z. Gan and G. Wu, *J. Phys. Chem. B*, 2013, **117**, 9643–9654.
74. J. R. Smith, W. Xu and D. Raftery, *J. Phys. Chem. B*, 2006, **110**, 7766–7776.
75. E. M. Heider, J. K. Harper and D. M. M. Grant, *Phys. Chem. Chem. Phys.*, 2007, **9**, 6083–6097.
76. D. V. Dudenko, J. R. Yates, K. D. M. Harris and S. P. Brown, *CrystEngComm*, 2013, **15**, 8797.
77. F. G. Vogt, L. M. M. Katrincic, S. T. Long, R. L. Mueller, R. A. A. Carlton, Y. T. Sun, M. N. Johnson, R. C. B. Copley and M. E. Light, *J. Pharm. Sci.*, 2008, **97**, 4756–4782.
78. X. Filip, I.-G. Grosu, M. Miclăuş and C. Filip, *CrystEngComm*, 2013, **15**, 4131–4142.
79. C. J. Pickard and F. Mauri, *Phys. Rev. B*, 2001, **63**, 245101.
80. R. K. Harris, P. Hodgkinson, C. J. Pickard, J. R. Yates and V. Zorin, *Magn. Reson. Chem.*, 2007, **45**, S174–S186.
81. J. R. Deschamps, *AAPS J.*, 2005, **7**, E813–E819.
82. T. R. Schneider, *HFSP J.*, 2008, **2**, 302–306.
83. K. T. Phuoc, F. Burgy, J.-P. Rousseau, V. Malka, A. Rousse, R. Shah, D. Umstadter, A. Pukhov and S. Kiselev, *Phys. Plasmas*, 2005, **12**, 23101-1–23101-8.
84. S. Kneip, C. McGuffey, J. L. Martins, S. F. Martins, C. Bellei, V. Chvykov, F. Dollar, R. Fonseca, C. Huntington, G. Kalintchenko, A. Maksimchuk, S. P. D. Mangles, T. Matsuoka, S. R. Nagel, C. A. J. Palmer, J. Schreiber, K. T. Phuoc, A. G. R. Thomas, V. Yanovsky, L. O. Silva, K. Krushelnick and Z. Najmudin, *Nat. Phys.*, 2010, **6**, 980–983.
85. G. H. Stout and L. H. Jensen, *X-ray structure determination: a practical guide*, Wiley, 1989.
86. M. F. C. Ladd and R. A. Palmer, *Structure determination by x-ray crystallography: analysis by x-rays and neutrons*, Springer, 2013.
87. L. Palatinus, in *Uniting Electron Crystallography and Powder Diffraction. NATO Science for Peace and Security Series B: Physics and Biophysics*, ed. U. Kolb, K. Shankland, L. Meshi, A. Avilov, W. David, Springer, Dordrecht, 2012, pp. 105–116.
88. L. Palatinus, *Acta Cryst*, 2013, **69**, 1–16.
89. G. Oszlányi and A. Sütő, *Acta Crystallogr. Sect. A Found. Crystallogr.*, 2008, **64**, 123–134.
90. G. M. Sheldrick, *Acta Crystallogr. Sect. A Found. Adv.*, 2015, **71**, 3–8.
91. N. Masciocchi and C. Vladiskovic, in *Crystallography for Health and Biosciences*, ed. A. Guagliardi and N. Masciocchi, Insubria University Press, Varese, 2012.

92. V. K. Pecharsky and P. Y. Zavalij, *Fundamentals of Powder Diffraction and Structural Characterization of Materials*, Springer, New York, 2009.
93. C. Vladiskovic and N. Masciocchi, *J. Pharm. Biomed. Anal.*, 2015, **107**, 394–402.
94. T. Egami and S. J. L. Billinge, *Underneath the Bragg peaks: structural analysis of complex materials*, Pergamon Press, Elsevier Ltd., Oxford, 2003.
95. P. Debye, *Ann. Phys.*, 1915, **351**, 809–823.
96. A. Cervellino, C. Giannini and A. Guagliardi, *J. Appl. Crystallogr.*, 2010, **43**, 1543–1547.
97. A. Cervellino, R. Frison, F. Bertolotti and A. Guagliardi, *J. Appl. Crystallogr.*, 2015, **48**, 2026–2032.
98. F. G. Vogt, *Futur. Med. Chem.*, 2010, **2**, 915–921.
99. B. E. Padden, M. T. Zell, Z. Dong, S. A. Schroeder, D. J. W. Grant and E. J. Munson, *Anal. Chem.*, 1999, **71**, 3325–3331.
100. R. K. Harris, P. Hodgkinson, T. Larsson, A. Muruganantham, I. Ymen, D. S. Yufit and V. Zorin, *Cryst. Growth Des.*, 2008, **8**, 80–90.
101. R. K. Harris, S. A. Joyce, C. J. J. Pickard, S. Cadars and L. Emsley, *Phys. Chem. Chem. Phys.*, 2005, **8**, 137–143.
102. N. Mifsud, B. Elena, C. J. J. Pickard, A. Lesage and L. Emsley, *Phys. Chem. Chem. Phys.*, 2006, **8**, 3418–3422.
103. E. Salager, R. S. Stein, C. J. J. Pickard, B. Elena and L. Emsley, *Phys. Chem. Chem. Phys.*, 2009, **11**, 2610–2621.
104. R. K. Harris, S. Cadars, L. Emsley, J. R. Yates, C. J. J. Pickard, R. K. R. Jetti and U. J. Griesser, *Phys. Chem. Chem. Phys.*, 2007, **9**, 360–368.
105. F. G. Vogt, J. Brum, L. M. M. Katrincic, A. Flach, J. M. Socha, R. M. Goodman and R. C. Haltiwanger, *Cryst. Growth Des.*, 2006, **6**, 2333–2354.
106. S. M. Reutzel-Edens, J. K. Bush, P. A. Magee, G. A. Stephenson and S. R. Byrn, *Cryst. Growth Des.*, 2003, **3**, 897–907.
107. D. A. Middleton, C. S. Le Duff, X. Peng, D. G. Reid and D. Saunders, *J. Am. Chem. Soc.*, 2000, **122**, 1161–1170.
108. A. Uldry, J. M. Griffin, J. R. Yates, M. Perez-Torralba, M. D. S. MAria, A. L. Webber, M. L. L. Beaumont, A. Samoson, R. M. M. Claramunt, C. J. J. Pickard and S. P. Brown, *J. Am. Chem. Soc.*, 2008, **130**, 945–954.
109. L. Mafra, S. M. Santos, R. Siegel, I. Alves, F. A. Almeida Paz, D. V. Dudenko and H. W. Spiess, *J. Am. Chem. Soc.*, 2012, **134**, 71–74.
110. R. Koike, K. Higashi, N. Liu, W. Limwikrant, K. Yamamoto and K. Moribe, *Cryst. Growth Des.*, 2014, **14**, 4510–4518.
111. R. K. Chandrappa, P. Ochsenbein, C. Martineau, M. Bonin, G. Althoff, F. Engelke, H. Malandrini, B. Castro, M. El Hajji and F. Taulelle, *Cryst. Growth Des.*, 2013, **13**, 4678–4687.
112. W. D. Wang, X. Gao, M. Strohmeier, W. Wang, S. Bai and C. Dybowski, *J. Phys. Chem. B*, 2012, **116**, 3641–3649.
113. A. C. Pinon, A. J. Rossini, C. M. Widdifield, D. Gajan and L. Emsley, *Mol. Pharm.*, 2015, **12**, 4146–4153.

114. A. Paudel, M. Geppi and G. Van Den Mooter, *J. Pharm. Sci.*, 2014, **103**, 2635–2662.
115. T. N. N. Pham, S. A. Watson, A. J. Edwards, M. Chavda, J. S. S. Clawson, M. Strohmeier and F. G. Vogt, *Mol. Pharm.*, 2010, **7**, 1667–1691.
116. X. Yuan, D. Sperger and E. J. Munson, *Mol Pharm*, 2014, **11**, 329–337.
117. F. G. Vogt, K. Roberts-Skilton and S. A. Kennedy-Gabb, *Pharm. Res.*, 2013, **30**, 2315–2331.
118. O. Policianova, J. Brus, M. Hruby, M. Urbanova, A. Zhigunov, J. Kredatusova and L. Kobera, *Mol. Pharm.*, 2014, **11**, 516–530.
119. T. G. Nunes, M. T. Viciosa, N. T. Correia, F. Danède, R. G. Nunes and H. P. Diogo, *Mol. Pharm.*, 2014, **11**, 727–737.
120. A. Newman, D. Engers, S. Bates, I. Ivanisevic, R. C. Kelly and G. Zografi, *J. Pharm. Sci.*, 2008, **97**, 4840–4856.
121. P. Kanaujia, P. Poovizhi, W. K. Ng and R. B. H. Tan, *Powder Technol.*, 2015, **285**, 2–15.
122. S. J. L. Billinge, T. Dykhne, P. Juhás, E. Božin, R. Taylor, A. J. Florence and K. Shankland, *CrystEngComm*, 2010, **12**, 1366–1368.
123. Y. Nguyen Thi, K. Rademann and F. Emmerling, *CrystEngComm*, 2015, **17**, 9029–9036.
124. M. G. Tucker, D. A. Keen, M. T. Dove, A. L. Goodwin and Q. Hui, *J. Phys. Condens. Matter*, 2007, **19**, 335218.
125. D. Hasa, C. Giacobbe, B. Perissutti, D. Voinovich, M. Grassi, A. Cervellino, N. Masciocchi and A. Guagliardi, *Mol. Pharm.*, 2016, **13**, 3034–3042.
126. J. A. Mason, J. Oktawiec, M. K. Taylor, M. R. Hudson, J. Rodriguez, J. E. Bachman, M. I. Gonzalez, A. Cervellino, A. Guagliardi, C. M. Brown, P. L. Llewellyn, N. Masciocchi and J. R. Long, *Nature*, 2015, **527**, 357–361.
127. A. Gavezzotti, *Molecular Aggregation*, Oxford University Press, NY, 2007.
128. E. Skorupska, A. Jeziorna, S. Kazmierski and M. J. Potrzebowski, *Solid State Nucl. Magn. Reson.*, 2014, **57–58**, 2–16.
129. T. Azaïs, C. Tourné-Péteilh, F. Aussenac, N. Baccile, C. Coelho, J. M. Devoisselle and F. Babonneau, *Chem. Mater.*, 2006, **18**, 6382–6390.
130. X.-P. Tang, N. C. Ng, H. Nguyen, G. Mogilevsky and Y. Wu, *Chem. Phys. Lett.*, 2008, **452**, 289–295.
131. R. Amorim, N. Vilaça, O. Martinho, R. M. Reis, M. Sardo, J. Rocha, A. M. Fonseca, F. Baltazar and I. C. Neves, *J. Phys. Chem. C*, 2012, **116**, 25642–25650.
132. N. Vilaça, F. Morais-Santos, A. F. Machado, A. Sirkecioğlu, M. F. R. Pereira, M. Sardo, J. Rocha, P. Parpot, A. M. Fonseca, F. Baltazar and I. C. Neves, *J. Phys. Chem. C*, 2015, **119**, 3589–3595.
133. N. Vilaça, R. Amorim, A. F. Machado, P. Parpot, M. F. R. Pereira, M. Sardo, J. Rocha, A. M. Fonseca, I. C. Neves and F. Baltazar, *Colloids Surf., B*, 2013, **112**, 237–244.

134. P. Horcajada, C. Serre, G. Maurin, N. A. Ramsahye, F. Balas, M. Vallet-Regí, M. Sebban, F. Taulelle and G. Férey, *J. Am. Chem. Soc.*, 2008, **130**, 6774–6780.
135. F. G. Vogt and M. Strohmeier, *Mol. Pharm.*, 2012, **9**, 3357–3374.
136. K. Anzai, H. Kono, J. Mizoguchi, T. Yanagi, F. Hirayama, H. Arima and K. Uekama, *Carbohydr. Res.*, 2006, **341**, 499–506.
137. M. Nonappa, S. Lahtinen, E. Ikonen, Kolehmainen and R. Kauppinen, *Cryst. Growth Des.*, 2009, **9**, 4710–4719.
138. M. O. Miclaus, X. Filip, C. Filip, F. A. Martin and I. G. Grosu, *J. Pharm. Biomed. Anal.*, 2016, **124**, 274–280.
139. J. M. Griffin, D. R. Martin and S. P. Brown, *Angew. Chem. Int. Ed.*, 2007, **46**, 8036–8038.
140. J. Brus, M. Urbanova, I. Sedenkova and H. Brusova, *Int. J. Pharm.*, 2011, **409**, 62–74.
141. I. C. B. Martins, M. Sardo, S. M. Santos, A. Fernandes, A. Antunes, V. André, L. Mafra and M. T. Duarte, *Cryst. Growth Des.*, 2015, 154–166.
142. J. S. Stevens, S. J. Byard, C. C. Seaton, G. Sadiq, R. J. Davey and S. L. M. Schroeder, *Phys. Chem. Chem. Phys.*, 2014, **16**, 1150–1160.
143. Z. J. Li, Y. Abramov, J. Bordner, J. Leonard, A. Medek and A. V. Trask, *J. Am. Chem. Soc.*, 2006, **128**, 8199–8210.
144. J. Mason, *eMagRes*, John Wiley & Sons, Ltd, Chichester, UK, 2007.
145. M. Witanowski, L. Stefaniak and G. A. Webb, 1982, pp. 1–486.
146. M. Khan, V. Enkelmann and G. Brunklaus, *CrystEngComm*, 2011, **13**, 3213.
147. M. Khan, V. Enkelmann and G. Brunklaus, *J. Am. Chem. Soc.*, 2010, **132**, 5254–5263.
148. E. Pindelska, I. D. Madura, A. Szeleszczuk, J. Żeszko, P. H. Jaśkowska, Marek and W. Kolodziejski, *Cryst. Growth Des.*, 2016, **16**, 6371–6380.
149. F. G. Vogt, J. S. S. Clawson, M. Strohmeier, A. J. Edwards, T. N. N. Pham and S. A. Watson, *Cryst. Growth Des.*, 2008, **9**, 921–937.
150. D. Sebastiani, *Chem. Phys. Chem.*, 2006, **7**, 164–175.
151. J. R. Yates, T. N. Pham, C. J. J. Pickard, F. Mauri, A. M. Amado, A. M. Gil and S. P. Brown, *J. Am. Chem. Soc.*, 2005, **127**, 10216–10220.
152. R. K. Harris, *J. Pharm. Pharmacol.*, 2007, **59**, 225–239.
153. J. A. Fernandes, M. Sardo, L. Mafra, D. Choquesillo-Lazarte and N. Masciocchi, *Cryst. Growth Des.*, 2015, 3674–3683.
154. A. E. Watts, K. Maruyoshi, C. E. Hughes, S. P. Brown and K. D. M. Harris, *Cryst. Growth Des.*, 2016, **16**, 1798–1804.
155. D. V. Dudenko, P. A. Williams, C. E. Hughes, O. N. N. Antzutkin, S. P. Velaga, S. P. Brown and K. D. M. Harris, *J. Phys. Chem. C*, 2013, **117**, 12258–12265.
156. E. D. L. Smith, R. B. Hammond, M. J. Jones, K. J. Roberts, J. B. O. Mitchell, S. L. Price, R. K. Harris, D. C. C. Apperley, J. C. Cherryman and R. Docherty, *J. Phys. Chem. B*, 2001, **105**, 5818–5826.
157. B. N. D. Elena and L. Emsley, *J. Am. Chem. Soc.*, 2005, **127**, 9140–9146.

158. E. Salager, G. M. Day, R. S. Stein, C. J. J. Pickard, B. N. D. Elena and L. Emsley, *J. Am. Chem. Soc.*, 2010, **132**, 2564–2566.
159. M. Baias, C. M. Widdifield, J. N. Dumez, H. P. G. Thompson, T. G. Cooper, E. Salager, S. Bassil, R. S. Stein, A. Lesage, G. M. Day and L. Emsley, *Phys. Chem. Chem. Phys.*, 2013, **15**, 8069.
160. S. M. Santos, J. Rocha and L. Mafra, *Cryst. Growth Des.*, 2013, **13**, 2390–2395.
161. C. A. Hunter, M. J. Packer and C. Zonta, *Prog. Nucl. Magn. Reson. Spectrosc.*, 2005, **47**, 27–39.
162. K. J. Cross and P. E. Wright, *J. Magn. Reson.*, 1985, **64**, 220–231.
163. K. Osapay and D. A. Case, *J. Am. Chem. Soc.*, 1991, **113**, 9436–9444.
164. R. Wolff and R. Radeglia, in *eMagRes*, ed. R. K. Harris and R. L. Wasylishen, John Wiley & Sons, Ltd, Chichester, UK, 2007.
165. A. D. Buckingham, *Can. J. Chem.*, 1960, **38**, 300–307.
166. M. Grayson and W. T. Raynes, *Magn. Reson. Chem.*, 1995, **33**, 138–143.
167. H. M. McConnell, *J. Chem. Phys.*, 1957, **27**, 226.

CHAPTER 7

Surface Plasmon Resonance for Identifying and Characterising Small Molecule Ligands

S. J. POLLACK

Bioscience Department, Sygnature Discovery, BioCity, Nottingham NG1 1GF, UK
Email: s.pollack@sygnaturediscovery.com

7.1 Introduction and Perspectives

7.1.1 Principles of SPR

Surface plasmon resonance (SPR) is a powerful biophysical technology that allows the interactions of virtually any two molecular species, from protein–protein to DNA-small molecule, to be observed in real time. It has been approximately 25 years since the first commercially available SPR instruments were introduced into the drug discovery scientific community by Biacore (now part of GE Healthcare). During this time, SPR instruments have evolved dramatically along with both the breadth and depth of their utility for applications in drug discovery and a host of other disciplines from environmental pathogen analysis to food processing quality control. The wide range of SPR applications in the realm of drug discovery will be discussed. This chapter first explores how SPR can accelerate the different stages of drug discovery, from initial hit finding (particularly the screening of fragment libraries) to mechanistic kinetic and thermodynamic profiling during hit validation and lead optimisation phases. It will become apparent that each of these stages of drug discovery involves somewhat different applications and hence different considerations for the optimal design and interpretation of SPR experiments.

There are many reviews that describe the technology behind SPR (see for example ref. 1), and its quantum mechanical principles are beyond the scope of this review. In essence, SPR works by acting as an extremely sensitive mass detector, quantifying subtle changes in mass (down to as low as 50 Da with some SPR instruments) as an analyte flows over and binds to a target immobilised on a gold-coated sensor surface (Figure 7.1, panel A). Plane-polarised light shines onto the opposite side of the prism-like surface under conditions of total internal reflection, generating surface plasmons at a critical angle of the incident light. The resulting absorption of the reflected light is observed as a decrease in intensity at this critical angle, which depends on the refractive index of the local environment. The increase in mass due to molecules binding to the sensor surface results in a linear change in this refractive index on the surface, thus changing the critical angle observed (Figure 7.1, panel B). SPR's strength lies in its ability to detect these mass changes with high sensitivity in real time and under a range of conditions as analytes are passed over the immobilised target surface, allowing the kinetics (association and dissociation rate constants) of the interactions to be assessed from the so-called sensorgram plot of binding response *vs.* time (Figure 7.1, panels C and D). In addition to kinetic-derived affinities, absolute binding levels and therefore steady state affinities can be determined routinely from the concentration dependence of binding (*i.e.*, using Languir binding isotherms). Moreover, temperature-dependent thermodynamic properties of the binding interactions can be readily assessed. In this way, SPR provides an extremely information-rich analysis of a biophysical interaction. As the target (often protein) of interest is immobilised on a surface over which the screening analytes are passed, very little protein (usually microgram amounts) is needed for a large SPR screen, unlike other biophysical techniques such as ITC that require large (often multi-milligram) amounts of protein. Furthermore, SPR binding measurements tend to be highly reproducible and correlate with data produced by other technologies (Figure 7.2). The affinity range in which SPR operates effectively is vast, from weak millimolar to very potent picomolar binding interactions.[2] This range surpasses those of most biochemical and biophysical technologies.

7.1.2 Overcoming the Challenges of SPR

For a balanced view of SPR, it is also important to consider its limitations and challenges. SPR is often discussed as one of many label-free biophysical techniques in that, unlike other biochemical and some biophysical approaches, it does not require the biochemical target species (typically a protein) or its ligand binding partner (*e.g.*, substrate, polypeptide interaction partner or small molecule reference inhibitor) to be labeled in order to detect a binding response. SPR can use the endogenous target and binding

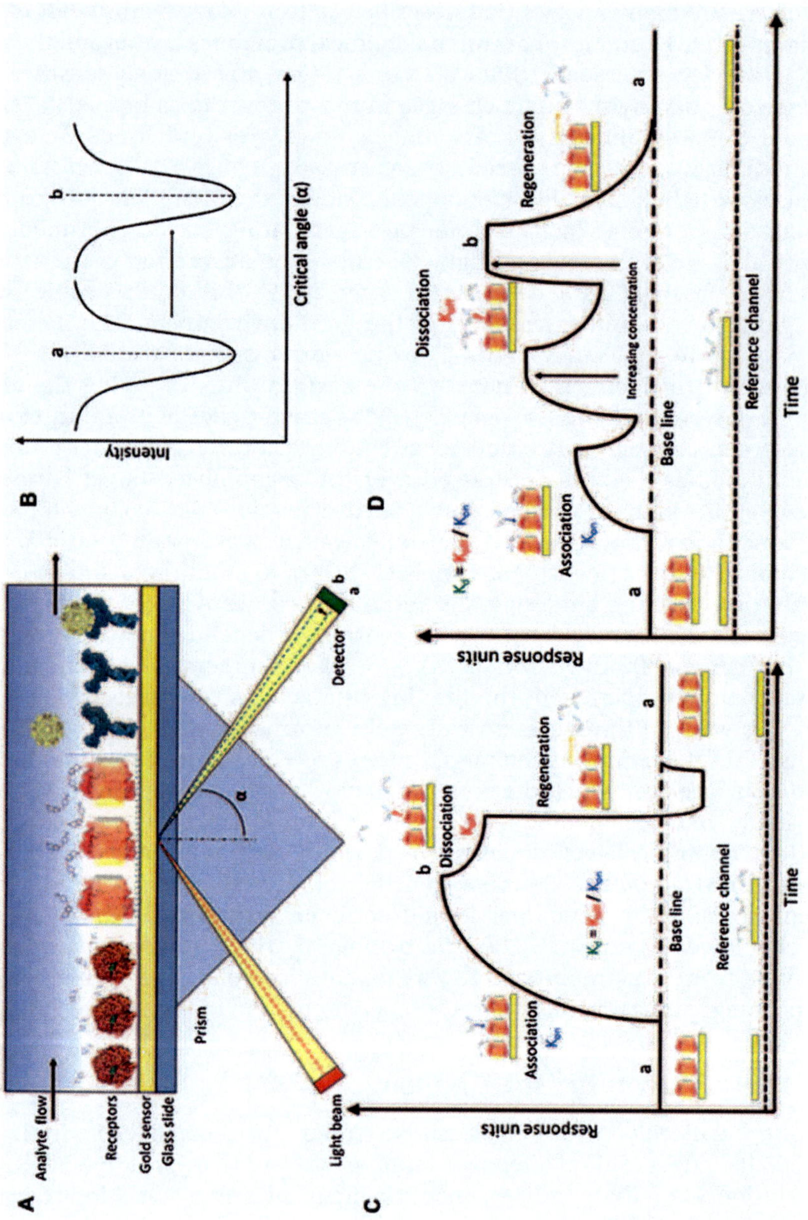
A
Analyte flow
Receptors
Gold sensor
Glass slide
Prism
Light beam
Detector
α
a
b
B
Intensity
Critical angle (α)
C
Response units
Association
Dissociation
Regeneration
Base line
Reference channel
Time
D
Increasing concentration

partners and therein lies one of its strengths. In this sense, SPR is label-free and is not subject to some of the interfering issues (*e.g.*, false positives) encountered with fluorescence-based, luminescence-based and other biochemical detection technologies. However, SPR is not a truly label-free technology because the protein (or non-protein) target must be immobilised at the surface of a gold sensor chip, either *via* covalent chemical coupling or *via* capture of a chemical feature of the target, such as biotinylation or an affinity tag. This immobilisation presents the target to potential analytes passing over the surface in a fashion that may not reflect its native conformation or orientation, as it might in a more homogeneous system. Nevertheless, a protein/target that is immobilised for SPR is more appropriately considered as tethered than immobilised, since it is only bound (ideally) at a remote site (often one of the termini in a protein target), leaving the functional portion of the molecule (*e.g.*, enzyme active site) free to adopt its native conformation. Of course, the physiological nature of the conformation(s) that can be adopted by a target tethered for SPR can and should be routinely ascertained through a number of controlled experiments, comparing the binding behavior of reference analytes with the properties expected based on other biophysical and biochemical techniques.

Figure 7.1 Illustration of the fundamental principles of SPR for drug discovery. Panel A: In SPR, plane-polarised light shines onto a prism-like gold-coated sensor surface under conditions of total internal reflection, generating surface plasmons observed as a decrease in intensity at a critical angle of the incident light (a) that is dependent on the diffraction index of the local environment near the surface. As an analyte flows over and binds to a target immobilised on the opposite side of the sensor surface, the increase in mass results in a linear change in this refractive index on the surface, thus changing the critical angle observed (b), also depicted in panel B. The schematic in panel A also illustrates representative target immobilisation strategies and binding interactions that can be assessed by SPR: a fragment binding to an immobilized soluble protein (left set); a small molecule binding to a membrane protein immobilized in a nanodisc (middle) and a protein binding to an immobilized antibody (right). Panel C: sensorgram of binding interactions observed by SPR. The mass changes detected in real time as analytes are passed over the immobilised target surface allow the kinetics (association and dissociation rate constants) of the interactions to be assessed from the so-called sensorgram plot of binding response *vs.* time. This panel illustrates a typical sensorgram for an analyte binding event, followed by dissociation and finally surface regeneration to prepare the surface for the next cycle. The bound (b) and unbound (a) forms of the immobilised target correspond to the critical angle changes depicted in panels A and B, quantified in so-called response units (RU), where 1 RU is defined as a shift in the critical angle of 10^{-4} of a degree. Panel D: a sensorgram depicting single-cycle kinetics, which can be used to assess a compound binding profile in a rapid fashion. A series of increasing concentrations of the test compound is injected in a single injection cycle, with complete dissociation (and regeneration if necessary) only at the end of the cycle.

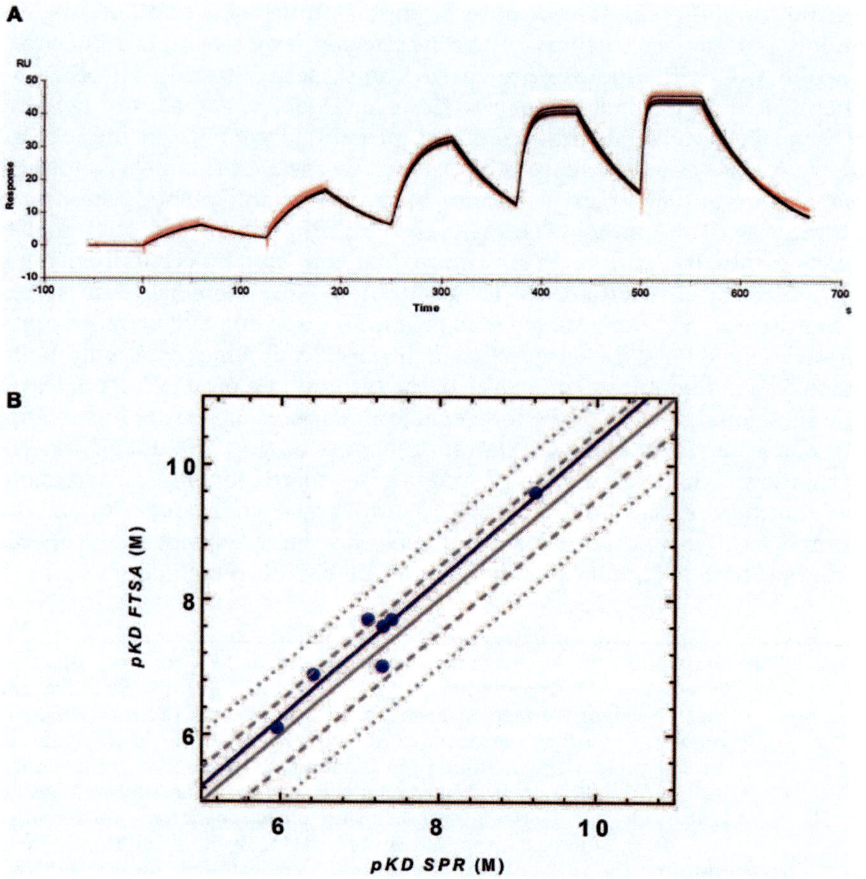

Figure 7.2 Panel A: Example data illustrating the reproducibility of SPR binding measurements. Single-cycle kinetic sensorgram data from 40 repeated injections of a compound five-concentration series, showing the high degree of overlap in the responses obtained. Panel B: Example of the consistency of SPR affinity data with analogous affinity data obtained from other biophysical technologies, in this case, a fluorescence thermal shift assay. Actual values are indicated by blue circles, with the black line representing the general trend. The solid grey line is the line of unity, and the dashed and dotted grey lines represent half an order of magnitude and an order of magnitude deviation in affinity from unity, respectively. Panel B is reprinted from *Analytical Biochemistry*, vol. 479, M. Redhead, R. Satchell, V. Morkūnaitė, D. Swift, V. Petrauskas, E. Golding, S. Onions, D. Matulis and J. Unitt, A combinatorial biophysical approach; FTSA and SPR for identifying small molecule ligands and PAINs, pp. 63–73, Copyright 2015, with permission from Elsevier.[13]

A related limitation that should be borne in mind with SPR *vis-à-vis* biochemical assays is that the immobilised protein is typically in the apo state, without associated substrates and cofactors. This format might fail to detect interactions with compounds that bind preferentially to ligand-bound conformations of the protein target. Although it is possible to include these components in the running buffer, this can complicate matters if enzymatic turnover takes place during the analysis. SPR, as with other direct binding biophysical approaches, does not assess the functionality of binding effects. For example, unless discernable from their different binding kinetics, SPR cannot readily distinguish receptor agonists from antagonists, or enzyme inhibitors from activators or indeed non-functional binders. Therefore, for hit identification, it is often advisable to run a functional activity screen initially or in parallel for these reasons.

On the other hand, in this regard, SPR can be viewed as a completely unbiased technology that does not rely on prior knowledge of the functions of the targets involved. For example, to characterise the interactions of a novel protein considered as a potential drug target with its binding partner, it is not necessary to understand or recapitulate the complete physiological context of the interaction, assembling other interacting partners, cofactors and conditions, as might be the case for functional and cell-based approaches. The natural ligand of a protein may not even be known, yet the protein could still be probed *via* an SPR screen for interacting analytes. Nevertheless, SPR experiments in the drug discovery context should generally be designed to be as (patho)physiologically relevant as possible.

Having a flow system in which analytes pass over a target surface naturally limits the throughput of SPR experiments as sufficient time is required to pass the analyte over the surface to (ideally) obtain a binding equilibrium and then to dissociate the analyte from the surface. This dissociation often requires a regeneration step with conditions that must be finely balanced to completely remove the analyte to allow subsequent cycles without damaging the target surface functionality in the process. Indeed, for some SPR protocols this end is only achievable by completely removing the target from the surface and re-immobilising it from scratch, further increasing the time and reagent costs of each test cycle and decreasing the throughput. Nevertheless, generally, SPR assays can be run with mild regeneration steps (if tolerated by the immobilised target) or without the need for regeneration at all (if analyte dissociation is rapid). Throughput is further enhanced by having several flow cells in parallel, as is the design of more recent SPR instruments such as the Biacore 4000 (four flow cells each with five surfaces), the Biacore 8K (16 flow cells in eight channels) and the BioRad ProteOn™ XPR36 (6×6 interaction array).[3] However, a related limitation of a flow-based system remains in that, despite the best possible regeneration conditions, samples pass over a surface that may be tainted by whatever has previous been passed over that surface. This could lead to fouling of the surface for subsequent cycles, particularly in the case of screening unknown compounds at high concentrations (*e.g.*, fragment screens) or screening antibodies present in crude cell supernatants. Much has been written about due care for SPR systems *via* careful sample preparation and buffer treatment to minimise the risk of affecting the experiment or instrument.[4] Nevertheless, it will always be important to monitor the functionality and stability of the target surface using the binding properties of known analytes throughout each SPR run to ensure that it is performing as expected. Some SPR-related systems have circumvented the issue of cross-contaminating flow cells between injection cycles or avoided microfluidics altogether. For example, FortéBio Octet instruments (which utilise biolayer interferometry technology related to SPR and are discussed later) have up to 96 independent channels with individual biosensing detectors in a plate-based format.[5] However, their sensitivity for detecting smaller analytes is more limited than that of the Biacore instruments with fewer channels. Overall, it is important to interpret SPR data appropriately to recognise false positives caused by artefactual responses (often due to aggregation or nonspecific binding). It is usually straightforward to distinguish real binding responses by using appropriate control surfaces and carefully examining sensorgrams and curve fits, facilitated by recent instrumentation software improvements.

Lastly, we return to the Achilles heel of SPR technology, which was alluded to in the sense that it makes SPR not a truly label-free technology: just as the repeated cry of "location" is the clichéd mantra of the estate agent, for SPR it is "immobilisation, immobilisation, immobilisation!" While immobilisation is what gives SPR its power to assess analyte interactions in real time, tethering the protein/target molecule at a suitable density and in a conformation that is at the same time functional, physiological and stable for the duration required, remains the biggest challenge of SPR. Physically getting a target immobilised onto the surface is the easy part. However, as a consequence of the principles underlying SPR, the magnitude of the signal arising from analyte binding is proportional to both the molecular mass of the analyte and the number of target molecules tethered to the surface, but inversely proportional to the molecular mass of the target. Hence, the ability to detect binding events diminishes with smaller analyte or larger target molecules. Achieving a suitably high immobilisation density can itself be a limitation of SPR when it comes to screening large protein targets and complexes for small molecule binding interactions. It will be shown how this situation cannot necessarily be overcome simply by maximising the amount of ligand immobilised on the surface, since this potentially introduces kinetic complications (*e.g.*, mass transport limitation and rebinding) that limit the ability to interpret the observed binding events.

An effective SPR immobilisation method requires the protein/target to be tethered in an active conformation for the duration and conditions of the experiment or screening run, often including relatively harsh regeneration conditions required to remove residual compounds bound to the surface. While it should be emphasised that a wide range of approaches can be and have been taken to immobilise proteins or other targets successfully, sometimes it is just not possible despite heroic efforts to find an immobilisation approach that keeps proteins functionally active when tethered to the SPR surface. This may be due to inherent instability of the protein or its requirement for a particular complexation, cofactors or buffering conditions. Membrane protein targets have their own particular challenges in this regard that have sometimes been overcome successfully, sometimes not. The next section explores how immobilisation methodologies can be applied appropriately to the design of SPR experiments for the desired application. The other elements of a successful SPR experiment, from surface validation to regeneration conditions to design of complicated competition-based formats will also be addressed but they all rely on that fundamental foundation of a successful immobilisation approach.

While it has been established that SPR cannot claim to be a truly label-free technique and that it is not homogeneous, with all the concomitant challenges that target immobilisation brings, the advantages of SPR over other biophysical and biochemical approaches will be evident when we consider how SPR can be used to its full potential to assesses interactions between virtually any two molecular species.

7.2 Applications of SPR

In this section, we will look at the design of SPR experiments to address a range of applications to support drug discovery. As alluded to in the Introduction, when describing how SPR is used successfully in drug discovery, it makes sense to break this down into how SPR is best employed at each stage. The approaches taken in initial hit finding will be rather different from those involved in the later phases of drug discovery, such as mechanistic kinetic and thermodynamic profiling during hit validation and lead optimisation phases. These considerations will be discussed chronologically in the context of a typical drug discovery project. Examples will then be provided of how SPR can be applied in each phase of drug discovery, illustrating the diverse approaches to tackling different target classes, immediate project remits and broader goals. Indeed, the range of target classes amenable to SPR is limited only by the imagination of the drug discovery scientists, covering everything from kinases, proteases, redox enzymes and ubiquitin-processing enzymes through non-enzymatic peptides and proteins such as bromodomains to non-protein targets such as carbohydrates, nucleic acids (DNA and RNA), lipids and even reconstituted cell membranes, virus particles and whole cells.[6,7] As there is such a wide range of SPR applications, emphasis will be given to more recent developments in the field.

7.2.1 SPR Considerations in the Different Phases of Drug Discovery

7.2.1.1 Hit-finding Approaches with SPR

In the initial hit identification process, SPR can be used either as a stand-alone hit-identification approach or in conjunction with other hit-finding technologies. It is often best to utilise more than one hit-finding approach in drug discovery, particularly when facing a challenging target such as a protein–protein interaction. This strategy ensures that a greater number of compound structures with different binding modes or mechanisms of action will be detected. In this regard, SPR provides an elegant orthogonal approach to complement functional biochemical assays. For example, screens for kinase inhibitors traditionally focus on identifying ATP competitive compounds through kinase assays or *via* displacement of an ATP site-directed labeled probe. In contrast, SPR provides an unbiased interrogation of the binding of small molecules to the target protein kinase, independent of their binding mode or mechanisms. As such, SPR is equally subject to the limitation of most biophysical analysis methods in that such binding events may not translate into effects on functional activity; hence the need to validate SPR hits in a functional biochemical assay. On the other hand, the hits identified using SPR often result in the identification of novel modes of binding and mechanisms of inhibition that would not have been discovered through the more traditional, directed approaches. As a pure biophysical approach, it is also possible to drive the detection of other binding modes by blocking common sites. Using the ATP binding example above, by including a non-hydrolysable ATP analogue the screen will effectively ignore ATP-binding modes and would be expected to only highlight novel binding modes. Alternatively, screening mutant forms of protein targets, that may not necessarily have functional activity, can illuminate a particular area in which to focus the drug design efforts.

SPR complements other hit-finding approaches that offer higher throughput, as it is ideal for screening small, focused libraries, and is particularly suitable for screening compounds at high (up to millimolar) concentrations. Together, these features allow SPR to lend itself nicely to fragment screening. Much has been written about fragment-based drug discovery and the advantages of fragment-based hit-finding approaches in covering a large amount of chemical space with a relatively small, diverse and optimisable library of small molecules (reviewed recently in ref. 8). Therefore, the emphasis here will be on applying SPR optimally to fragment-based drug discovery approaches in the context of other equally valid hit-finding strategies.

7.2.1.2 Immobilisation Approaches

As emphasised in the Introduction, immobilisation is the most critical feature in the design of a successful SPR experiment, and this applies as much (if not more) to the initial hit-finding strategy as it is does to other phases of drug discovery. An additional consideration that was alluded to in the Introduction is that most SPR instrumentation can assess the binding interactions of an analyte at several target surfaces in parallel. These multiple parallel flow cells allow experiments to be designed to either increase the throughput (with several identically-immobilised target surfaces addressed with different compounds) or to assess interactions with several targets, anti-targets and selectivity targets in parallel. For example, for an enzyme target, the target of interest can be immobilised on one surface, a second surface used to assess the binding of analytes to a completely unrelated protein (specificity determination), and others to assess related members of the same family, mutant proteins, different activation states of the enzyme or enzymes from a different species (selectivity determination) or the target protein with a tool compound irreversibly bound (binding site determination). Additionally, with some SPR instrumentation (*e.g.*, the Biacore 4000), it is possible to run different buffers through different flow cells in parallel, providing the option to compare different conditions (buffers, pH or co-solvent concentration) or the presence *vs.* absence of competing analytes in the buffer (to assess binding competition). In other recent instrumentation (*e.g.*, the Biacore S200), a so-called ABA autoinjection system streamlines the analysis of ligand competition in a single flow cell by assessing additivity of responses. This system, which facilitates mode of action assessment in fragment screening, is equally useful for rapid buffer scouting, for example, when optimising conditions for compound screening or orthogonal assays, including X-ray crystallography. The combined data obtainable from these different formats add to the information richness obtainable with SPR.

As mentioned in the Introduction, binding sensitivity is a consideration and may be limited by the mass of the immobilised protein/target. Generally, for small molecule screening, protein targets with high molecular masses will be difficult to analyse by SPR, particularly if the analytes are very small fragment-like compounds. However the upper limit has been extended substantially with the recent advent of more sensitive Biacore instruments (*e.g.*, the Biacore S200 and Biacore 8000). Nevertheless, higher molecular weight proteins and complexes are generally better suited to a mass-independent methodology such as thermal shift or microscale thermophoresis. Another consideration is the purity of the target being assessed. While the immobilisation method itself may filter out impurities in the preparation, generally proteins immobilised for SPR should be highly pure and well characterised to eliminate aggregation effects and other non-specific artefacts in the data interpretation.

In terms of the immobilisation strategy, proteins and other targets immobilised for SPR need to be sufficiently stable both for the immobilisation itself and for flowing buffer across their immobilised surface for an extensive time period (usually hours or days). The immobilisation approach will also be partly dictated by whether the active site of the protein needs to be accessible in preference to other parts of the protein, or whether the goal is an unbiased screen for binders anywhere on the target molecule (*e.g.*, allosteric as well as orthosteric binders).

Choosing an immobilisation method and suitable conditions is critical to the successful outcome of an SPR experiment. Sensor chip surfaces are covered in a carboxymethylated dextran matrix, which allows greater immobilisation levels by introducing a third dimension, and provides functional groups that can be covalently modified. Traditionally, direct amine coupling is the most commonly used approach, chemically derivatising lysine residues of the target protein onto the activated carboxyl functionalities of the carboxymethylated dextran surface. The net negative charge on the dextran surface acts as both an electrostatic attraction focal point for the protein and at the same time limits nonspecific adsorption of analytes. Direct amine coupling can be used with any purified protein, without the need for addition of a tag, and has the added advantage of usually providing a high (but controllable) level of surface density of the protein on the surface, which is desirable for small molecule (*e.g.*, fragment) screening. On the other hand, the protein is modified chemically and is coupled heterogeneously in this methodology, possibly with more than one lysine residue bound to the dextran matrix. Consequently, the capacity of the tethered protein to bind analyte may be decreased or abolished, particularly if the modification occurs near putative binding or conformation-critical sites. There is little control over where the chemical modification occurs, though the inclusion of a ligand during the immobilisation step can physically occlude the binding site and protect it from chemical derivatisation. A less commonly used variation of covalent coupling is thiol coupling, which requires a similar chemical modification but is more restricted to reactive cysteine residues of the protein, which could provide a degree of selectivity if the cysteines are suitably located. If the chemical linkage of the protein to the surface is a disulfide, produced in a thiol–disulfide interchange coupling reaction, the immobilization can be easily reversed, if necessary, with mild reducing conditions.[6]

Streptavidin capture of biotinylated proteins is an essentially irreversible derivatisation method often used for SPR immobilisation. It still involves chemically modifying the protein by introducing one or more biotin groups, but they can be placed strategically on one of the termini (*via* an Avi-tag system) as an alternative to the random biotinylation of lysine residues. Nevertheless, the latter, when carried out minimally using well-established techniques and optimal linkers between the protein and the biotin, can often lead to successful immobilisation when direct amine coupling fails. However, in addition to the lack of control over the locations of the derivatised sites for both direct amine coupling and chemical biotinylation, the actual chemical process in both cases involves a low pH step (in the case of direct amine coupling, typically one pH unit below the pI of the protein to electrostatically attract the protein to the negatively-charged carboxymethylated dextran surface), which can be prohibitive for proteins with low pI and/or sensitivity to low pH.

A popular alternative is to combine mild amine coupling at neutral pH with the polyhistidine tag capture of a protein onto a nitrilotriacetic acid (NTA)-derivatised chip surface *via* chelated nickel. Many proteins can be conveniently tagged on their N- or C-termini, providing an orientated capture onto the surface. However, the 6xHis tag itself (the most commonly one employed) is usually not sufficiently tightly bound to nickel–NTA to provide a stable linkage for the duration of an SPR run. Therefore, when combined with mild amine coupling at neutral pH to chemically crosslink the captured target to the surface, the polyhistidine tag capture provides a directed immobilisation approach without the harshness of standard amine coupling. Alternatively, longer or repeated polyhistidine tags provide much more stable tethering in the complete absence of covalent amine coupling.[9,10] The BioRad ProteOn™ SPR system provides access to a trimeric NTA surface that increases affinity for the standard 6xHis tag by an order of magnitude.[11] An advantage of these noncovalent approaches is that the captured protein can be removed by injecting EDTA to dissociate the nickel from the NTA chip.

Other tag approaches are available based on antibody capture and provide additional options for immobilisation if a tagged protein is available. These include haemagglutinin (HA), glutathione *S*-transferase (GST), maltose binding protein (MBP) and Fc tags. Again, the tags offer the advantage of providing control over the orientation of the protein tethered to the SPR surface away from its active site. However, the capture of a tagged protein with an antibody creates a much larger total mass on the surface, reducing the sensitivity for small molecule detection. Moreover, it also introduces additional sites of binding (the tag or antibody itself) that can lead to false positives unless controlled for with antibody and tag only surfaces. On the other hand, the tag can also cause steric hindrance and interfere with compound binding, leading to false negatives, or even protein instability. In the domain of biologic drugs, antibody capture is widely used for antibody screening approaches: secondary antibodies are immobilised on the surface and then the test antibodies passed over the surface followed by their desired ligand to assess binding potential. One strategy to better control the orientation of immobilized ligands on the SPR surface is to covalently couple cysteine thiols directly to the gold surface.[6] Proteins can be engineered to incorporate a key cysteine residue for this purpose. This approach has been used to immobilise protein G to enable antibody binding in a controlled conformation.

7.2.1.3 Surface Validation and Regeneration

Once a target has been immobilised, it is essential to validate the activity and stability of the SPR surface. This is usually carried out using reference compounds to monitor both the level of binding (which indicates the proportion of immobilised target that is active) and affinity (injecting a concentration range), with repeated injections over a run time similar to that anticipated for a typical experimental run. If no tool compound is available, a screen can proceed at risk but it is advisable to try to identify a suitable compound from an initial pilot screen. Protein DMSO tolerance should be ascertained if the small molecules being analyzed are dissolved in DMSO, since some proteins are much more tolerant to DMSO than others, and high concentration fragment screens may require substantial DMSO concentrations [up to 5% (v/v)]. At this point it should be noted that co-solvents, particularly DMSO with its high refractive index, require careful matching between test and reference flow cells, and corrections to account for slight inevitable mismatches between samples to avoid their bulk shift and volume exclusion effects on binding responses.[4,12] Fortunately, in recent years, most SPR software has integrated these corrections into its analysis to make the overall process fairly immune to solvent mismatch issues.

As part of the assessment of the immobilised surface, the sensorgrams should be examined to determine whether typical analytes passing over the surface dissociate cleanly in a reasonable time frame (usually seconds to minutes). If not, a regeneration step may be required between injection cycles. It is important that the regeneration step is optimised so that it is sufficiently mild to retain subsequent protein functionality while cleanly removing all of the analyte from the surface in between injection cycles. There are several options for regeneration conditions that can be explored, including mild acid, mild base, detergents and mild denaturants. While not as tricky overall as immobilisation, consistent regeneration can be difficult to achieve satisfactorily. One method to get around the issues with optimising regeneration conditions that can be applied to biotinylated targets is to use a Biotin CAPture kit (GE Healthcare). In this system, the biotinylated target is captured using a special streptavidin–DNA oligonucleotide linker that is bound to a standard SPR chip *via* hybridisation with a covalently coupled DNA antisense strand. After an analyte is bound, the whole analyte–target complex can be dissociated by disrupting the DNA hybridisation that links the complex to the surface using a standard set of conditions (high concentration of guanidine), leaving the surface free for another injection cycle. However, this system removes the target protein between each cycle, leading to its relatively high reagent consumption. DNA-based sensor chips have also been used to direct the immobilisation of proteins tagged with a complementary DNA strand or proteins with a DNA binding domain, again with the caveats that covalent modification or docking of the protein target may interfere with its binding functionality.[6]

7.2.1.4 Considerations for Low Affinity Binders

An emerging theme in this section will be how the expected potency range of the interactions being probed in an SPR binding analysis dictates how the experiment is designed and interpreted. We begin with the early stages of drug discovery, when low affinity compounds are identified in initial hit-finding campaigns. At the extreme (but also most common) range of potencies for screening, this includes fragments, which have affinities typically in the high micromolar to millimolar range. With their low expected potencies, fragments pose challenges of requiring high concentrations for screening to produce measurable binding, giving rise to potential solubility issues. In addition, as alluded to earlier, for small molecule and especially fragment screening, due to the low molecular masses being detected, it is important to maximise detection sensitivity to ensure that binding responses are above the background Gaussian noise generated by

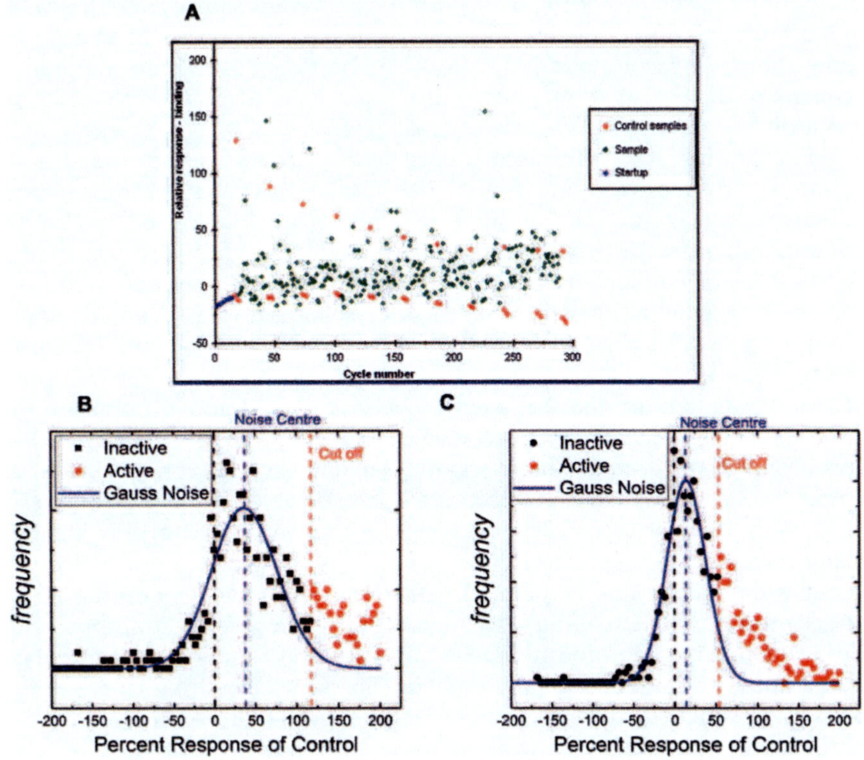

Figure 7.3 Panel A: Typical SPR data from a fragment screening run. Startup injection responses are shown in blue, control sample responses (calibrator compound and buffer blanks) in red, and test compound responses in green. Panels B and C: Distribution of responses from an SPR fragment screening run, expressed as a percentage of the control compound response. A higher surface density of the immobilised target (Panel C, *vs.* lower density surface in Panel B) improves the ability to distinguish active compounds from the Gaussian noise generated by injections. The larger compound responses trend the noise centre towards a zero response.

the injection events themselves (Figure 7.3). Therefore, the surface density of the target protein should be as high as possible. Often this can be achieved best *via* direct amine coupling (provided it is not detrimental to the protein). Biacore provides a higher density CM7 chip that allows even greater coupling levels than the standard density carboxymethylated dextran surfaces. Fragment and other compound libraries should be checked for solubility in the appropriate buffer conditions before screening using methods such as turbidity (absorbance readout) to avoid issues with fouling the Biacore surface. Likewise, a clean screen of the library should be carried out by SPR before the compounds are used for screening campaigns. This entails screening the library against a selection of diverse protein targets and assessing compounds for poor binding behaviour (*i.e.*, aberrant sensorgrams), typified by slow dissociation or effects on subsequent injection cycles. So-called pan-assay interfering compounds (PAINs) can be flagged or removed from a screening compound library using this approach together with chemical intuition.[13]

Upon binding, low affinity fragments (and other hits) are expected to associate and dissociate rapidly from the surface, manifested by "square wave" sensorgrams (Figure 7.4). For such compounds, a slow association or dissociation phase may be indicative of a pathological binding mechanism (*e.g.*, aggregation or nonspecific binding). However, there are exceptions to this rule, with fragments known to have low dissociation rates.[14] Report points are used to mark specific events during the compound injection cycle, indicating the binding response at a particular time point. Ratios of appropriate report points can be used to assess the shape of sensorgrams quantitatively. Flags can be given for non-ideal binding behavior that deviates from the expected square-shaped sensorgrams (Figure 7.4). For example, compounds that exhibit slow association, superstoichiometric binding, or slow or incomplete dissociation can be flagged in this manner.

It is advisable to assess the binding of the screening compounds to related and unrelated proteins in parallel with the assessment of binding to the target protein. The use of an unrelated protein can give an additional indication of nonspecific binding. However, initial screening hits, particularly low affinity fragments, are not expected to bind selectively to the target protein by SPR (Figure 7.5). Comparisons of well-behaved fragments that bind a target protein typically show that some legitimate binders are non-selective against an unrelated protein. Fragments may bind different proteins in distinct low affinity binding modes. Therefore, the selectivity filter should be used with caution to help sort and prioritise hits rather than to eliminate them. Indeed, even within the possible modes of binding to the target protein itself, secondary interactions unrelated to the desired binding site may complicate the evaluation of a fragment screen output.

Following the primary screen, at a single concentration and repeat confirmation screen in a similar format, concentration–response titrations are undertaken. Hits are further prioritised by their ability to reach saturated binding with reasonable binding isotherms (Figure 7.6). Prioritisation can be given to compounds that bind with reasonable (near 1 : 1) stoichiometry. Nonspecific binders tend to exhibit linear, nonsaturable concentration–response relationships. The equilibrium dissociation constant (K_D) is determined from a fit to the steady-state binding response against compound concentration. Where measurable, the association rate constant should be independent of compound concentration for legitimate binders. It is also useful to compare the SPR findings with those obtained from an orthogonal biophysical or biochemical assay format at this stage. In order to elucidate the mode of compound binding, it may be desirable to show competition with an active site probe or other known binding ligand. This type of experiment can be used to help prioritise hits (*e.g.*, for crystallography). In SPR

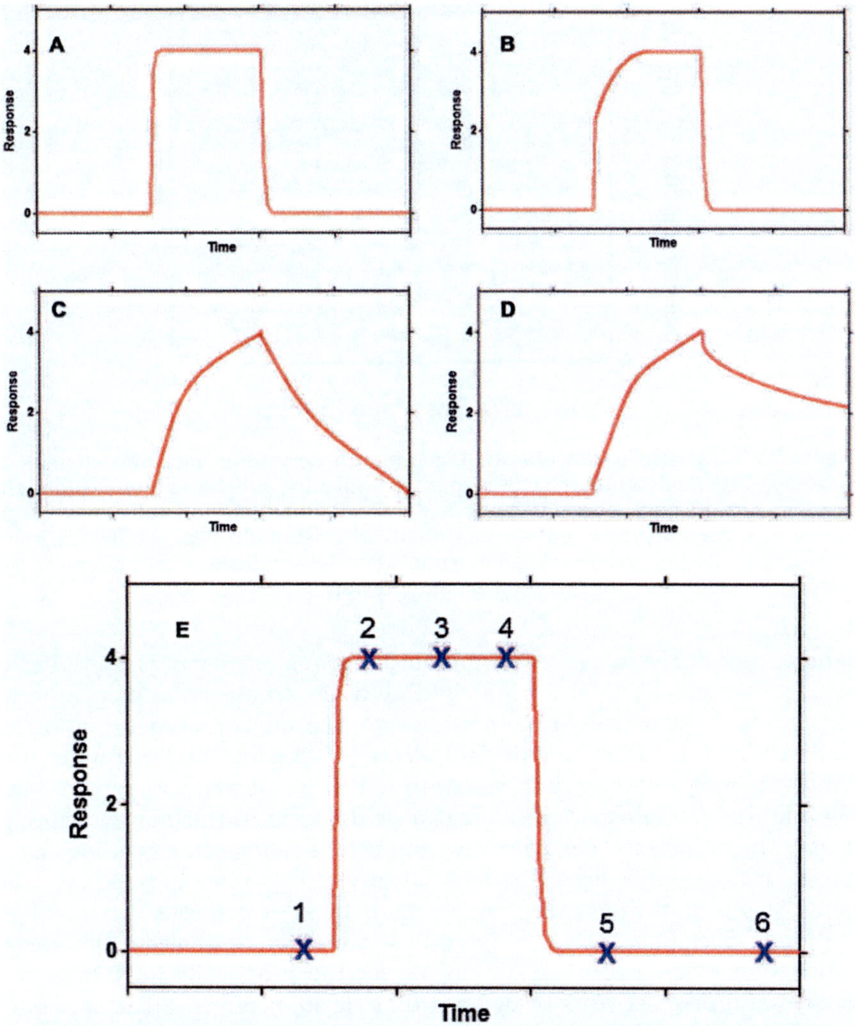

Figure 7.4 Examples of sensorgrams obtained in an SPR fragment screen. Panel A illustrates a well-behaved square-shaped senorgram indicative of rapid association and dissociation events. Panels B, C, and D illustrate progressively worse binding behaviour, with a slow association and/or dissociation phase that may indicate a pathological binding mechanism. Panel E: Report points are used to mark specific events during the compound injection cycle, indicating the binding response at a particular time point. 1: Baseline; 2: Early binding; 3: Mid-binding; 4: Late binding; 5: Early dissociation; 6: Late dissociation. Ratios of appropriate report points can be used to quantitatively assess the shape of the sensorgrams.

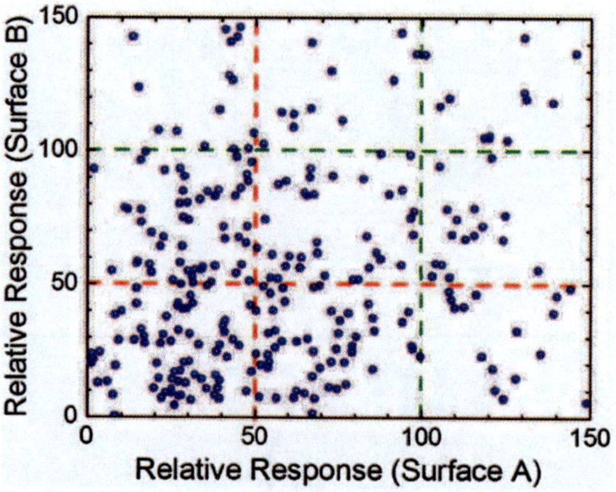

Figure 7.5 SPR single concentration compound screening data against two unrelated proteins. The responses indicated are background subtracted and normalised to molecular weight. Only data for well-behaved fragments (based on sensorgrams) are included. 47% of the fragments showed some selectivity, while 23% showed little or no preference and 30% remained in the background noise.

competition analysis, the test compound and reference compound binding responses are measured separately and when co-injected. As mentioned earlier, recent Biacore instrumentation with ABA injection programming has facilitated this experimental design. While competitive binders cannot bind simultaneously, additive responses are indicative of non-competitive binding (Figure 7.7). Allosteric binders can be detected in this manner, though pure non-competitive behavior may indicate a nonspecific binding mechanism. It is useful to use other technologies at this stage to weed out compounds with undesirable binding mechanisms. For example, dynamic light scattering can help assess compound or compound-induced target aggregation behaviour. Redox assays can assess the propensity for hits to self-oxidise and/or react oxidatively with their protein target. Structural information (*e.g.*, that obtained from X-ray crystallography) is highly desirable at this point to shed further light on the binding mode and mechanism of hits obtained *via* SPR.

As hits progress and improve in potency, the next stage of analysis takes advantage of the ability of SPR to characterise the kinetics of binding interactions in real time. Both the association and dissociation phases of the binding sensorgram provide useful information that allows the determination of rate constants for the forward (k_{on} or k_a) and reverse (k_{off} or k_d) binding events. The inverse ratio of these rates k_{off}/k_{on} provides the kinetic affinity K_D. It is useful to compare this value to the K_D obtained from steady state equilibrium binding analysis of the concentration–response titrations (Languir binding isotherm) as an additional validation of the results, as these values should be reasonably consistent (Figure 7.8). An alternative

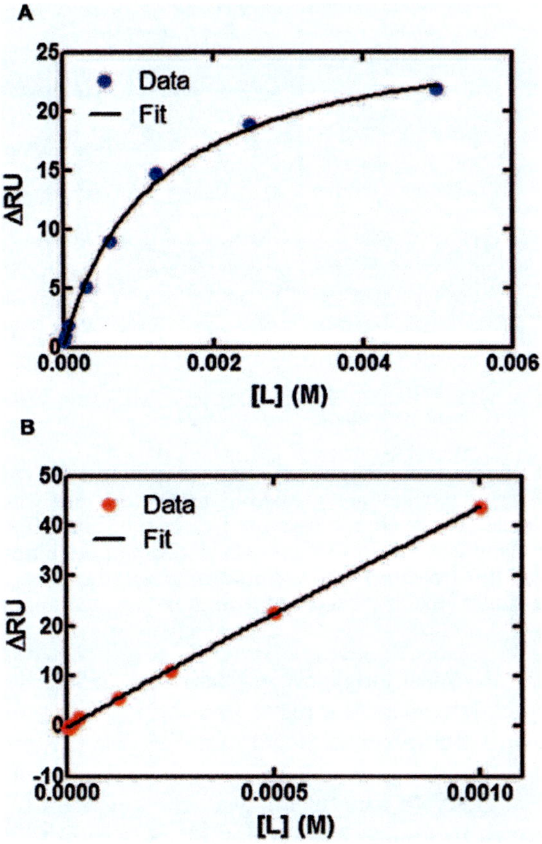

Figure 7.6 Fitting SPR concentration–response titration data. Panel A illustrates data for a compound that reaches saturated binding, enabling the determination of a steady state K_D, whereas the compound in panel B exhibits a linear, nonsaturable concentration–response indicative of nonspecific binding.

approach to the traditional concentration–response titration produced Languir binding isotherm covering a set of several compound concentrations is the OneStep™ method developed recently by SensiQ. As the name implies, this technology utilises a concentration gradient injection method based on a Taylor dispersion, producing a high resolution analyte titration in a single injection.[15]

At this stage, SPR can be used to profile compounds for the hit-to-lead screening phase of a drug discovery programme. SPR screening methods can be highly reproducible and efficient, with as many aspects of the process as possible automated, from compound handling to data analysis. Classical lower throughput multi-cycle kinetic analysis, in which separate injection cycles are used for each compound concentration, can be replaced by single-cycle kinetics for higher throughput affinity and kinetic screening without

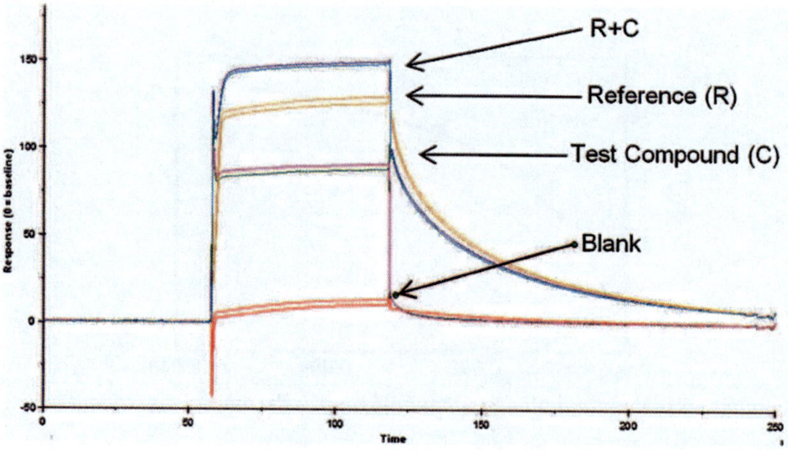

Figure 7.7 SPR competition analysis. The test compound and reference compound binding responses are measured separately and when co-injected as indicated. While competitive binders cannot bind simultaneously, additive responses are indicative of non-competitive binding. The example shown illustrates competitive binding since the additive response is only marginally greater than that for the reference compound alone.

loss of information. With single-cycle kinetics, a series of increasing concentrations (typically five) of the test compound is injected in a single injection cycle, with complete dissociation only at the end of the cycle (Figure 7.1, panel D and Figure 7.8). An example of the typical reproducibility possible with single-cycle kinetic analysis during a hit-to-lead screening campaign is shown in Figure 7.9. While SPR data tend to be highly reproducible, the largest source of variability among measurements is usually the compound preparation and/or dilution process. In particular, inconsistencies in concentration will affect the fit of k_{on}, while the determination of k_{off} is independent of concentration.

7.2.1.5 *Considerations for the Hit-to-lead Stage*

As compounds are optimised and their potencies improve further in the hit-to-lead phase, SPR can be used to fine-tune both the kinetic and thermodynamic properties of the hits progressing to leads (Figure 7.10). If possible, it is useful to know the required kinetic or thermodynamic profile beforehand. For example, long residence times or slowly dissociating compounds may have greater therapeutic benefit and have often (but not always) been linked to greater clinical efficacy.[16] Generally, slowly dissociating compounds will be relatively potent. Moreover, depending on the physiological turnover of the protein target, slow dissociation may be able to overcome pharmacokinetic and pharmacodynamics challenges, which could otherwise limit a compound's bioavailability. As part of a bromodomain inhibitor

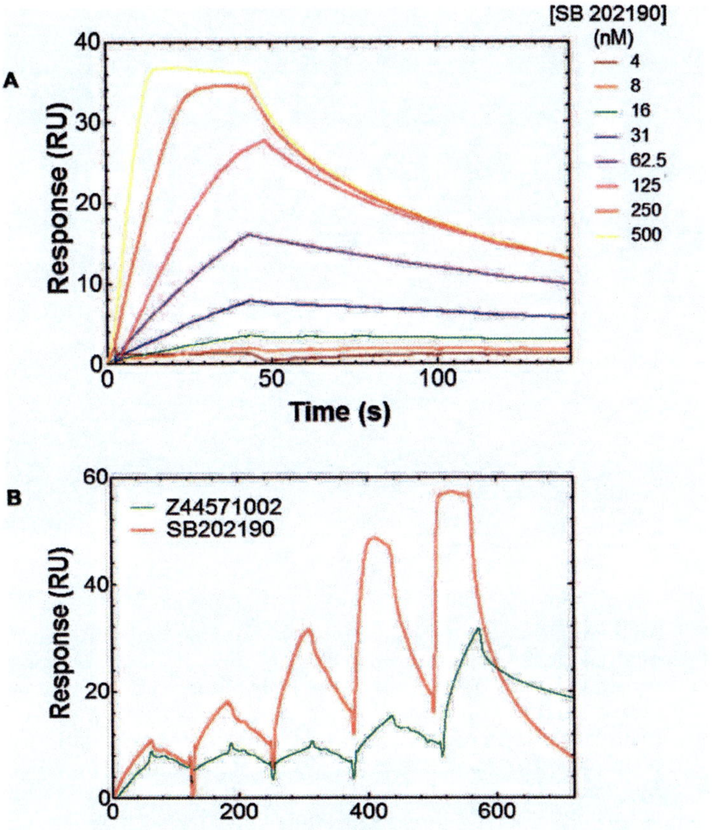

C

	k_{on}	k_{off}	K_D
SB202190	6.5E+5	0.027	4.094E-8
Z44571002	2.3E+4	0.001	5.168E-8

Figure 7.8 Typical SPR concentration–response titration data from multicycle kinetics (Panel A) or single-cycle kinetics (Panel B), using the kinase p38 as an example target. Panel C illustrates the kinetic data obtained by fitting the data from the single-cycle kinetics experiment to a one-site binding model.

project, over 300 compounds were screened against the target and a related bromodomain by SPR (unpublished results). Compound potency and selectivity were found to track primarily with decreased dissociation rate constant (k_{off}) and also tracked well with cellular potency (Figure 7.11). Since dissociation rates are independent of compound concentration, yet are the most important determinant of potency, compound libraries and even crude

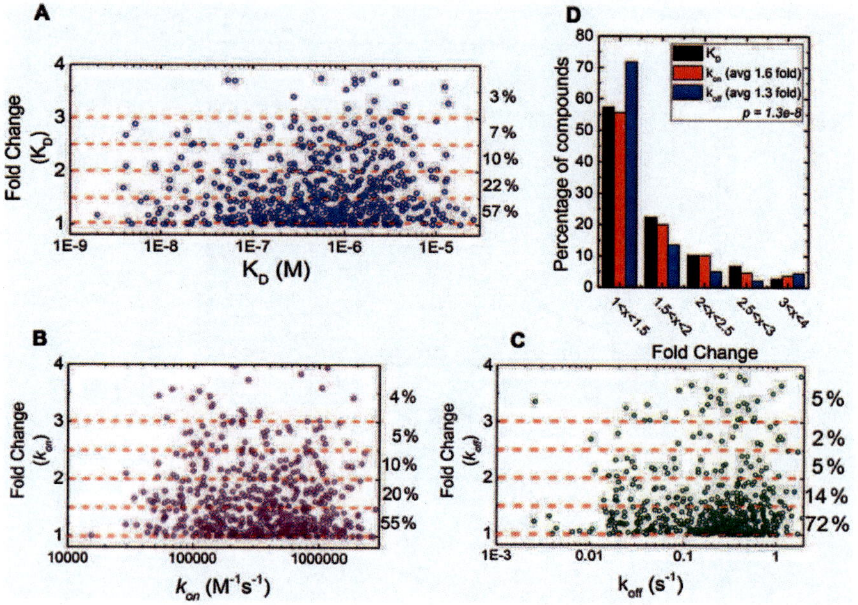

Figure 7.9 An example of the reproducibility obtainable with single-cycle kinetic analysis during a hit-to-lead screening campaign. 400 compounds were screened by SPR against two identically-prepared bromodomain protein target surfaces on two separate occasions, thereby providing 1600 measurements in total. The $n=1$ data are plotted against the fold change observed with the $n=2$ determination for K_D (Panel A), k_{on} (Panel B) and k_{off} (Panel C). Panel D illustrates that most of the measurements are consistent within 1.5-fold when comparing the two data sets. Overall, the k_{off} values were slightly more consistent than k_{on} values, which depend on compound concentration.

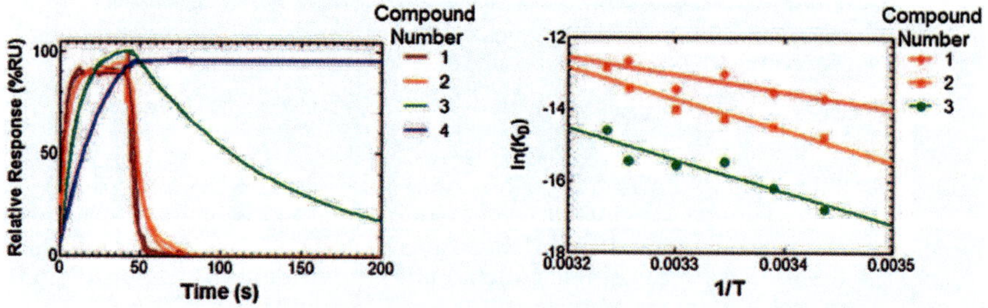

Figure 7.10 Panel A: Example SPR data used to determine kinetic parameters. The differently-shaped sensorgrams illustrate different dissociation rates which can be optimised during hit progression. Panel B: Van't Hoff plot of ln(K_D) as a function of reciprocal temperature for a set of compounds screened by SPR. This plot can be used to derive thermodynamic properties, enthalpy ($\Delta H = -R \times \text{slope}$) and entropy ($\Delta S = R \times \text{intercept}$), and overall Gibbs free energy ($\Delta G = \Delta H - T\Delta S$), for the binding of hits progressing to leads.

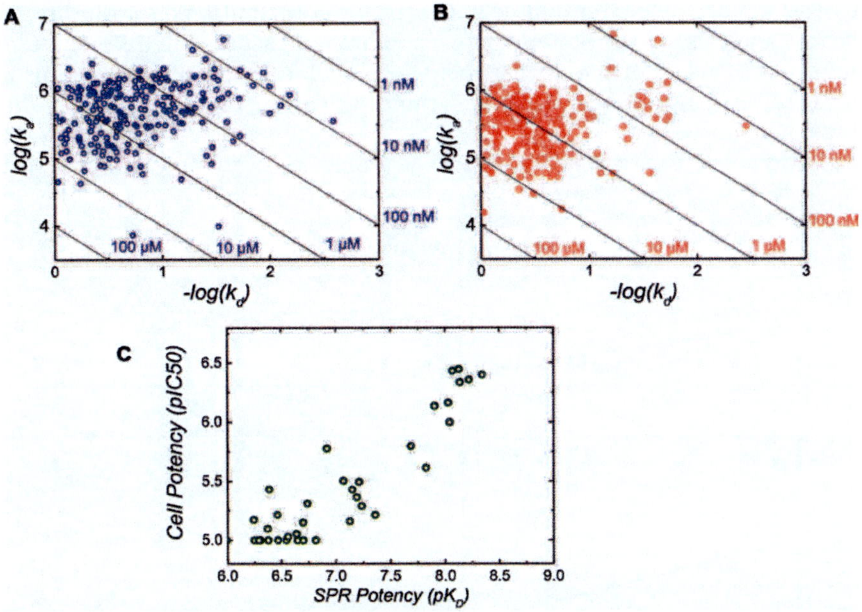

Figure 7.11 Representative data from the screening of over 300 compounds against a bromodomain target and a related protein by SPR. Panel A: plot of the kinetic SPR data, with log(k_a) plotted as a function of log(k_d). Panel B: A similar plot for the related selectivity of the protein surface. Compound potency and selectivity track primarily with decreased dissociation rate constant (k_d or k_{off}). Panel C: SPR affinity determinations also tracked well with cellular potency in a functional (anti-proliferation) assay.

synthesis reaction mixtures can be screened efficiently and rapidly by SPR to optimise dissociation rates during hit-to-lead progression.[17] However, it is not always as simple as optimising off rates; the rate of association can also be relevant, and depends on both intrinsic k_{on} and the concentrations achievable. SPR can therefore be used to optimise association and dissociation rate constants independently for a given lead series to drive potency in a medicinal chemistry programme.

In addition to optimising kinetic parameters, though less obvious, SPR is also used to characterise and optimise the thermodynamic properties of ligands (Figures 7.10 and 7.12). Measurement of thermodynamic parameters allows an interrogation of structure–activity relationship (SAR) hypotheses. By assessing binding properties at different temperatures and using Van't Hoff analysis, enthalpic and entropic contributions to binding can be determined, providing greater insight into compound binding mechanisms.[18,19] Results support crystallographic and computational binding studies and usually correlate well with those obtained by isothermal calorimetry, often considered the gold standard for thermodynamic parameter assessment, but a technology which is limited by its heavy use of protein, particularly in comparison with SPR. SPR can also be used to determine activation parameters (enthalpic and entropic contributions) for binding *via* Eyring analysis of the kinetic rate constants measured at different temperatures. This information can help to elucidate and optimise the properties of lead compounds involved in their high affinity binding interactions.

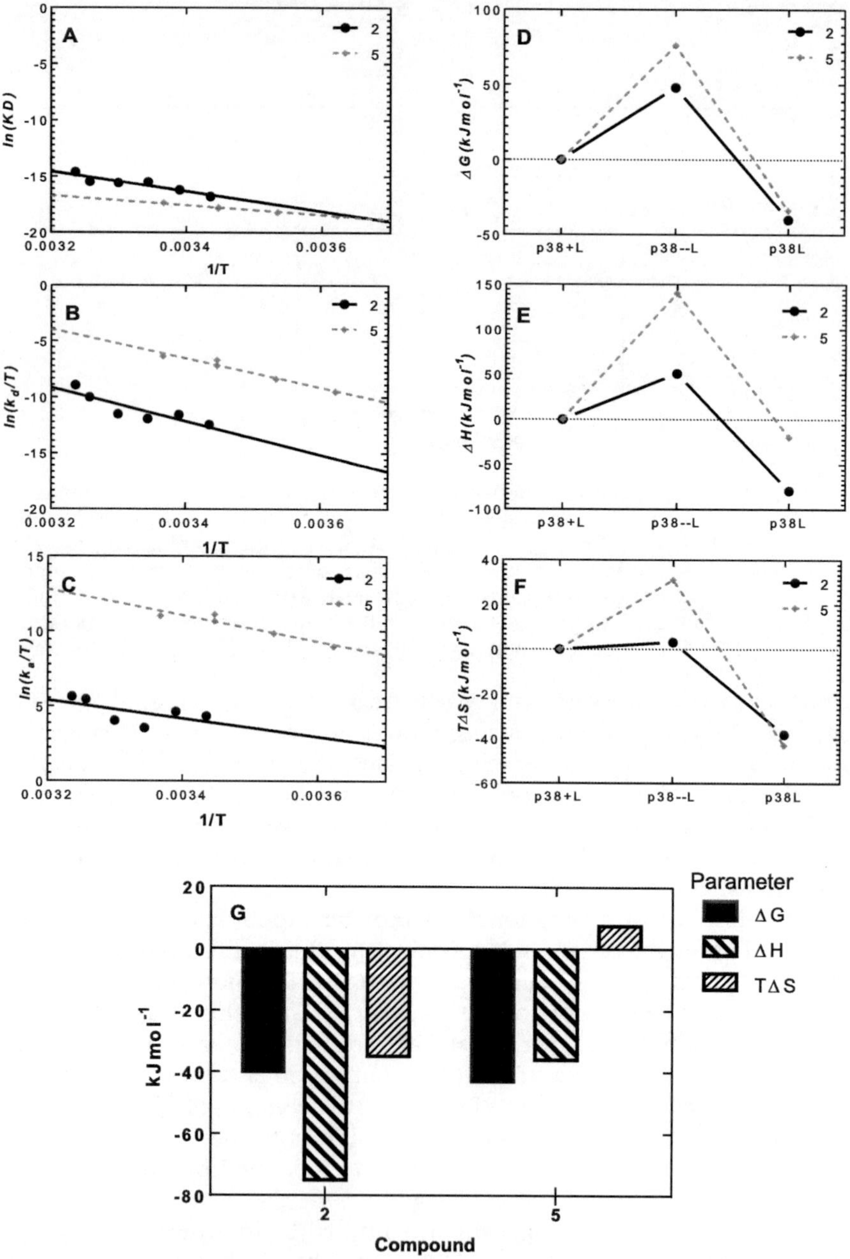

Mass transport limitations must be borne in mind in the design of effective SPR experiments. This phenomenon occurs when the capture of an analyte by the immobilised target surface is faster than the rate it is delivered by the flow system.[20] Under these conditions, mass transport is rate-limiting, which leads to an underestimation of k_{on} values. Moreover, the analyte may rebind after dissociating from the target, resulting in an underestimation of the k_{off} value. Mass transport effects create a concentration gradient between the flow buffer and the surface and are characterised by linear association and higher order dissociation phases. For very high affinity compounds at low concentrations, mass transport becomes unavoidable but can be minimised with high flow rates and low surface densities.

7.2.1.6 Further Considerations for Lead Optimisation

With the very high affinity interactions (*e.g.*, nanomolar) encountered during lead optimisation, extremely slow dissociation rates cannot be quantifiable by SPR. In these cases, it is still useful to analyse the SPR data qualitatively and simulate the interactions with modelled sensorgrams. In one example, slowly dissociating non-nucleoside reverse transcriptase inhibitors were assessed using simulated sensorgrams that fit best with an induced fit mechanism, in which an initial enzyme–inhibitor complex converted to a higher affinity stabilised complex.[21] In this sense, SPR can also indirectly provide information about conformational changes involved in binding. Slow association events detected by SPR can likewise be indicative of a two-step mechanism caused by protein conformational changes. Although SPR technology only detects mass changes and as such does not assess protein conformation directly, the structural changes that take place in a protein can

Figure 7.12 Measurement of thermodynamic parameters by SPR. By assessing binding properties at different temperatures and using Van't Hoff analysis (see also Figure 7.10) or Eyring analysis, enthalpic and entropic contributions to binding and activation parameters, respectively, can be determined. Panel A depicts the Van't Hoff plot obtained for two p38 inhibitors (compounds 2 and 5). Panels B and C show Eyring plots of the activation parameters for dissociation and association, respectively. Panels D, E and F depict the free energy diagrams obtained from the Eyring analysis in terms of the Gibbs free energy change (D), enthalpy change (E), and entropy change (F). In panels D, E and F, p38 + L represents the unbound ground state. The difference between p38 + L and p38 − L is the change in association and the difference between p38 + L and p38L is the change in dissociation. Therefore, the difference between p38 − L and p38L is the equilibrium. Panel G indicates the free energy, enthalpy and entropy changes determined from the Van't Hoff analysis in panel A. Panels A to F are reprinted from *Analytical Biochemistry*, vol. 479, M. Redhead, R. Satchell, V. Morkūnaitė, D. Swift, V. Petrauskas, E. Golding, S. Onions, D. Matulis and J. Unitt, A combinatorial biophysical approach; FTSA and SPR for identifying small molecule ligands and PAINs, pp. 63–73, Copyright 2015, with permission from Elsevier.[13]

also affect the SPR signal at the surface to a degree.[6] Moreover, kinetic modelling of complex, multistage binding events can be indicative of conformational changes that take place in the formation of higher affinity binding pockets.[22] Such ligand-induced protein conformational changes have been further resolved into fast and slow changes.[23] In another approach, conformation-specific antibodies can be used to assess the different conformations of proteins bound by SPR. This strategy was used successfully to distinguish between normal and TRAIL-induced conformations of the pro-apoptotic Bax protein.[24]

Although already discussed in the broader context of poorly behaved compounds, appropriate regeneration conditions are particularly relevant as compound potencies increase and dissociation rates decrease. Baseline drift is detrimental to subsequent injection cycles. Typical regeneration solutions include mild basic (NaOH) or acidic (glycine) conditions, high ionic strength buffers or detergents. While experience and regeneration scouting experiments can often lead to optimised regeneration conditions suitable for slowly dissociating compounds, sometimes it is more straightforward (though also more costly on protein) to regenerate the whole protein surface between cycles.

7.2.2 Recent Applications of SPR in Drug Discovery

Having overviewed the design of SPR experiments to suit the type of analyses being undertaken, we will return to the applications of which SPR is capable. In terms of target classes, SPR has traditionally been used mostly with soluble protein targets, such as cytosolic kinases and proteases, which are readily amenable to immobilisation *via* the various methods described. This discussion will focus on the more recent and often less straightforward applications of SPR. These have included the immobilisation of integral membrane proteins and the use of antibodies in biologic drug development. Furthermore, while we have mainly been describing protein targets, nucleic acids have also been immobilised successfully to enable the screening of DNA- and RNA-binding drug candidates or the characterization of nucleic acid–protein interactions by SPR, as reviewed recently.[6,7,25,26] In an extension of this approach, SPR has also been used for RNA aptamer selection, in which the target protein is immobilised on the SPR chip and an RNA pool injected over the surface to select for high affinity binders.[6,26] Chemically modified nucleic acids have also been analyzed by SPR in the development of antisense drugs.[6,7] Finally, SPR has been used in combination with the principles of the polymerase chain reaction (PCR) in hybrid systems in which PCRs are carried out directly on an SPR sensor chip and the PCR product probed for DNA complementarity *via* binding interactions.[6] Through a multi-step amplification scheme combining ATP-binding aptamers bound to magnetic beads, a hybridization chain reaction on an SPR surface and target DNA strand displacement, the lower limits of DNA and ATP detection, were recently extended to the sub-femtomolar range.[6,27]

SPR assay designs have also diversified to satisfy the range of demands of drug discovery projects. In addition to the straightforward binding of small molecules to a tethered target protein surface, several competition formats have been devised to assess the ability of analytes to compete with a reference compound for a known binding site. This could involve co-injecting the test and reference analytes to assess the additivity of responses, as described earlier, or including the reference analyte in the binding buffer. The reverse orientation has also been used, whereby the protein in solution is passed over an immobilised small molecule- or peptide-bound surface. Test compounds can then be assessed for their ability to compete for this interaction.[28] This format has the advantage of not requiring protein immobilisation, although it does require much greater protein usage than protein immobilisation approaches. Finally, a sandwich-based approach can be used, analogous to a sandwich ELISA, in which analyte binding to an immobilized target (antibody or other protein) surface is interrogated with a secondary analyte (*e.g.*, an anti-analyte antibody or other analyte binder) subsequently passed over the surface. This format allows an additional degree of specificity (as well as sensitivity) in the binding assessment.[7]

A variant of SPR used at the hit-identification stage extends the principle of immobilising the compounds and passing the protein target over the surface: Small molecule microarray SPR, used particularly for fragment screening, requires a diverse screening library to be chemically derivatised and coupled to a microarray surface. In a so-called SPR imaging (SPRI) system, a charge-coupled device camera is then used to assess binding events at the immobilised surface, allowing the measurement of thousands of binding interactions in parallel.[6] Unlike traditional SPR, SPRI measurements are performed at a constant wavelength and angle of the incident light, so it is the intensity rather than angle-dependent profile of the reflected light that is monitored. This technology can be applied to both high throughput screening and biomarker profiling studies and, for example, was used to evaluate over 9000 ligand–target interactions in a single overnight experiment.[7] Graffinity Pharmaceuticals (now part of Novalix) pioneered the approach of small molecule libraries micro-arrayed on SPR biosensor surfaces and now uses it routinely for screening a 116 000 compound library (3rd Novalix Conference, Biophysics in Drug Discovery, Strasbourg, France, 2016). This approach has also been used for applications ranging from estrogenic hormone detection to a quadruplex DNA ligand screen.[7] Since the small molecule rather than the protein target is immobilised, small molecule microarray SPR also allows the interrogation of different protein complexes, *e.g.*, individual components *vs.* paired subunits *vs.* higher order complexes, for their binding to the immobilised library of compounds being screened. This approach is also useful for proteins that naturally aggregate or are otherwise difficult to immobilise in a functionally active conformation. Another variation on higher throughput SPR is the FlexCHIP system, which utilises a diffraction grating in place of a prism to generate an SPR effect, allowing the simultaneous kinetic measurements of

400 micro-sample surfaces.[29] However, both this approach and SPRI still have limited resolution and fields of view.

A plethora of formats has also been used to assess protein–protein interactions by SPR. Immobilisation of each protein partner in the interaction can be used to ascertain the kinetics of binding of the other partner and the effects of competing analytes (proteins or small molecules) on these interactions. Applications ranging from antibody screening to identifying disrupters of a multimeric complex have been enabled through this type of design. A key consideration in such experiments is the surface immobilisation density, which should be low for protein–protein interactions to avoid steric interference and mass transport limitation effects, in contrast to the sensitivity considerations for typical small molecule analytes.

7.2.3 Advances in Membrane Protein Capabilities for SPR

Membrane proteins represent at least 50% of validated molecular drug targets.[29,30] Recent membrane protein capabilities have expanded the possibilities of these target classes for SPR dramatically, allowing binding affinities and kinetics for drug target classes such as GPCRs, ion channels and other transmembrane cellular receptors to be assessed directly (as reviewed recently).[30] These structurally complex proteins have posed a particular set of challenges for biophysical experimentation, and SPR is no exception here. They require demanding detergent solubilisation or lipid reconstitution conditions, and often complex immobilisation methods, typically necessitating screening a wide range of reagents and parameters in order to obtain surfaces on which the proteins retain both an appropriate surface orientation and (as close as possible) their native physiologically-active properties (as reviewed recently).[31,32] Surfactant screening has been used extensively, but not always successfully, to identify appropriate conditions for retaining membrane proteins in their native conformations.[29] Detergent/lipid-solubilised membrane proteins have then been tethered for SPR *via* antibody capture or biotinylated ligand-directed immobilisation.[29] Tagged receptors have also been solubilised and captured for SPR-based screening. Both antibody and tagged receptor coupling methods have an advantage over the non-specific surface attachment of membrane vesicles, which often give variable, unstable responses.[32] In one such example, the GPCR (in this case CCR5) could be captured directly and selectively from a solubilised cell pellet onto an SPR sensorchip *via* an antibody binding to its C-terminal peptide tag.[33] Another recent method combined His tag capture of the membrane protein (CXCR5) with mild amine coupling, chemically crosslinking the receptor to stabilise the chip surface.[34] In the same report, this so-called capture–stabilise approach was subsequently extended to the GPI-anchored protein CD52 in its native virus-like particle environment.

Membrane associated proteins have also been captured using wheat germ agglutinin-coated or other specially-designed lipophilic sensor chip surfaces.[29,30,35] For example, hydrophobic sensor chip surfaces containing long-chain alkanethiols linked directly to the gold surface or carboxymethyl dextran chips modified with hydrophobic alkane groups allowed membrane protein-containing vesicles to be adsorbed to form lipid monolayer or bilayer surfaces, respectively. These chips can also be used to immobilise detergent-solubilised membrane proteins with lipids replacing the detergent in the immobilised complex. In one study, using the tagged GPCR CXCR4 captured with a specific antibody, detergent solubilisation and alkane chip approaches were compared directly.[29,36] The latter approach involved reconstitution of the lipid bilayer around the receptor to enable a more native-like environment on the SPR surface. Lipid bilayer-immobilised GPCR functionality was also addressed directly by measuring the large decrease in SPR response from ligand-induced G-protein alpha subunit dissociation,[29,37] obviating the need to look at downstream signalling responses. However, successful approaches for immobilisation, while still retaining the physiological activity of the target membrane protein, have been limited to a few examples and likely involve extensive trial and error. Indeed, until recently, approximately half of the published studies on membrane proteins for SPR centred on rhodopsin and chemokine receptors.[32] This repertoire has since expanded in conjunction with improved membrane protein expression, purification, solubilisation and stabilisation methods.

Several other methods to improve the probability of success for addressing membrane proteins by SPR have been developed in recent years. One such approach is to introduce artificial mutations to thermostabilise membrane-bound GPCRs in active (agonist or antagonist binding) conformations for immobilisation (Hepatares' stabilised receptor or StaR approach).[38] This technology, which involves systematic mutagenesis of the GPCRs to generate variants with the desired properties, has the obvious caveat that it utilises non-native protein sequences, which have pharmacological properties consistent with a single conformation. Nevertheless, the validity of the resulting StaRs for biophysical applications is usually justified by comparing functional behavior with their wild-type equivalents, and has the advantage that compounds can be screened for their ability to bind a specific receptor conformation. It is a less suitable approach if the screening of compounds has a more agnostic view to their potential mode of action or if the goal is to identify allosteric modulators.

Nanodisc and nanopore array approaches represent other recent strategies that obviate the requirement for detergents to reconstitute membrane proteins in a native-like environment for SPR.[29] Nanodiscs are essentially artificial lipid bilayer membranes encompassed by recombinant membrane scaffolding protein (*e.g.*, apolipoprotein A-1) to encapsulate the target membrane protein. They have been demonstrated to be useful for immobilising a wide range of membrane-associated proteins for SPR and have the added advantage of allowing access of ligands to both sides of the membrane interface. Nanodiscs have also been used as the injected analyte passed over an immobilised target surface.[30] In one such example, nanodiscs incorporating an integral CD4 receptor fused to a ubiquitin/polyhistidine tag were analysed for their binding to an antibody-coated surface.[39] The nanopore array format (to be discussed further in Future Perspectives) incorporates a network of nanopores embedded in the gold film layer, allowing free-standing lipid bilayers arrayed on the gold surface to present a physiologically relevant membrane orientation accessible from both sides.[29] This orientation allows the receptor to be attached to one side of the lipid bilayer and its ligand to bind at the other side. In addition to utilising native protein sequences, both the nanodisc and nanopore methods eliminate the need for detergent solubilisation, providing a more stable protein that behaves much like it would in a physiological environment. Overall, SPR capability for membrane protein analysis now sits nicely between traditional radiometric ligand binding assays and functional cellular signalling assays, in its ability to provide real-time kinetic information on ligand binding.

7.2.4 SPR in Drug Metabolism and Pharmacokinetics

SPR has also made the leap from the domain of hit-finding and characterisation to wider applications in drug metabolism and pharmacokinetics (DMPK). In addition to providing target engagement information from direct binding analysis, SPR can be used to determine plasma protein and lipid binding, for example, assessing drug candidates for their degree of binding to common plasma protein constituents (*e.g.*, human serum albumin, α1-glycoprotein), and thereby providing an information-rich alternative to the traditional equilibrium dialysis method.[40,41] The binding affinities determined by SPR agreed well with those obtained by other techniques, with the added advantages SPR provides in richer information content and higher throughput.[7,42] Cytochrome P450s, which are often implicated in drug–drug interactions, can be immobilised and profiled for their binding to potential drug candidates. It may even be possible to assess the toxicity of potential drug candidates by SPR through appropriate but indirect readouts such as key signalling protein binding or mitochondrial membrane potential.[7,43]

Drug permeability can also be modelled by SPR, using lipid layer biosensor surfaces as a mimic of the membrane microenvironment to evaluate candidate compounds.[44,45] Cell membranes or liposomes are immobilised rapidly and reversibly onto the SPR sensor surface and interrogated with potential drug analytes to monitor binding interactions directly in real time to obtain liposome binding affinities and dissociation kinetic data complementary to those obtained from traditional methods such as octanol/water partitioning and parallel artificial membrane permeability assays (PAMPA).[7] Oral drug permeability could even be predicted indirectly by immobilising rodent-derived brush border membranes to an SPR sensor chip and assessing the binding responses of drug candidates.[7,46]

7.2.5 Biomarker Characterization by SPR

Biomarker identification and profiling is another growing application of SPR in the later stages of drug discovery (recently reviewed).[6] An even more sensitive variation of SPR, termed localised surface plasmon resonance (LSPR; discussed further below), is particularly suited to biomarker research. LSPR is an optical detection method based on the localised electromagnetic field around gold nanoparticles, which is much smaller than the wavelength of the incident light,[7,47,48] providing intrinsic signal amplification at the SPR surface for the sensitive detection of low concentrations of analytes under physiological conditions. For example, LSPR was used with nanoparticles coupled to a prostate specific antigen (PSA) antibody, allowing detection of PSA in human serum at concentrations well below the threshold level for prostate cancer diagnosis.[6] LSPR has also been used for efficiently screening compounds for cytochrome P450 binding[48] and profiling invasive cancer cells for biomarkers.[6,49,50]

7.3 Future Perspectives

It has been shown how SPR clearly contributes to drug discovery at all stages. The greatest impact of this biophysical technology lies in (1) its ability to assess kinetic and, increasingly, thermodynamic information about compounds binding to their target to support their optimisation, (2) its utility for fragment-based drug discovery, and (3) its ability to address more challenging molecular targets such as protein–protein interactions and membrane-bound proteins. As with other biophysical techniques, SPR provides useful complementary information to that obtained from biochemical and cellular assay technologies. In terms of future trends, optimising SARs through the tuning of kinetic rates, as opposed to equilibrium affinity (potency), although on the increase, is still in its infancy and somewhat empirical.[51] Apart from trying to recognise rules that govern trends in dissociation (and to a lesser extent, association) rate constants, often limitations of compound and target protein stability (pharmacokinetics and turnover, respectively) will dictate the maximal residence times of compounds *in vivo*. Therefore, off-rate screening tends to be carried out as a data gathering exercise along with all of the other SAR data obtained during hit characterisation. Nevertheless, some early guidelines have been established. For example, k_{on} optimisation is often based on charge–charge complementation, while k_{off} optimisation can be enhanced by π–π stacking interactions.[52] SPR will continue to provide a robust tool to enable kinetic-based optimisation to be further refined in the future.

Likewise, the value to drug discovery programmes of obtaining thermodynamic parameters (enthalpy and entropy) for candidate compound binding interactions is also a developing and probably underutilised capability of SPR. Until recently, it was generalised that negative enthalpies and positive entropies should be sought in this process to ensure binding potency and selectivity. However, it is now understood to be much more complicated than that. Active site water molecule networks and enthalpy–entropy compensation are among the phenomena that make it difficult to predict thermodynamic trends, much less how they should be used.[53–55] Moreover, equilibrium thermodynamics only tells half the story. The potential for activation parameters to be determined by SPR will undoubtedly be used more in the future to enhance drug discovery programmes through a greater understanding of the events required for a ligand to engage with its binding site.

While SPR has traditionally been undertaken with soluble, well-behaved protein targets, its capabilities for addressing membrane-bound proteins and other difficult targets will undoubtedly continue to expand. In addition to the approaches already described, one potential future strategy might involve the use of styrene maleic acid lipid particles (SMALPs), which capture membrane proteins in a highly physiological polymer-encapsulated background that is amenable to a range of biophysical and biochemical approaches.[56]

SPR instrumentation has come a long way since it was first introduced 25 years ago. It is anticipated that SPR will evolve further with innovative sensor chip chemistries, improved flow cell systems that limit the impact of detrimental effects of samples, and other instrumentation design modifications to provide higher throughput and even greater sensitivity. SPR sensitivity is desirable for fragment screening and evaluating low level binding responses while increased capacity array-based approaches will extend the utility of SPR to medium or even high throughput screening, from drug candidate starting points to antibody–antigen interactions in biomarker profiling. SPR instruments such as the Biacore S200 used currently for fragment screening are approximately 20 times more sensitive than those used 20 years ago when fragment screening was in its infancy. SPR sensitivity can also be vastly improved by more widely implementing known signal amplification methods. In the LSPR nanoparticle-based approach described above, colloidal gold nanoparticles in solution but in proximity to the SPR surface can magnify the sensing surface area, refractive index changes and localised electromagnetic field-induced SPR effects to enhance sensitivity. The colloidal gold nanoparticles can be conjugated with an appropriate probe (*e.g.*, antibody) to direct them to the surface-bound analyte, leading to signal amplifications of as much as 25-fold.[6,47] Another strategy reported to improve SPR sensitivity by as much as 100-fold, termed label-enhanced SPR, was developed recently by Episantic.[57] As the name suggests, this is not a label-free method (apart from the SPR chip surface coupling), in that the sensitivity enhancement comes from a dye labelling system. Selective visualisation of the binding responses of a dye-labelled anti-analyte antibody (or secondary antibody) *vs.* the contribution of other non-labelled analytes can be compared using specialised SPR software to enhance the signal of the specific binding interactions of the analyte and eliminate non-specific binding responses. "Label-free" binding data can still be obtained with this technology by using a competition format with an unlabeled analyte. Pharma Diagnostics has extended the LSPR principle by exploiting the properties of the colloidal gold nanoparticles themselves, which can be coupled with ligands and probed in bulk suspension for the SPR-induced absorbance changes indicative of binding interactions without the need for an SPR instrument.[58] This SoPRano™ technology has the advantage of a much higher throughput than traditional SPR but is also much more limited in sensitivity and information content. In addition to gold-based nanoparticles, other elemental metal nanoparticles, and magnetic and carbon-based nanoparticles have also been developed (sometimes in combination) to increase SPR sensitivity, specificity and other properties for particular biomolecular applications, as reviewed recently.[6] Like LSPR, nanopore array technology (introduced earlier in the context of membrane protein analysis) also provides both vastly increased throughput and enhanced SPR signalling through proximity effects. Unlike conventional SPR, which utilises a prism to focus the incident light beam and LSPR, in which gold nanoparticles produce strong localised SPR effects, nanopore arrays produce amplified SPR signals through the combined light energy of individual nanopores arrayed on a gold surface.[29] A phenomenon termed extraordinary optical transmission results from plasmon generation within the gold surface when light of specific wavelengths shines on a large array of such nanopores configured on the surface. Their additive plasmon resonance causes a funneling of light energy through the surface. Binding interactions at the surface lead to changes in the optimally transmitted wavelength of light. Nanopore array technology is particularly suitable for three dimensionally complex targets such as membrane proteins, which can be suspended in bilayers for unimpeded access to ligand binding in a near physiological environment.[30]

Alongside improvements to SPR materials and instrumentation, SPR software is continuously developing to improve the quality and ease of experimental set up and data analysis. SPR data analysis of large screening data sets has often been considered a bottleneck, with several manual steps, subjective quality control assessments and the necessity for visual inspection of individual sensorgram data. Biologic developed Scrubber™ as a programme designed to "clean up" raw senorgram data systematically (*e.g.*, align injection times, eliminate noise and correct for DMSO effects) for subsequent analysis, and now has a second version of the software that can also fit kinetic data to different models. Recently, AstraZeneca and GeneData joined forces to develop a browser-based software solution that not only is transferrable across different types of SPR instruments but also streamlines the whole data analysis process, formatting the data appropriately for experimental documentation, presentations and project databases.[59] It is anticipated that SPR data analysis methods will evolve further, with bespoke software suitable for particular applications including fragment screening and software to support structure–activity relationship studies that can complement (or perhaps even substitute for) protein X-ray or NMR structural data that is not always available for the desired molecular target.

Related technologies have also expanded the capabilities beyond traditional SPR. In Forté Bio's Octet systems, the use of bio-layer interferometry (BLI), a technology related to SPR, together with 96- or 384-well plate-based systems with parallel biosensors, allows the independent analysis of samples with minimal potential for cross-sample contamination issues, while improving throughput at the same time.[5] In addition, samples can be assayed in crude cell media and lysates since BLI is less susceptible to interference from components in the media. BLI involves the analysis of an interference pattern of light reflected from two surfaces on a fiber optic biosensor. A layer of immobilised target molecules resides on the first surface on the biosensor, while the second layer provides an internal reference surface. Similar to SPR, a change in the density of matter bound to the biosensor tip produces changes in the interference pattern that can be measured in real time.

Along with BLI, several other emerging biophysical techniques have adopted similar principles to SPR in the sense that an immobilised target protein surface is "interrogated" in real time with an analyte passed over the surface in a flow system and binding events are detected through a photometric effect on the surface. Creoptix' proprietary grating-coupled interferometry (GCI) technology extends the principles of waveguide interferometry, an optical method which, like SPR, detects refractive index changes within an evanescent field proximal to a biosensor surface caused by mass changes upon binding interactions at the surface.[60] Unlike SPR, however, the molecular interactions in waveguide interferometry are detected as refractive index changes within an evanescent field that cause a phase shift of the light beam in an optical waveguide, giving rise to an interference pattern in a parallel reference beam. Light travels over a longer surface than with SPR, providing greater sensitivity through cumulative effects of binding events, while minimising bulk refractive index changes through the shallow penetration of the evanescent field. GCI technology further enhances waveguide interferometry to eliminate alignment issues between the measurement and reference beams, making the technique less sensitive to temperature changes and vibrations.

We have mentioned that SPR can be used to assess protein conformational changes indirectly. Based on recent developments, it is anticipated that variations in SPR and related technologies will enable more direct observation of protein conformational changes that take place upon compound binding. For example, surface acoustic wave (SAW) technology monitors phase shifts in acoustic waves traveling along a biosensor, enabling the real time detection of conformational changes induced by binding. SAW technology has recently been used in combination with a membrane protein capture methodology to characterise GPCR binding in a comparative study with SPR.[35] SwitchSense so-called nanolevers utilise a different technology: target proteins or other biomolecules are tethered to short DNA nanolevers immobilised to the surface.[61] Upon compound binding, conformational changes alter the amplitude of waves produced by the electro-switchable DNA nanolever surface, measured in real time by fluorescence energy transfer. Second-harmonic generation incorporates a dye into the protein tethered to a lipid bilayer surface, and as such, is not a label-free technique.[62] Upon compound binding, conformational changes that affect the orientation of the dye cause changes to the evanescent wave of light reflected from the sensor surface. Finally, back-scattering interferometry (BSI) is a label-free, conformation-sensitive technology that can be used to assess binding interactions and conformational changes.[63] Similar to SPR, BSI detects changes in the refractive index in a microfluidic channel, but, unlike SPR, BSI does not involve protein immobilisation as the detection occurs in solution *via* a microscale interferometer. As such, BSI is a truly label-free technology that is not limited in sensitivity by target molecular mass and is therefore particularly useful for membrane proteins and other complex and difficult targets.[64] Although the technology has been available for approximately a decade, the means by which conformation-changing binding events in solution produce signals detectable by BSI are only just becoming elucidated[65] and its ability to provide kinetic information still requires target immobilisation. As instrumentation evolves to a similar extent to that which has already occurred with SPR, these technologies have the potential to make equally significant contributions to drug discovery platforms in the future.

In terms of increasing throughput and reduced sample consumption, taken to the extreme, there is even the promise of miniaturisation of microfluidic systems for single-molecule studies that could be applied to SPR.[66] In the other extreme of scale, SPR and related immobilised surface/refraction index-based technologies will find more applications for larger, more complex target systems, even including living cells. Cells have recently been used in SPR as the analytes passed over the immobilised target surface. In one such example, a suspension of TNF-α-expressing Jurkat cells was passed over surfaces immobilised with different anti-TNF agents (*e.g.*, adalimumab, etanercept and infliximab) and the binding responses analyzed in the absence or presence of competing soluble receptor or other anti-TNF agents.[67] Surface plasmon resonance microscopy (SPRM) is a recently-developed technology that allows the measurement of binding kinetics of surface protein interactions in single cells.[30,68,69] SPRM imaging is carried out on cells cultured on a gold-coated surface using an inverted microscope to perform optical and fluorescent image analysis simultaneously. Binding interactions can be measured with microsecond temporal precision and micrometer spatial resolution.[30] SPRM also enables the analysis of thousands of sample surfaces in parallel, thus providing both higher throughput and enhanced sensitivity for kinetic analysis compared with traditional SPR. The technology has already been applied to examine nicotinic acetylcholine receptor distribution and glycoprotein–lectin interactions.[30,68,69] Cellular responses from biochemical interactions and signalling events also cause specific protein movements that translate into dynamic mass redistributions which, with immobilised cells in SPR imaging (SPRI) systems, can be detected as refractive index changes in reflected light. Ion channel activity and

GPCR functional effects such as cAMP inhibition, calcium release and β-arrestin redistribution can all be detected in this manner. Other related technologies such as BLI, impedance and optical resonant waveguide grating (*e.g.*, Corning Epic[70]) are now being used for similar purposes. These non-invasive real time observations of live cellular responses allow SPR (and related technologies) to interrogate the functional cellular effects of compounds in an informative yet agonistic manner: Unlike other cellular functional assays that measure a specific signal transduction phenomenon, SPR and related approaches examine the integral of all phenomena occurring in response to a compound. As such, the non-characterised effects on alternative pathways or of orphan receptors will not be missed. While these capabilities of SPR are far removed from its traditional applications in screening and characterising small molecule binding interactions, they are equally instrumental to drug discovery.

In conclusion, while the fundamental principles of SPR have been established for decades, its utility for drug discovery continues to evolve on multiple fronts. In particular, instrument sensitivity and throughput continue to improve. Immobilisation and capture techniques and experimental designs are being refined to enable more complex targets and binding models to be addressed. For these reasons, SPR, used alongside complementary biophysical techniques, will continue to play an ever greater role in all phases of drug discovery, from initial hit finding to clinical candidate.

Acknowledgements

I would like to thank Rupert Satchell, David Allcock and John Unitt for critically reviewing the manuscript and Jordan Lane and Martin Redhead for assistance with the figures.

References

1. W. Huber and F. Mueller, *Curr. Pharm. Des.*, 2006, **12**, 3999–4021.
2. R. L. Rich and D. G. Myszka, *J. Mol. Recognit.*, 2001, **14**, 223–228.
3. Y. N. Abdiche, K. C. Lindquist, A. Pinkerton, J. Pons and A. Rajpal, *Anal. Biochem.*, 2011, **411**, 139–151.
4. A. M. Giannetti, Fragment-Based Drug DesignTools, Practical Approaches, and Examples, in *Enzymology*, ed. L. C. Kuo, Academic Press, 2011, vol. 493, pp. 169–218.
5. J. Concepcion, K. Witte, C. Wartchow, S. Choo, D. Yao, H. Persson, J. Wei, P. Li, B. Heidecker, W. Ma, R. Varma, L.-S. Zhao, D. Perillat, G. Carricato, M. Recknor, K. Du, H. Ho, T. Ellis, J. Gamez, M. Howes, J. Phi-Wilson, S. Lockard, R. Zuk and H. Tan, *Comb. Chem. High Throughput Screening*, 2009, **12**, 791–800.
6. H. H. Nguyen, J. Park, S. Kang and M. Kim, *Sensors*, 2015, **15**, 10481–10510.
7. A. Olaru, C. Bala, N. Jaffrezic-Renault and H. Y. Aboul-Enein, *Crit. Rev. Anal. Chem./CRC*, 2015, **45**, 97–105.

8. G. M. Keseru, D. A. Erlanson, G. G. Ferenczy, M. M. Hann, C. W. Murray and S. D. Pickett, *J. Med. Chem.*, 2016, DOI: 10.1021/acs.jmedchem.6b00197.
9. F. Khan, M. He and M. J. Taussig, 2006, *Data in Brief*, **78**, 3072–3079.
10. M. Fischer, A. P. Leech and R. E. Hubbard, *Anal. Chem.*, 2010, **83**, 1800–1807.
11. S. Lata and J. Piehler, *Anal. Chem.*, 2005, **77**, 1096–1105.
12. G. A. Papalia, S. Leavitt, M. A. Bynum, P. S. Katsamba, R. Wilton, H. Qiu, M. Steukers, S. Wang, L. Bindu, S. Phogat, A. M. Giannetti, T. E. Ryan, V. A. Pudlak, K. Matusiewicz, K. M. Michelson, A. Nowakowski, A. Pham-Baginski, J. Brooks, B. C. Tieman, B. D. Bruce, M. Vaughn, M. Baksh, Y. H. Cho, M. De Wit, A. Smets, J. Vandersmissen, L. Michiels and D. G. Myszka, *Anal. Biochem.*, 2006, **359**, 94–105.
13. M. Redhead, R. Satchell, V. Morkūnaitė, D. Swift, V. Petrauskas, E. Golding, S. Onions, D. Matulis and J. Unitt, *Anal. Biochem.*, 2015, **479**, 63–73.
14. I. Navratilova and A. L. Hopkins, *ACS Med. Chem. Lett.*, 2010, **1**, 44–48.
15. J. G. Quinn, *Anal. Biochem.*, 2012, **421**, 391–400.
16. H. Lu and P. J. Tonge, *Chem. Biol.*, 2011, **14**, 467–474.
17. J. B. Murray, S. D. Roughley, N. Matassova and P. A. Brough, *J. Med. Chem.*, 2014, **57**, 2845–2850.
18. I. Navratilova, G. A. Papalia, R. L. Rich, D. Bedinger, S. Brophy, B. Condon, T. Deng, A. W. Emerick, H.-W. Guan, T. Hayden, T. Heutmekers, B. Hoorelbeke, M. C. McCroskey, M. M. Murphy, T. Nakagawa, F. Parmeggiani, X. Qin, S. Rebe, N. Tomasevic, T. Tsang, M. B. Waddell, F. F. Zhang, S. Leavitt and D. G. Myszka, *Anal. Biochem.*, 2007, **364**, 67–77.
19. D. G. Myszka, in *Energetics of Biological Macromolecules, Part C*, Academic Press, 2000, vol. 323, pp. 325–340.
20. D. G. Myszka, X. He, M. Dembo, T. A. Morton and B. Goldstein, *Biophys. J.*, 1998, **75**, 583–594.
21. M. Elinder, P. Selhorst, G. Vanham, B. Öberg, L. Vrang and U. H. Danielson, *Biochem. Pharmacol.*, 2010, **80**, 1133–1140.
22. N. J. de Mol and M. J. E. Fischer, *Handbook of Surface Plasmon Resonance*, The Royal Society of Chemistry, 2008, pp. 123–172.
23. M. Geitmann, K. Retra, G. E. De Kloe, E. Homan, A. B. Smit, I. J. P. De Esch and U. H. Danielson, *Biochemistry*, 2010, **49**, 8143–8154.
24. M. Kim, S. O. Jung, K. Park, E.-J. Jeong, H.-A. Joung, T.-H. Kim, D.-W. Seol and B. H. Chung, *Biochem. Biophys. Res. Commun.*, 2005, **338**, 1834–1838.
25. H. Šípová and J. Homola, *Anal. Chim. Acta*, 2013, **773**, 9–23.
26. S. C. B. Gopinath, *Sens. Actuators B Chem.*, 2010, **150**, 722–733.
27. X. Li, Y. Wang, L. Wang and Q. Wei, *Chem. Commun.*, 2014, **50**, 5049–5052.
28. R. Karlsson, M. Kullman-Magnusson, M. D. Hämäläinen, A. Remaeus, K. Andersson, P. Borg, E. Gyzander and J. Deinum, *Anal. Biochem.*, 2000, **278**, 1–13.

29. J. Maynard and N. Lindquist, *Biotechnol. J.*, 2009, **4**, 1542–1558.
30. S. G. Patching, *Biochim. Biophys. Acta, Biomembr.*, 2014, **1838**, 43–55.
31. I. Navratilova and A. L. Hopkins, *Future Med. Chem.*, 2011, **3**, 1809–1820.
32. S. Locatelli-Hoops, A. A. Yeliseeva, K. Gawrischa and I. Gorshkova, *Biomed. Spectrosc. Imaging*, 2013, **2**, 155–181.
33. I. Navratilova, J. Besnard and A. L. Hopkins, *ACS Med. Chem. Lett.*, 2011, **2**, 549–554.
34. R. Chu, D. Reczek and W. Brondyk, *Sci. Rep.*, 2014, **4**, 7360.
35. R. G. Heym, W. B. Hornberger, V. Lakics and G. C. Terstappen, *Biochim. Biophys. Acta, Proteins Proteomics*, 2015, **1854**, 979–986.
36. P. Stenlund, G. J. Babcock, J. Sodroski and D. G. Myszka, *Anal. Biochem.*, 2003, **316**, 243–250.
37. C. Bieri, O. P. Ernst, S. Heyse, K. P. Hofmann and H. Vogel, *Nat. Biotechnol.*, 1999, **17**, 1105–1108.
38. N. Robertson, A. Jazayeri, J. Errey, A. Baig, E. Hurrell, A. Zhukov, C. J. Langmead, M. Weir and F. H. Marshall, *Neuropharmacology*, 2011, **60**, 36–44.
39. J. M. Glück, B. W. Koenig and D. Willbold, *Anal. Biochem.*, 2011, **408**, 46–52.
40. R. L. Rich, Y. S. N. Day, T. A. Morton and D. G. Myszka, *Anal. Biochem.*, 2001, **296**, 197–207.
41. Å. Frostell-Karlsson, A. Remaeus, H. Roos, K. Andersson, P. Borg, M. Hämäläinen and R. Karlsson, *J. Med. Chem.*, 2000, **43**, 1986–1992.
42. K. Vuignier, J.-L. Veuthey, P.-A. Carrupt and J. Schappler, *Drug Discovery Today*, 2013, **18**, 1030–1034.
43. H. Nishijima, A. Kosaihara, J. Shibata and T. Ona, *Anal. Sci.*, 2010, **26**, 529–534.
44. C. L. Baird, E. S. Courtenay and D. G. Myszka, *Anal. Biochem.*, 2002, **310**, 93–99.
45. E. Danelian, A. Karle, R. Karlsson, S. Winiwarter, A. Hansson, S. Lo and H. Lennerna, *J. Med. Chem.*, 2000, **43**, 0–3.
46. K. Kim, S. Cho, J. H. Park, Y. Byun, H. Chung, I. C. Kwon and S. Y. Jeong, *Pharm. Res.*, 2004, **21**, 1233–1239.
47. S. Szunerits, J. Spadavecchia and R. Boukherroub, *Rev. Anal. Chem.*, 2014, **33**, 153–164.
48. A. Das, J. Zhao, G. C. Schatz, S. G. Sligar and R. P. Van Duyne, *Anal. Chem.*, 2009, **81**, 3754–3759.
49. Y. Hong, M. Ku, E. Lee, J.-S. Suh, Y.-M. Huh, D. S. Yoon and J. Yang, *J. Biomed. Opt.*, 2014, **19**, 051202.
50. T. Jiang, M. Minunni, P. Wilson, J. Zhang, A. P. F. Turner and M. Mascini, *Biosens. Bioelectron.*, 2005, **20**, 1939–1945.
51. A. C. Pan, D. W. Borhani, R. O. Dror and D. E. Shaw, *Drug Discovery Today*, 2013, **18**, 667–673.
52. Z. Yu, J. P. D. Van Veldhoven, J. Louvel, I. M. E. T'Hart, M. B. Rook, M. A. G. Van Der Heyden, L. H. Heitman and A. P. IJzerman, *J. Med. Chem.*, 2015, **58**, 5916–5929.

53. J. D. Chodera and D. L. Mobley, *Annu. Rev. Biophys.*, 2014, **42**, 121–142.
54. G. Klebe, *Nat. Rev. Drug Discovery*, 2015, **14**, 95–110.
55. G. Klebe, *ChemMedChem*, 2015, **10**, 229–231.
56. M. Jamshad, J. Charlton, Y.-P. Lin, S. J. Routledge, Z. Bawa, T. J. Knowles, M. Overduin, N. Dekker, T. R. Dafforn, R. M. Bill, D. R. Poyner and M. Wheatley, *Biosci. Rep.*, 2015, **35**, 1–10.
57. L. Eng, L. Nygren-Babol and A. Hanning, *Anal. Biochem.*, 2016, **510**, 79–87.
58. P. Englebienne, M. Verhas and A. Van Hoonacker, *Analyst*, 2001, **126**, 1645–1651.
59. G. Dahl, S. Steigele, P. Hillertz, A. Tigerström, A. Egnéus, A. Mehrle, M. Ginkel, F. Edfeldt, G. Holdgate, N. O'Connell, B. Kappler, A. Brodte, P. B. Rawlins, G. Davies, E.-L. Westberg, R. H. A. Folmer and S. Heyse, *J. Biomol. Screening*, 2016, DOI: 10.1177/1087057116675316.
60. P. Kozma, F. Kehl, E. Ehrentreich-Förster, C. Stamm and F. F. Bier, *Biosens. Bioelectron.*, 2014, **58**, 287–307.
61. J. Knezevic, A. Langer, P. A. Hampel, W. Kaiser, R. Strasser and U. Rant, *J. Am. Chem. Soc.*, 2012, **134**, 15225–15228.
62. B. Moree, K. Connell, R. B. Mortensen, C. T. Liu, S. J. Benkovic and J. Salafsky, *Biophys. J.*, 2015, **109**, 806–815.
63. D. J. Bornhop, J. C. Latham, A. Kussrow, D. A. Markov, R. D. Jones and H. S. Sørensen, *Science*, 2007, **317**, 1732–1736.
64. M. M. Baksh, A. K. Kussrow, M. Mileni, M. G. Finn and D. J. Bornhop, *Nat. Biotechnol.*, 2011, **29**, 357–360.
65. D. J. Bornhop, M. N. Kammer, A. Kussrow, R. A. Flowers and J. Meiler, *Proc. Natl. Acad. Sci. U. S. A.*, 2016, **113**, E1595–604.
66. Y. Zhao, D. Chen, H. Yue, J. B. French, J. Rufo, S. J. Benkovic and T. Jun Huang, *Lab Chip.*, 2013, **13**, 2183–2198.
67. T. Ogura, Y. Tanaka and H. Toyoda, *Anal. Biochem.*, 2016, **508**, 73–77.
68. C. T. Campbell and G. Kim, *Biomaterials*, 2007, **28**, 2380–2392.
69. W. Wang, Y. Yang, S. Wang, V. J. Nagaraj, Q. Liu, J. Wu and N. Tao, *Nat. Chem.*, 2012, **4**, 846–853.
70. D. Verzijl, T. Riedl, P. W. H. I. Parren and A. F. Gerritsen, *Biosens. Bioelectron.*, 2017, **87**, 388–395.

CHAPTER 8

Fluorescent Thermal Shift Assays for Identifying Small Molecule Ligands

ANDREW D. SCOTT

Bioscience Department, Sygnature Discovery, BioCity, Nottingham NG1 1GF, UK
Email: a.scott@sygnaturediscovery.com

8.1 Introduction

Ligand-induced protein stabilisation refers to the phenomenon whereby the binding of inhibitors, substrates, metal ions, cofactors and other proteins (ligands) enhances the stability of a protein (receptor) due to the energetic coupling of the ligand-binding and receptor-melting reactions. This biophysical property has been exploited in the drug discovery process by measuring the thermal stability of target proteins in the presence and absence of small compounds or protein therapeutics. This thermal stability had traditionally been investigated using biochemical activity assays, circular dichroism (CD), or differential scanning calorimetry (DSC). However, for the biochemical assays, there is a requirement for a specific assay to be developed along with a catalytically active protein. Both CD and DSC require large amounts of protein and are relatively low-throughput methods that only measure single protein–ligand interactions at a time. To address the need for a higher throughput, simple, universal, automated assay to measure the thermal stability of proteins in the presence and absence of ligands, FTSA was developed. FTSA, uses the well-established thermodynamic principle that the thermal stability of a protein target can be altered by a binding ligand in a concentration- and potency-dependent manner.[1,2]

FTSA, variously known as ThermoFluor,[3] differential scanning fluorimetry (DSF),[4] or temperature-dependent fluorescence (TdF),[5] is one of the simplest, cheapest and most widely used biophysical methods for identifying and characterising small molecule ligand binding to a target protein. Unlike other biophysical methods such as nuclear magnetic resonance (NMR), mass spectrometry (MS), surface plasmon resonance (SPR), microscale thermophoresis (MST) or isothermal titration calorimetry (ITC), FTSA does not require a significant investment in instrumentation and is relatively straightforward in terms of method development, optimisation and data interpretation. FTSA can also be used in a much higher throughput mode than other biophysical techniques, which lends itself to an early-stage triage assay. Indeed FTSA has been used as both an orthogonal technique for hit confirmation following HTS[4] and as a primary assay for fragment-based drug discovery.[5]

8.2 FTSA Principle

FTSA quantifies the change in thermal denaturation temperature of a protein under varying conditions using a fluorescent probe molecule. The probe binds non-specifically to hydrophobic surfaces and has a low quantum yield in a polar environment (*e.g.*, an aqueous solution) but is highly fluorescent in nonpolar environments with low dielectric constants (*e.g.*, attached to hydrophobic patches on a protein).[3] This compound is added to a protein in solution and the fluorescence is continually monitored while the solution is gradually heated. For the majority of proteins, this gradual increase of temperature has little effect until a temperature is reached where the protein chain quickly unfolds. When this unfolding begins, the hydrophobic core becomes exposed and the dye attaches to these sites and the fluorescent signal increases. As heating continues, the protein becomes more denatured and the fluorescence continues to rise until full denaturation occurs. At this point, the maximum fluorescent signal is reached, as after this, the protein will start to degrade or aggregate, the complex of the probe and protein will start to precipitate, and the fluorescent signal will drop. The stability of a protein is related to its temperature-dependent Gibbs free energy of unfolding (ΔG_u). Typically, the stability of a protein decreases as the temperature increases and eventually the ΔG_u is equal to zero at the equilibrium where concentrations of folded and unfolded protein are equal. The free energy contribution of ligand binding usually results in an increase in ΔG_u. A larger (more positive) ΔG_u relates to a more stable protein. Therefore, ligand binding frequently causes an increase in the temperature at which half of the protein population is unfolded, which corresponds to the melting temperature (T_m).

Real-time monitoring of these fluorescent changes provides a thermal melting curve, or thermogram, an example of which is shown in Figure 8.1.

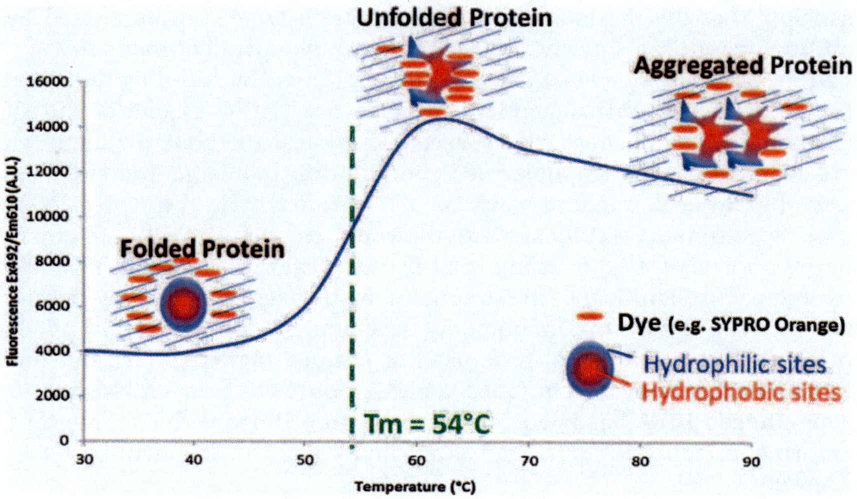

Figure 8.1 Typical FTSA thermal denaturation graph. The melting curve shows the protein being heated in the presence of a hydrophobic dye (*e.g.*, SYPRO Orange) and upon denaturation, the dye binding to the internal hydrophobic core of the protein, increasing the fluorescence. When unfolding is complete, maximal fluorescence is reached, followed by a decrease in fluorescence as the dye dissociates from the aggregating protein. The inflection point corresponds to the melting temperature (T_m), at which 50% of the protein is unfolded. In this instance, a T_m of 54 °C has been calculated.

The value most often reported from FTSA is the temperature of the midpoint of the sigmoidal transition region of the fluorescence curve, *i.e.*, the T_m of the protein, and this is a measure of its inherent thermal stability.[4] Other attributes of the curves also contain information, such as the steepness of the transition or the flatness of the initial baseline. The melting temperature can also be derived by calculating the peak of the first derivative (Figure 8.2), as long as the curve does not show significant asymmetry. It has been shown that the stabilising effect of compounds upon binding is proportional to the concentration and affinity of that ligand. Therefore, in most cases it is possible to also determine an estimate of the affinity (K_d).

Due to the highly reproducible, sharp melting curves produced by FTSA, it can be used to study proteins under various conditions, for example differing pH, salts, additives, ligands, oxidised/reduced states or mutations.[6] A positive shift in the melting temperature can be directly linked to an increase in structural order. This chapter will concentrate on the use of FTSA in drug discovery and, specifically, the use of the technique for screening compounds and fragments. Alternative uses of the method in both the drug discovery process and other environments will be highlighted but not discussed in great detail. For example, FTSA can be used to determine optimal protein storage conditions, perform functional characterisation or even improve the success rates of protein purification and crystallisation.

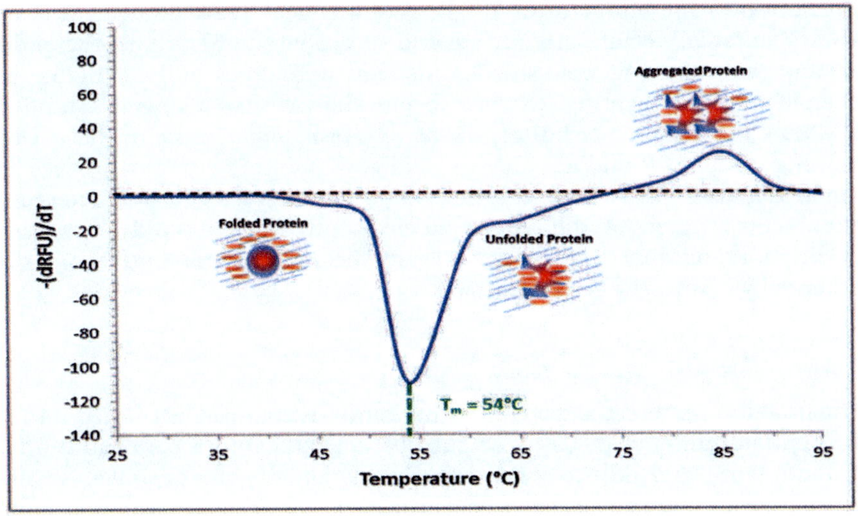

Figure 8.2 An alternative representation of the melting curve using the first derivative – (dRFU)/dT of the raw data. In this case, the T_m corresponds to the apex of the negative curve.

8.3 Optimal Experimental Set-up

As stated above, the experimental set-up for FTSA is extremely simple.[4] The typical set-up involves five steps: (i) initial optimisation of buffer, protein concentration and speed of denaturation (°C per min); (ii) mixing the protein with the fluorophore and compound; (iii) a brief centrifugation (*e.g.*, ~1000 × g, for 1 min) of the assay plate to mix the compounds into the protein solution; (iv) melting the solution in an optically clear sealed plate whilst monitoring the fluorescent changes in real time, typically using a quantitative PCR instrument (refer to Section 8.3.7). For SYPRO orange, an excitation range of 510–530 nm and an emission signal at 567–596 nm is used for data analysis (for other dyes, refer to 8.3.5); (iv) calculating the melting point changes and, if appropriate, estimating the affinity constants.

A typical mixing protocol would be in the following order:

(1) Water to make up to volume, (2) buffer, (3) salt or additive if required, (4) protein, (5) ligand, and (6) dye, with the total volume generally being between 5–20 µl, depending on the plate used. The main aim of the order of protein preceding the diluted dye and then compound is to avoid damage to the protein by contact with high concentrations of DMSO.

8.3.1 Buffer

The best buffer systems for FTSA tend to be relatively simple, for example, HEPES or phosphate at near-neutral pH, containing only a few necessary salts and additives. It should be noted that the pH of Tris is not always constant over the temperature gradient of a typical FTSA assay, so it is not always suitable. Acetate, cacodylate and phosphate buffers have the added advantages of having very small ionisation enthalpies and negligible pH changes during heating. It has been shown that using a standard HEPES + 150 mM NaCl buffer, 58 of 61 *Plasmodium* proteins have clear melting transitions and are able to yield good melting curves.[7] It is recommended that stabilising reagents such as glycerol, DTT, TCEP and detergents should be omitted or diluted as much as possible to avoid nonspecific fluorophore binding that would elevate the background and could also potentially distort the melting curves.

8.3.2 Protein

The protein of interest is typically either buffer-exchanged or diluted into an appropriate buffer (*e.g.*, assay or purification buffer) to a concentration of around 1 μM and adjusted accordingly, depending on response. Smaller proteins may require higher concentrations, whereas for proteins above 50 kDa, concentrations as low as 100 nM can be used. For a 96-well assay, approximately 100 μg of a protein with a molecular weight of around 30 kDa will be required. Protein with a purity of at least 75% is required in order to obtain a reliable melting point in FTSA experiments. Additionally, the protein T_m should be between 30 and 90 °C, as, if the melting point is higher or lower, no melt curve will be observed. In the case of an extremely low T_m, a high initial fluorescence will be measured due to the protein being partially or fully unfolded already, and this fluorescence will only decrease with increasing temperature due to aggregation. In these cases, the use of another method to screen for conditions that either increase or decrease the thermal stability of the protein (*e.g.,* change buffer, add ligand/substrate/cofactor or mutate the protein) may be required.

It is good practise to ensure protein is folded and not aggregated using CD, dynamic light scattering (DLS) or size exclusion chromatography (SEC). If there are a significant number of exposed hydrophobic sites on the protein, then you are likely to get a high background signal, whereas, if there are insufficient hydrophobic sites exposed upon unfolding, then the signal after melting will be low.

It is generally desirable to remove any tags, whether they be solubility or affinity based, from the protein before running FTSA. For example, a His6-TEV site tag results in a dye-mediated fluorescence signal at about 45 °C and can also decrease the protein thermal stability.[6]

8.3.3 Dyes

FTSA makes use of environmentally sensitive dyes that have low quantum yields (low QY) in solvents with high dielectric constants, such as water, but are highly fluorescent (high QY) in solvents with low dielectric constants, such as the hydrophobic sites of an unfolded protein. Various fluorescent dyes have been used for FTSA-type studies and the choice as to which to use is primarily down to the protein being studied. The dyes differ with respect to their optical properties, particularly in the fluorescence quantum yield caused by binding to the hydrophobic sites of a denatured protein.[8] It is recommended that a variety of dyes are tested, especially when no unfolding transition is observed with SYPRO orange.

The first dyes to be used in FTSA type-assays were 1-anilinonaphthalene-8-sulphonic acid (1,8-ANS) and 2,6-ANS, 2-(*p*-toluidinyl)-napthalene-6-sulphonic acid (2,6-TNS),[3,9] the former of which is particularly useful when detergent is present.[10] However, for these dyes to work, they require UV excitation and quartz cuvettes, and typically the excitation and emission maxima are not accessible by commercial filter sets for RT-PCR machines.

The most common fluorescent compound used in FTSA is SYPRO orange dye, which is favoured due to its large intensity change upon protein binding. Also, there is minimal interference from intrinsic small molecule fluorescence due to the properties of the Sypro orange–protein complex, which has an excitation maximum around 480 nm and maximal emission at 569 nm. This high excitation wavelength is matched by Nile Red (554 nm), which is a common alternative but generally has a weaker fluorescent gain. Typically, a small aliquot of concentrated (*e.g.*, ×5000) Sypro orange or Nile red in DMSO is added to the protein solution to give a final concentration of less than ×20, as higher concentrations can actually have a destabilising effect on the protein.[11] It should be noted that the physical properties of both dyes preclude their use in the presence of surface-active excipients, due to the binding of the dyes that occurs in the hydrophobic micelle core. This limitation, especially in the study of protein formulation, has led to other probes being investigated.

A specialised variation of FTSA, which was introduced by Alexandrov *et al.* (2008),[12] replaced Sypro orange with *N*[4-(7-diethylamino-4-methyl-3-coumarinyl)phenyl]maleimide (CPM), a compound that only fluoresces after reacting with a nucleophile. CPM has a high preference for thiols over other typical biological nucleophiles and therefore will react specifically with cysteine side chains to form fluorescent adducts. Hydrophobic cysteines are typically buried in the interior of a folded protein so that as the protein denatures, the thiols become available and a fluorescence signal obtained. This is particularly useful for membrane-bound proteins that require detergent to stabilise them, as there will be no background noise from the dye binding to the hydrophobic micelle core. One drawback to this approach is that the excitation and emission wavelengths for reacted CPM are 387 nm and 463 nm, respectively, and this potentially requires specialised filters. Another alternative is the Proteostat® dye from Enzo Life Sciences, which monitors protein aggregation rather than folding and is also compatible with detergents[13] and 4-(dicyanovinyl)julolidine (DCVJ), a molecular rotor probe with fluorescence that is strongly dependent on the rigidity of its environment.[14] In this case, when protein denatures, the more ordered environment provided by the exposed hydrophobic regions produces DCVJ fluorescence. This has been used extensively in formulation development as there is little interference from a micelle environment.

8.3.4 Compounds

Compound concentrations should be within the kinetic solubility limits and preferably in excess of the protein concentration, as ΔT_m is driven by the free ligand, rather than ligand bound to the protein, and significant ligand depletion will prevent accurate ΔT_m measurements. To ensure full occupancy of the binding sites, the ratio between compound and protein should be maintained at around 10 : 1 or higher and with high affinity compounds of sub-µM, it is usually best to keep the concentration at about 10 µM. Ligands are generally prepared to a 50- to 100-fold concentrated solution, typically in the 10–100 mM range. For titration studies, any number of different concentrations of a test compound can be employed, along with a no-compound negative control. Compounds in DMSO typically require diluting in final screening conditions before addition to avoid excessive DMSO (cut-off is usually 2–5% v/v DMSO final concentration).

8.3.5 Controls

No protein/ligand control (NPLC) wells should contain only buffer (including volume to match protein sample), ligand solvent (*e.g.*, DMSO) and dye. High fluorescence would indicate contamination in the wells or dye interaction with buffer components or plate material. Ligand only control (LOC) wells should contain only ligand, buffer (including volume to match protein sample) and dye. High fluorescence or a melt curve might indicate ligand–dye interactions or the presence of a fluorescent ligand. No dye control (NDC) wells should contain only protein, buffer, ligand and dye solvent. High fluorescence might indicate a fluorescent ligand and a melt curve might indicate the potentially interfering auto-fluorescent nature of the protein. However, if consistent melt curves without dye present are achieved, this suggests that the assay could be run using the intrinsic protein fluorescence and therefore eliminate the possibility of dye–ligand competition (refer to Section 8.10). Ideally, a positive control compound should be included on each plate. This could be a known binder, a natural substrate (*e.g.*, ATP) or a known cofactor.

8.3.6 Temperature

The speed of the temperature ramp is typically about 1 °C min^{-1}, but it can be altered dependent on the need for resolution *versus* throughput. Senisterra *et al.*[15] measured the T_m values of nine proteins under various heating rate conditions and observed that the faster the melting speed, the higher the apparent T_m of the protein. However, it was also observed that the net melting point shifts as a result of ligand binding appear to be affected only minimally by the melting speeds.[16] The maximum rate of heating so far used for screening compound libraries against a protein is 4 °C min^{-1} but it can be as low as 0.1 °C min^{-1} to obtain highly accurate T_m. The temperature range can be set anywhere between 5 °C (if a cooling system is available) and 99 °C, and can be adjusted to focus around the expected T_m (based on the initial melting experiments). Obviously, the range must go to high enough temperatures to allow the T_m of the protein to be measured in the presence of high affinity binders.

8.3.7 Instrumentation

FTSA can be performed in either a fluorescent plate reader with a heated stage or a standard quantitative real-time PCR instrument, as long as the latter has an optical system compatible with the fluorescent properties of the dye. Some commonly used instruments include the LightCycler 486 from Roche,[17] iCycler/iQ from BioRad,[18] the Stratagene Mx3005P,[19] the Rotorgene from Corbet Research,[20] the FluoDia T70 from PTI and various real-time PCR instruments from Applied Biosystems (*e.g.*, StepOnePlus). The choice of instrument will depend mainly on the required capacity and/or temperature range, but could also be influenced by the dye used and the fluorescent filters available.

8.4 Data Analysis

8.4.1 T_m Determination

The output of FTSA is a series of melt curves that need to be interpreted in order to get value from the assay. As well as reliable detection of the thermal denaturation process, an accurate measurement of the melting point is a prerequisite for obtaining a binding affinity between the ligand and the target. A detailed description of the analysis of protein folding curves to determine the melting temperature is given in the literature.[3,21] When the fluorescence intensity of the probe is reported as a function of the temperature of the solution, a sigmoidal curve is generated (Figure 8.1). As long as the thermogram is symmetrical, the point of inflection can be calculated from the eqn (1.1) below:

$$Y(T) = \text{Max} + \frac{(\text{Min} - \text{Max})}{1 + e^{\left(T_m - \frac{x}{a}\right)}} \tag{8.1}$$

where Max signifies the maximum fluorescence intensity of the melting transition, Min is the minimum intensity and a is the slope of the curve at the T_m.

Alternatively, the T_m value can be derived from a variety of available RT-PCR instruments using the publically available analytical tool from the SGC,[4] or the instrument vendors themselves, as they are now providing the means to measure accurate T_m as a standard part of their software.

Another way to calculate T_m values is to determine the maximum of the first derivative (Figure 8.2), which is offered by most PCR software packages or more specialised programmes, *e.g.*, 'Meltdown',[22] MTSA[23] and ThermoQ.[24] One of the advantages of these specialised programmes is the ability to reject curves that are inappropriate for use in T_m estimations.

Care should be taken when ranking compounds based on their ΔT_m values, especially if either the protein is not the same or the compounds are substantially different from each other. This is due to the thermodynamic influence on the magnitude of the T_m shift (see Section 8.6), so that a ΔT_m is not unique to a certain binding affinity, because a range of K_d values with different entropic and enthalpic contributions can give rise to very different changes in melting temperatures. There is generally a good correlation between melting point shifts and binding affinities when the enthalpy changes of binding are comparable for a related series of compounds,[21] but the correlation is poor when the enthalpy changes of binding are substantially dissimilar.[25,26]

8.4.2 K_d Determination

The increase of a protein's thermal stability due to ligand binding is generally concentration-dependent and proportional to the affinity of the ligand.[27] Affinity constants at the T_m can be calculated from the resulting curves (Figure 8.3), however, an affinity at another temperature (*e.g.*, 25 °C) cannot be directly determined from a thermal shift assay without knowing the relative contributions of enthalpy and entropy of ligand binding. In addition, binding constants determined at the melting point must be extrapolated to a common temperature in order to make direct comparisons between compounds.

Global fitting of all melting curves for a common protein target in the absence and presence of serial dilutions of a given compound is only possible provided that individual melting curves are uncomplicated and do not have distortions. It is therefore necessary that all FTSA data are accompanied by some degree of manual examination to confirm simple monophasic profiles.

Various models and equations are available to determine affinity at a particular temperature from FTSA data. They range from very simple equations, where zero enthalpy is assumed, to highly complex equations which require additional parameters to be measured by other techniques (*e.g.*, enthalpy from ITC and heat capacity from DSC). Certainly for initial screening of compounds and even basic mechanism of action studies where speed and robustness are more important than absolute precision, the simple equations are most suitable. Indeed, by the time that highly accurate affinities are required for a drug discovery programme, it is likely that other assays will be available to accurately measure either K_d or IC_{50}, meaning the

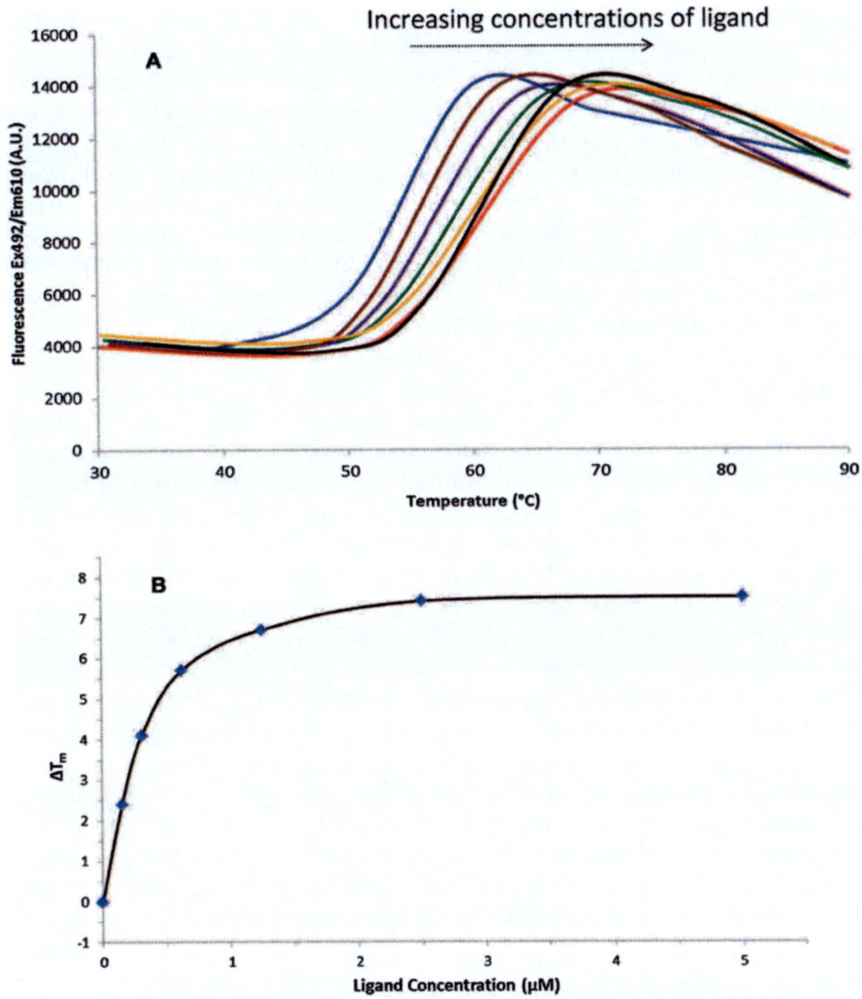

Figure 8.3 Detection of ligand binding by FTSA. (A) Effect of ligand binding on the stability of a protein. (B) The ΔT_m values at different concentrations of protein illustrating a concentration-dependent stabilisation effect. T_m values should be calculated from fitting the transition curves in (A) by using the Boltzmann sigmoid function as described.[28]

complex equations that could be used for FTSA data may be redundant. However, there is still a need in drug discovery to ascertain both accurate affinities and mechanisms of action as early in the process as possible. The first analyses of FTSA data used a modified version of the equations of Brandts and Lin, removing the sloped asymptotes of the FTSA curves to improve the fitting. However, the K_d values obtained depended on ligand concentration and this cannot be the case as an affinity constant must be independent of the concentration of ligand tested. Further to this, the equations require an 'estimated enthalpy of binding' of about 80 kJ mol^{-1}, which would mean any extrapolation of the data would be highly error prone. One cause of this is that the equations do not account for the conservation of mass, and the ligand concentration is assumed to be unaffected by the concentration of bound ligand, which may be true when the ratio of ligand to protein is high, but means that application of the equations is not universal.

The models developed in Matulis' lab at the University of Vilnius account for the sloped asymptotes of the FTSA curves, so they allow for more accurate fitting to the data and show a strong correlation with DSC and ITC.[21] The authors encourage using the solvent accessible area of the protein to estimate the specific heat capacity of the protein and fixing this within the fitting of the raw data. Instead of calculating the heat capacity from fitting, they instead float the reference temperature term in a non-linear Van't Hoff equation, and, due to a quirk of the non-linear fitting algorithm, this caused the term to trend to the value where it has the least effect on the calculated parameters, the midpoint of the transition. Although this method of calculating T_m was unusual, it was effective, and, as a consequence, allowed the entropy of unfolding to be determined during the fitting process for the first time. Having determined the thermodynamic parameters of denaturation, the authors used the conservation of mass (converted to several non-linear Van't Hoff equations) to determine the amount of ligand required to raise the melting temperature to a particular value. This implicit method was shown to be accurate in determining the affinity of ligands to a protein, although it still required an assumed enthalpy of binding and an estimated enthalpy of unfolding *via* fitting. As had previously been observed, there was not an upper limit to the binding energy that this technique could measure, and modified equations, which were able to describe biphasic melting transitions, demonstrated the ability to determine picomolar binding affinities.[16] These curves showed characteristic biphasic features in the FTSA assays which demonstrated the ligand depletion.

It is also useful to understand how interfering compounds behave in these assays and the first study to address this consideration was by Redhead *et al.*[29] Additional to this, the team have also derived a number of equations for the various mechanisms of action that can be studied using FTSA. Equations for competitive, uncompetitive and cooperative allosteric binding are to be published in the near future.

It should also be noted that several related algorithms have been published for data analysis to derive affinity constants based on thermal shift assays.[3,18,30,31] Although these algorithms share core formulae,[25] they have subtle differences in relation to the handling of the melting curves and formula simplifications.

8.5 Advantages

FTSA is high-throughput, inexpensive, robust, easy to set up and requires only a small amount of protein at low concentration. The assay requires only one additional reagent on top of the protein and compound, that being the fluorescent dye. Moreover, it does not involve any target modification or additional separation steps. Due to the fact that FTSA uses a fundamental biophysical property, there is no need to design new assays for each new target and no inherent properties of the protein, such as enzymatic activity, are needed. Indeed, no prior knowledge of a specific therapeutic target's function is required. The universal nature of the assay was illustrated in a couple of systematic case studies, where 65–75% of soluble proteins examined by FTSA using SYPRO Orange produced typical denaturation curves, with clear temperature transitions, enabling an accurate measurement of the melting points.[28,32] The method can also be applied to samples for which aggregation or low stability hinders purification efforts and can be used to screen for inhibitors of protein–protein interactions, a rich source of potential targets for drug action.

FTSA provides a good initial estimate of binding constants, which can facilitate the experimental setup for orthogonal approaches. Indeed, when properly estimated, FTSA can yield reliable dissociation constants over a far wider range than what is possible to obtain with other assays. It is one of the few methods that is sensitive enough to examine binding affinities in the millimolar range while at the same time it has no upper limit on the affinity that can be measured.[33] Most assays designed to measure IC_{50} values have an upper limit at half the active target concentration. This constraint often leads to compounds hitting the tight-binding limit during the lead discovery and lead optimisation stages of drug discovery and therefore limiting SAR (structure–activity relationship) analysis. Other biophysical methods such as ITC or SPR can extend this limit to low nanomolar binders, or even high picomolar binders with the use of well-designed competition experiments. However, the success of these assays relies on known mid-affinity binders, non-zero enthalpy changes upon binding and the ability to measure very slow off-rates. Along with DSC, FTSA is the only method that can theoretically determine affinities at or even below femtomolar levels. However, unlike DSC, FTSA can achieve these measurements at medium to high throughput and therefore is much more cost effective and only uses <10 pmol of protein per assay point. Practically, the upper limit is reached only when the thermal shift caused by the high affinity ligand takes the T_m of the protein above the boiling point of the aqueous solvent.

Like other biophysical techniques, FTSA has the potential to identify allosteric ligands that may not necessarily inhibit the primary activity of the target. This could be especially useful for families of proteins (*e.g.*, kinases, bromodomains *etc.*) where selectivity within the active site is difficult due to common features (*e.g.*, ATP and acetyl–lysine bind sites). Targeting a less conserved area of the protein could give selectivity and such a ligand could be further engineered to produce the desired effect on the activity of the target.

8.6 Limitations

The artefacts in FTSA measurements have been well documented, especially in fragment-based lead discovery.[34] When compared to other binding and activity assays used in fragment screening, FTSA gives a relatively high rate of false negative and false positive results.[35] However, other papers highlighting the applicability of FTSA to the high-throughput screening of small molecules have shown little or no false positives.[36]

False positives can be caused by experimental errors or by the melting transition of a fluorescent compound being mistaken for a shift in the protein T_m. As with any binding assay, a compound that associates with the target but has no functional consequence may be classed as a false positive. However, this does not take into account the opportunity of taking a non-active binding compound and, through structural studies and SAR analysis, producing a compound that has the desired function. Other causes for false positive results include promiscuous binders that associate with multiple sites on the protein surface. Compounds that can bind to more than one site on a protein could appear more potent due to the additive effect on the ΔT_m. Although there may be advantages to taking these compounds further in the drug discovery process, they might also wrongly bias the structure–activity relationship of the ligands and lead to sub-optimal lead compounds. In these cases, it would be best practice to obtain stoichiometry from another method (*e.g.*, ITC) for all compounds showing an unusually large T_m shift.

False negatives are also present in any FTSA screen, and these can be caused by a number of factors. If the protein is particularly unstable and is partially unfolded even at the starting temperature, then large shifts of T_m following compound binding might not be observed. Similarly, if the compound has a comparable affinity for both the native and unbound states of the protein, then even though the affinity might be quite high, the shift in T_m will not be present. Indeed, this could be exacerbated if a compound binds tightly and enthalpically to the native state (true positive), but also binds weakly and entropically to the denatured state, and therefore the binding is masked in FTSA. It is also important to understand the solubility of the compounds as screening at concentrations around or above the solubility concentration will result in false negatives as the true (soluble) concentration of the compound will be much lower than expected. A compound that is competitive with the fluorophore itself might cause K_d determinations that are much weaker than the true affinity. Similarly, a compound that requires the presence of a cofactor, another ligand or a substrate to bind, which is not present in the FTSA buffer, could be incorrectly excluded from a screening cascade. It has been noted previously that many multidomain proteins show a sharp two-state melting transition. This may be caused by a cooperative unfolding of the attached proteins due to energetic coupling between the domains or by a strong binding preference of the dye to one of the domains. In either case, a ligand binding to a domain that does not cause the dominant signal could result in a false negative. Alternatively, a weak binder to the dominant domain might be preferred to a stronger binder of another domain, even though either or both could be the active compounds.

Both the kinetics and thermodynamics of binding can affect FTSA results. Kinetically, an extremely slow association rate might lead to the assay starting before the compound has a chance to bind to the target. However, more importantly, it is the thermodynamics of the binding event that makes a significant difference to the extent of stabilisation of a particular compound, and subsequently the affinity estimate. For example, high enthalpic binders, associated with the occurrence of specific interactions such as H-bonds, typically have their binding affinities underestimated when measured by FTSA, whereas entropic binders, associated with non-specific contacts such as hydrophobic interactions, are measured with higher affinity than other assays. This is mainly due to the assumption of an enthalpy change of around $-7\ \text{kcal mol}^{-1}$ and no heat capacity changes to allow affinities measured at higher temperatures to be extrapolated to 25 °C. However, as enthalpic contributions to binding range from 0 to $-14\ \text{kcal mol}^{-1}$,[37] the measured and calculated K_d from FTSA will be between 2.5- and 5-fold different from the true K_d for the majority of interactions. For entropically-driven binders, the enthalpic contribution will be positive, resulting in a large deviation (>10-fold) from the estimated to the true affinity. In this case, the affinities will be overestimated by FTSA (*i.e.*, the ligands will appear to be much stronger binders than they truly are). It is possible to use this discrepancy between assays to investigate the mechanism of action, as large differences between the estimated K_d from FTSA and measured K_d from an alternative method performed at 25 °C will indicate either enthalpically- or entropically-driven interactions. In addition, the heat capacity change of binding is rarely zero, so care must be taken when extrapolating the affinity measured at high temperatures to that at physiological temperature. If the T_m of the protein is high (>50 °C), then estimated affinities can deviate from the true K_d by up to 10-fold.

Although FTSA works with most soluble proteins under most conditions, there are occasions when a strong, clear fluorescent signal is not achieved. Non-globular, large multi-domain proteins or proteins with intrinsically disordered regions tend not to exhibit distinct phase transitions during thermal denaturation. Due to the complexity and irreproducibility of the melting curves, it is almost impossible to interpret them. Small proteins can sometimes have a weak hydrophobic core resulting in the lack of a fluorescent transition from folded to unfolded states. Partially unfolded proteins or proteins containing exposed hydrophobic patches in their native state can display very high initial fluorescence which can subsequently mask any changes due to unfolding at higher temperatures. For this reason, it is important to ensure that all aggregated material is removed from the samples by either centrifugation or filtration. In most of these cases where FTSA gives either ambiguous results or no measurable curve, it will be necessary to measure the thermostability of the protein by other methods such as CD[38] or DLS.[39]

One important aspect of FTSA that has been discussed extensively in the literature is the treatment of destabilising compounds that result in a negative T_m shift. Typically, these have been disregarded as compounds that bind exclusively to the unfolded state of the protein, most probably due to their hydrophobic nature or their promiscuity of binding. However, it is possible that some of these destabilising binders may also bind to the native state of the protein and that the reduction in T_m is caused only by a preference for the unfolded state.[40] Other causes for destabilising true binders include target-disruptive allosteric binding,[41] covalent modification,[42] detergent-like denaturation and changes in the ionic strength leading to a depletion of ions that stabilise the protein.[43] Since a ligand's overall effect may be masked by binding to both the native and the denatured protein states, with a net result being only a minor shift in the equilibrium, it is recommended that some destabilising compounds should be further investigated using either other binding assays or functional assays before dismissing them. One recent example showed that when the hits following a fragment screen of a bacterial O-GlcNAc hydrolase were triaged by FTSA, they fell into two camps, stabilisers and destabilisers.[44] When these compounds were further tested in an activity assay, the stabilisers inhibited the enzyme; however, the destabilisers caused an increase in activity. Structurally the inhibitors stabilised a closed 'active' form of the enzyme which may have been less structurally stable than the open 'inactive' form.

The effect of various pan-assay interference compounds (PAINs) on FTSA data was investigated by Redhead *et al.*[29] They found that different PAINs had very different effects on FTSA results, depending on their mechanism of interference. For example, promiscuous aggregators caused sharp decreases in fluorescence, similar to known fluorescence quenchers. Redox cycling compounds had various effects, but those that generated H_2O_2 in solution and potentially caused oxidation of the dyes gave reduced fluorescence with increased concentration. It was concluded that compounds that lower the overall fluorescence without causing a shift in melting temperature can be discarded as they are likely to be aggregators, fluorescent quenchers or fluorophore bleachers.

8.7 FTSA in Drug Discovery

Recent increases in the sensitivity and throughput of many biophysical assays have resulted in these techniques being used earlier in the drug discovery process. As FTSA is able to screen thousands of compounds for protein binding at a much quicker rate, and for a much lower cost, than other biophysical methods, it has primarily been utilised in primary and secondary (orthogonal) screening. This has been true for both high-throughput screening and fragment-based drug discovery programmes (refer to Section 8.8). Other binding techniques tend to be more information rich, so, until recently, FTSA was not employed in the hit-to-lead or lead optimisation stages. However, with clever experimental design and more sophisticated data analysis, mechanism of action studies can be performed by FTSA.

8.7.1 Screening for Ligand Binding

As discussed previously, the binding of low molecular weight ligands can increase the thermal stability of a protein[45] (Figure 8.4). This has been utilised in the drug discovery process to conduct high-throughput screening for small molecule and/or fragment binding to numerous targets. FTSA has also served as a secondary validation assay following screening for either activity or binding by biochemical and other biophysical assays, respectively.

As discussed, the main application of FTSA is in ligand affinity ranking by either medium or high-throughput screening or identification of novel ligands (including fragments). It is important to realise that a shift in T_m upon ligand binding has a simple thermodynamic meaning only when the target is saturated. It is therefore important to use ligands at high concentrations ($\gg K_d$) or perform a full concentration–response type of experiment to ensure that the binding site has indeed been fully saturated.

As FTSA can be applied to any protein target that has a discernible melting curve, it is possible to rapidly screen for target specificity by profiling compound binding to a set of related or even unrelated proteins.[46] Fedorov *et al.*[47] evaluated the specificity of 156 validated kinase inhibitors against 60 active and inactive human Ser/Thr kinases using FTSA. The choice of off-target proteins could include post-translationally modified variants (*e.g.*, differing phospho-forms), site directed mutants,[48] truncated proteins or domain-segregated targets. Typically, it is the cost-effectiveness of FTSA in terms of both money and time and the relatively low false-positive rate that makes it the assay of choice to use for such complex studies. For FTSA, these variants do not always need to be functionally active, therefore the range of variables that can be monitored can be extended.

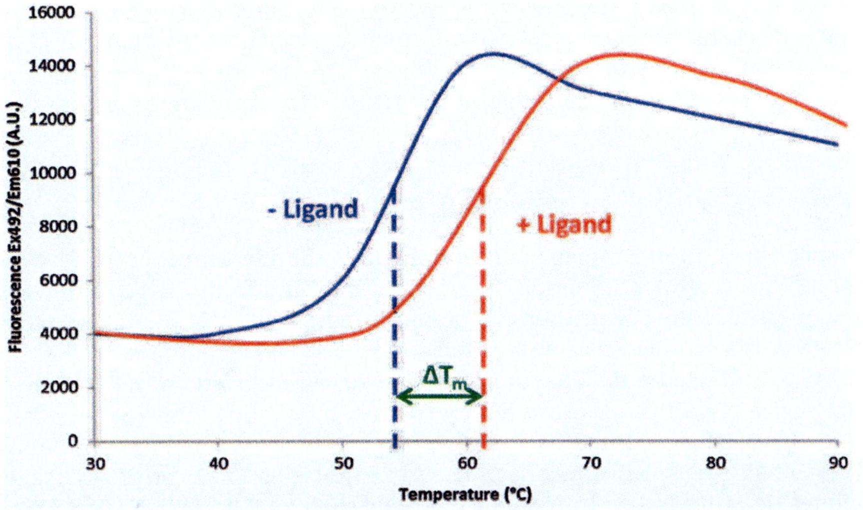

Figure 8.4 Example of an increase in the T_m of a protein when the ligand is bound.

For increased throughput, FTSA can be used to screen against a panel of ligand mixes, *i.e.*, high-density screening. In this case, pooled ligands are used as an initial step to assign protein binders. If a significant thermal shift is detected with a ligand pool (hit pool), the protein is then screened against all individual ligands in the pool to determine specific binding ligands.

Another growth area of small molecule drug discovery where FTSA is being applied more frequently is in the identification and validation of pharmacological chaperones.[49] These molecules bind to protein mutants and stabilise them and, even though they have only been identified relatively recently as potential drugs, a few are already in clinical trial for lysosomal storage diseases such as Fabry,[50] Gaucher[51] and Pompe disease.[52] In these cases, the chaperones bind and stabilise mutant lysosomal enzymes, helping them to fold properly and ultimately get transported across appropriate cellular compartments to exert their glycolytic degradation activity.[53] Here, the use of FTSA is particularly relevant due to the potential advantages of non-inhibitory chaperones, which, of course, could not be screened for using activity assays. An example of FTSA being used to characterise these chaperones can be found in Patnaik *et al.*, 2012, where pyrazolopyrimidines were discovered to act as noninhibitory chaperones of glucocerebrosidase.[54]

When FTSA or other binding assays are used alongside, or following, a high-throughput activity assay, it is important that the target protein in both is as near to identical as possible. A study by Rudolf *et al.*[55] found that the two types of assays (activity and binding) were usually performed using different expression constructs, expression systems and purification procedures. To investigate the influence these factors, they expressed and purified 15 kinases from *E. coli* and screened them against 244 known kinase inhibitors. A threshold of 4 °C was set for the T_m shifts and of those compounds showing this shift, 71% exhibited IC_{50} values of less than 0.5 μM. This was a significant improvement over earlier studies in which only 49% of compounds that showed a ΔT_m of 4 °C had IC_{50} values of less than 0.5 μM.[56] Additionally, it was found that by using constructs with multiple domains, and not just the catalytic domain, the fraction of compounds passing the criteria improved to 86%.

8.7.2 Screening for Fragment Binding

A specialised form of ligand binding involves the identification of very low molecular weight compounds (fragments) that bind to the target of interest and are subsequently further elaborated into high affinity small molecules. Fragment-based lead discovery (FBLD) has been widely adopted in both industrial and academic labs and has been reviewed extensively.[57–60] Due to the very weak affinities exhibited by fragments, various biophysical techniques have been adapted as initial screening methods in FBLD. Amongst these techniques, FTSA has been used in numerous FBLD projects, usually alongside other lower throughput methods such as SPR, ITC, NMR or X-ray crystallography.

8.7.3 Mechanism of Small Molecule Inhibition

Recently, there has been a growing use of FTSA to go beyond simple ligand–protein binding studies to further probe the compound mechanism of action (MOA). Chemodynamic analysis, which examines the qualitative details of any particular interaction, can be obtained by analysing identical FTSA experiments that differ only in buffer pH, ionic strength or the presence of other substances that may influence the interaction. This type of analysis will give valuable information about the dominant forces behind the interaction. Determining the mechanism of action of a compound early in the screening cascade is of growing importance in drug discovery. In this regard, FTSA can help to distinguish the mode of binding between two ligands, for example between a compound and an endogenous ligand, substrate or cofactor. Depending on the mutual influence, the mode of binding, *i.e.*, competitive, non-competitive or uncompetitive, can be revealed. The most straightforward of these approaches is the use of competition/co-dependency type studies where the compound of interest is tested in the absence and presence of another binder (*e.g.*, known inhibitor, endogenous ligand, substrate/product or cofactor). One simple case would be kinases, where compound binding in the presence or absence of ATP (or an ATP analogue) can determine whether the compound is ATP-competitive, -non-competitive or -uncompetitive. This kind of information is important for the hit-confirmation and lead-identification stages of drug discovery, as one mode of binding may be known to be more desirable, or different modes could be progressed in parallel until obvious advantages become apparent. It should be noted that competitive binding cannot only be caused by ligands binding in the substrate pocket but also by conformational changes due to orthosteric attachment. Verification of the mode of binding and exact site of interaction can only be achieved using either NMR or X-ray crystallography.

In some cases, complex transition curves can be used to extract additional information on the MOA. For example, if more than one transition is detected in the presence of a ligand, then there may be heterogeneous solutions of bound and unbound protein. If this is the case, then using higher ligand concentrations should result in fewer transitions until saturation is reached and only one transition remains. However, if increasing ligand concentration leads to more distinct multiple transitions, then it may be the case that the ligand is binding to a distinct domain in a multi-domain protein. Substrates or inhibitors with known binding sites on particular domains could then be used to ascertain whether a ligand with an unknown binding site is binding to the same domain as the substrate or a different one. For example, human GDP-D-mannose 4,6-dehydratase (GMD) is a multi-domain protein where NADP+ and GDP bind to distinct domains, and the shifts caused by these two compounds can be easily distinguished on the double transition curves.[15]

Lea and Simeonov[61] investigated the utility of FTSA in MOA studies using glutathione S-transferase (GST) as a model enzyme that has a range of inhibition modes due to its two distinct substrates. A FTSA study looking at small molecule stabilisation of the protein in the presence of the enzyme's native co-substrate glutathione (GSH) indicated that compounds with different inhibition modes displayed distinct thermal shift signatures. GSH competitive inhibitors tended to have a thermal profile for which the shifts of T_m with and without GSH converged at higher concentrations of compound. Compounds that worked by product inhibition (conjugate formation) showed a profile for which the T_m shifts in the presence of GSH became gradually greater with increasing concentration, whereas in the absence of GSH, the temperature shifts remained relatively constant over the concentration range. Finally, the 'ligandin-type inhibitors', which likely dock into the H-site of the GST, without interference with the G-site, have thermal profiles in which the + and − GSH shifts increase in parallel over the compound concentration range.

Another example of co-dependency experiments can be found in a study by Niesen *et al.*[62] Here, 50 hits from a screen of >160 000 compounds against 15-PGDH were investigated by FTSA in the absence and presence of the cofactor NAD^+ in both its oxidised and reduced states. A combination of assays was then able to identify and confirm both competitive and non-competitive nanomolar inhibitors of 15-PGDH.

A variation in the cofactor/co-substrate dependency analysis was shown by Auld *et al.*[63] Here, an excessively large increase in T_m by 3,5-diaryl oxadiazole inhibitors of firefly luciferase in the presence of the ATP substrate indicated the formation of a covalent adduct that spanned an extended binding pocket and thus provided exceptional stabilisation of the protein target. This was confirmed by subsequent LC-MS and X-ray crystallography studies.

FTSA-based ligand screening has also been instrumental in helping to explain the biology of certain proteins. Screening proteins with unknown functions against physiologically relevant compounds is a way to decrypt their biological function. For example, Carver *et al.*[64] screened 3000 ligands against an unknown but essential protein from *Streptococcus pneumoniae.* The identification of both pyridoxal phosphate and pyridoxamine phosphate as high affinity ligands helped to identify the protein as a nucleoside diphospho-keto-sugar aminotransferase.

Protein–protein interactions are critical to many biological processes and include enzyme–substrate binding, protein–antibody binding and protein–receptor interactions. Understanding these interactions can provide insight into how the proteins function in a biological system. FTSA has recently been used to measure the stabilisation of a protein by another protein.[39] Interestingly, the majority of protein complexes studied showed a two-state model of thermal denaturation usually observed for single proteins. This finding suggests that complex formation provides additional stabilisation and the single melting point was the temperature of dissociation. This provides a potential use for FTSA in screening small molecules that disrupt protein–protein interactions by the appearance of two distinct transitions.

8.8 Successful Applications of FTSA for Ligand Binding Screens

There is an extensive and ever growing list of examples where FTSA has been utilised successfully in the drug discovery process. These examples range from primary screens of small molecules or fragments to detailed MOA studies. Below are a few examples from the literature that highlight the depth and breadth of FTSA.

High-throughput FTSA has been used as a primary screen to identify hits against mitogen-activated protein kinase kinase 4 (MAP2K4),[65] an activator of pro-invasion signalling pathways in human prostate cancer. The group used automated systems for both compound transfer and protein dispensing, and initially used pools of three compounds to increase throughput. Hits that showed concentration-dependent increases in the T_m of MAP2K4 were further validated using an *in vitro* kinase assay, with a good correlation between T_m shifts and IC_{50} values.

Similarly, FTSA was used as the primary assay in an HTS campaign to discover inhibitors of the epigenetic target DOT1L histone methyltransferase.[66] The team from the SGC screened 3120 kinase inhibitors by FTSA and identified 5-iodotubercidine (5ITC) as a stabiliser (2.5 °C at 50 μM). An IC_{50} of 18.2 μM was confirmed in an enzyme assay, and through extension and merging of this compound with known cofactors, a new molecule (BrSAH) with an IC_{50} of 77 nM was found.

A dual binding analysis of small molecules to poly-ADP-ribose polymerases using FTSA and SPR was undertaken by Wahlberg *et al.*[67] After showing an excellent correlation between the data from both the FTSA and SPR methods, FTSA was then used to evaluate 185 inhibitors against 13 different members of the PARP family. It was concluded that the most widely used inhibitors bound to multiple members of the PARP family.

An early example of FTSA's ability to discover inhibitors of protein–protein interactions was shown by Grasberger *et al.*[68] HDM2, a human ubiquitin E3 ligase that binds to p53 and inhibits its tumour suppressive function, was screened against almost 350 000 compounds by FTSA, which resulted in a hit rate of <0.35% (1216 compounds). Follow-up screening by a fluorescent peptide displacement assay and hit expansion based on crystal structures resulted in potent HDM2 antagonists that activated p53 in cells.

The ability of FTSA to screen for ligands binding to enzymes in an inactive state was highlighted in its use to help identify and optimise small molecule Akt1 inhibitors that act through a mechanism distinct from that observed for kinase domain ATP-competitive inhibitors.[69] Akt is part of the PI3K/mTOR pathway and plays a central role in cell migration, proliferation and angiogenesis. To gain a more selective inhibitor, the authors targeted the hydrophobic clusters in the ATP-binding cleft and used a combination of FTSA, MS and SPR to screen compounds for binding against the inactive kinase. FTSA identified 80 stabilisers from the 1200 binders identified by MS. A full hit-to-lead and lead optimisation programme was then run, utilising FTSA throughout the process, to eventually obtain a highly selective, 2.7 nM Akt inhibitor.

The applicability of FTSA to a wide range of target types was highlighted in DeSantis *et al.*,[36] who used FTSA to screen a large number of commercially-available compounds against a partially purified human estrogen receptor α ligand binding domain (ERαLBD). The outcome of the screen identified three compounds, all of which had been previously confirmed as ERα interactors. The team also showed that FTSA produced no false positives in this instance and is therefore well suited to screen larger compound libraries for binders of this important class of proteins. In addition, it is hoped that by using FTSA to identify ligands for nuclear receptor ligand binding domains that are renowned for being difficult to purify and concentrate, the added stability could lead to these proteins being studied by different techniques including crystallography.

The extensive use of FTSA in fragment-based lead discovery has been highlighted earlier and an excellent example of its use was shown by Silvestre *et al.*[70] A screen using 1250 fragments against *Mycobacterium tuberculosis* pantothenate synthetase was performed. The primary FTSA screen, which identified 39 hits (3.1% hit rate), was followed by secondary screening by 1-D NMR, ITC and X-ray crystallography. This multi-biophysical approach identified three distinct binding sites for various fragments, suggesting both elaboration and/or linking of the fragments might lead to potent antitubercular compounds.

FTSA was used successfully for the initial screening of the bacterial orphan protein CFE97.[64] This essential bacterial protein from *S. pneumoniae* was screened against both a 'drug discovery' compound library and a functional probe library, the latter containing known enzyme substrates, inhibitors, sugars and nucleotides used to decrypt the function of unknown proteins. From the initial FTSA screen, the enzyme cofactor pyridoxal phosphate (PLP) was found to cause a large thermal shift. Subsequent BLAST searching against various databases confirmed the existence of a PLP binding site within the CFE97 protein. Additional FTSA screening was then used to identify which amino acids were the preferred substrates in the conversion of CFE97:PLP to CFE97:PMP. This is an excellent example of the power of FTSA to probe targets of unknown function and, with the help of other technologies, to quickly assign a role for these proteins.

8.9 Other Uses of FTSA

Stability profiling has been used to identify buffer conditions,[71] mutations[72] and additives that help to stabilise proteins (including antibodies). Increasing the thermal stability of a protein has many applications including, potentially, making it more suitable for crystallography,[73] NMR[74] and/or FTSA ligand screening itself.[7] Indeed, identifying conditions under which a protein is most stable and least likely to aggregate will help in any analytical or biophysical experiment that requires high protein concentrations or is sensitive to protein aggregation.

FTSA can be particularly useful in the search for stable protein variants for therapeutic use as there is currently no reliable method to predict stabilising mutations and other biophysical methods such as circular dichroism, calorimetry and fluorescence spectroscopy are low-throughput techniques. Crowther *et al.*[7] ran a large profiling study comparing the T_m of 31 recombinant enzymes from *Plasmodium* parasites to their activity. They found that in general, the T_m and quality of the melting profile correlated well with the activity of the enzyme, and could therefore be used to bin protein stocks into 'properly-folded' or 'possibly-denatured'. Lavinder *et al.*[72] used FTSA in a high-throughput manner, which they termed high-throughput thermal screening (HTTS), to search for conservative, stable mutants of biotechnologically important proteins. This enabled them to build a sequence–stability relationship and identify the stabilising mutations in the four-helix bundle protein, Rop, and the yeast triosephosphate isomerase (TIM) protein. The HTTS apparent T_m values were roughly related linearly to the CD T_m values.

Production of proteins appropriate for successful structural studies can be difficult to achieve in a timely and cost-effective manner. A number of properties can determine the likelihood of the successful crystallisation of a protein including solubility, homogeneity and stability. In addition, the introduction of a stabilising ligand can greatly enhance the chances of obtaining crystals. Ericsson *et al.*[73] applied a high-throughput FTSA method to identify optimal *E. coli* protein formulations for crystallisation. The screens included variations in pH, ionic strength, additives, precipitants and protein concentrations. Dupeux *et al.*[32] conducted a systematic study showing that proteins with a melting point greater than 45 °C had a significantly higher chance of successful crystallisation. They also showed that the best incubation temperature for successful crystallisation should be at least 25 degrees below the T_m.

It should be noted that with all of the above examples of monitoring T_m to determine stability, increased T_m does not always translate into improved resistance of the protein preparation to proteolytic or other types of degradation upon storage at room temperature or at 4 °C and below.

8.10 Non-fluorescent Dye Thermal Shift Assays

Alternative thermal denaturation-based assays have also been utilised for screening proteins against libraries of compounds. These assays can measure the temperature of aggregation (T_{agg}) by light scattering (known as differential static light scattering – DSLS) or measure the temperature at which protein can no longer bind to a high-affinity radiolabelled ligand.

DSLS is a label-free method that relies on the immediate aggregation of proteins upon unfolding.[75] The method is applicable to auto-fluorescing proteins and ligands and is also insensitive to the presence of detergents or hydrophobic proteins. However, even though for most proteins the T_{agg} and

T_m values are reasonable close, DSLS is an indirect measurement of unfolding and not all proteins aggregate immediately after denaturation. It is also insensitive to unfolding mechanisms that do not affect the size of the protein.

The high-affinity radiolabelled ligand method is performed by incubating protein with a saturating amount of labelled ligand for 30 minutes at a particular temperature, then quenching on ice.[76] The mixture is then passed through a gel filtration mini column and the radiation levels within the protein are quantified. If the protein is unfolded, then no radiation will be present due to the lack of binding of the ligand. This approach has been used successfully to study both agonist and antagonist binding to GPCRs.[77] The advantage of radiometric assays is the small amount of protein required, with the obvious caveats of safety issues around radioactivity. The assay set-up described above also requires a high-affinity ligand for the target protein.

Other techniques such as CD, infrared (IR) spectroscopy and DSC have advantages over FTSA as they require no reporter dye, but all have a much lower throughput than FTSA. DSC detects phase transitions, such as the melting of a protein, by measuring the difference in energy required to keep a cell containing sample at the same temperature as a control cell containing only buffer or water. Since the absorption of circularly polarised light is weak and decreases upon protein precipitation, CD typically requires concentrations greater than 1 $mg\,ml^{-1}$. For IR spectroscopy, the strong absorption of water and its temperature dependency limit the applicability of this technique for thermodynamic studies.

It is also possible to run FTSA without dyes if the proteins either contain an endogenous fluorescent cofactor[78] or contain sufficient intrinsic fluorescence from buried aromatic residues (tryptophan, tyrosine or phenylalanine).[79] The excitation and emission wavelengths of tryptophan are dependent on the immediate environment and therefore differ between folded and unfolded protein. Similarly, the lifetime of tryptophan fluorescence differs between folded and unfolded protein and by measuring the lifetime of UV-excited fluorescence over increasing temperature, an estimate of the melting point of the protein is obtainable. These techniques have the obvious advantage of not requiring reporter dyes and are therefore accessible for membrane protein analysis. However, intrinsic fluorescence techniques are limited to proteins containing sufficient tryptophans to obtain a reasonable signal.

8.11 Cellular Thermal Shift Assays

The ability of a compound to bind to a purified protein does not always mean that it will bind to that target in cells. Indeed, the measured affinity *in vitro* can be significantly different to that found in cellular assays, mostly due to issues around cell permeability, off-target engagement and the presence of co-factors, natural substrates *etc.* To address cellular target engagement directly, 'in-cell' thermal shift assays have been developed through a variety of approaches to assess drug–protein interactions in a physiologically relevant setting.

8.11.1 CETSA

Cellular thermal shift assay (CETSA)[80] is a relatively new technique that is the in-cell equivalent to FTSA. This method allows studies of the target engagement of drug candidates in a cellular context (*e.g.*, cell lysates, intact cells or even tissues) and gives an indication of ligand permeability to a relevant cellular compartment. The assay involves treating cells with compounds of interest, heating to both denature and aggregate the proteins, lysing the cells and separating cell debris and aggregates from the soluble protein fraction by centrifugation. The supernatant is then run on Western blots and if a protein is stabilised in the presence of a compound, then the protein will still be present in the soluble fraction at higher temperatures. Alternative analyses to Western blots include using SDS-PAGE, especially if the protein is tagged and specific staining is used (*e.g.*, His-tagged staining), AlphaScreen[81] and FRET[82] assays. A typical CETSA assay involves the comparison between the temperature-induced aggregation curves of a target protein in the presence and absence of various ligands. However, the technique can also be used to generate isothermal dose response curves, where protein stabilisation with increasing ligand concentration is monitored. This has been termed an isothermal dose–response (ITDR) experiment and is used extensively in drug discovery programmes to ascertain whether the drug candidates actually bind to their intended target in a biologically relevant setting, and subsequently, at what concentration they exert their effect.

Although the strengths and weaknesses of CETSA have yet to be fully elucidated, certain limitations are known. For example, not all proteins aggregate upon unfolding but might instead populate highly soluble, compact conformations. In addition, it is possible that proteins will co-precipitate with their less stable protein interaction partners and therefore show lower apparent stability than their actual thermodynamic stability. CETSA also requires generation of sufficiently selective antibodies suitable for western blotting and pairs of conformational selective antibodies with non-overlapping epitopes in the case of AlphaScreen or FRET.

A recent example of the successful use of CETSA in small molecule screening is described by Almqvist *et al.*[83] The assay not only detected all of the known drugs acting on human thymidylate synthase (TS), but it also found inhibitors with novel chemistry and identified marketed drugs that were not previously known to target TS. In addition, the time-dependent nature of the target engagement could be linked to both the cellular uptake and the enzymatic production of active metabolites. This highlights the utility of CETSA to be able to associate drug uptake and activation with target binding during drug discovery.

8.11.1.1 FASTpp

Fast parallel proteolysis (FASTpp) is a variation of CETSA whereby a thermostable protease (thermolysin) is added to cell supernatants, which are subsequently heated.[84] FASTpp exploits the facts that the more unfolded the protein is, the more susceptible it is to proteolysis, and that thermolysin cleaves at the hydrophobic residues Phe, Leu, Ile and Val, which are typically found in the core of the proteins. The seminal paper describes the use of FASTpp to monitor the thermal unfolding of a wide range of protein targets from 10 to 240 kDa. Additionally, the assay was used to detect the effects of ligand binding on Maltose Binding Protein (MBP).

Limitations include the fact that thermolysin is dependent on calcium ions for activity and therefore calcium must be present in the buffer. Similarly, calcium chelators must be absent. Compounds and detergents that may interfere with the function of the protease could also cause problems in FastPP.

8.11.1.2 TPP

An extension of CETSA that enables the monitoring of changes in protein thermal stability across the cellular proteome using quantitative mass spectrometry was developed by Savitski *et al.*[85] and was termed thermal proteome profiling (TPP). The team compared the intracellular thermal profiles of almost 7000 proteins from human K562 cells in the presence and absence of various drugs and were able to identify direct and indirect targets. Initial hits were further ranked for binding affinities by determining stability profiles across a range of compound concentrations at a defined temperature. The ability to monitor so many targets was attributed to the use of isobaric tandem mass tag 10-plex (TMT10) reagents, which allowed multiplexing on the mass spectrometer. A detailed description of the protocol, including data analysis with freely available software, is available in Franken *et al.*[86] This method could be particularly valuable in the deconvolution of phenotypic screening results, to identify the proteins and pathways targeted by the bioactive molecules.

8.11.2 Future Perspectives

With an ever greater need to deconvolute phenotypic screens and to discover druggable targets with indistinct activity, HTS by biophysical assays is expected to become an ever more important tool to find selective ligands of proteins, especially where functional assays have not yet been configured. As FTSA is a powerful, high-throughput assay used for studying biomolecular–ligand interactions and is straightforward to adapt to any set of conditions for a particular protein, it is perfectly placed to be used in such cases.

The successful measurement of binding affinity for assorted ligands binding to highly diverse classes of protein targets demonstrates that FTSA has a wide utility and is particularly useful in situations where detailed enzymological and spectroscopic studies are difficult to implement. Data analysis is relatively simple using any standard commercial software and can easily be used to investigate specific protein–ligand interactions. Screening enzymes against libraries of compounds in the presence and absence of cofactors also provides additional information on the mechanism of binding. Indeed, this is an example of the true power of thermal shift assays as they are increasingly being used not only in the high-throughput affinity ranking of compounds during hit finding, hit optimisation, and lead identification stages but also alongside other assays, such as NMR, ITC and SPR, to provide MOA-informed SARs. As shown, FTSA can contribute to several aspects of drug-binding MOA studies including specificity, stoichiometry, kinetics and thermodynamics. These important advances in small molecule MOA studies are partly due to the increase in ligand binding assays in a concentration–response format. This is important not only because dose-range assay formats allow for a better assessment of the actual interaction trends, but also because single-concentration tests carry a much greater risk of both false positive and false negative results, especially in cases where the compound is insoluble at the concentration utilised.

FTSA has found a particular relevance in the primary screening of fragment libraries. Due to the nature of fragment interactions, biophysical techniques are ideally suited as primary and secondary screening tools for FBLD and as detailed, FTSA has been used in numerous FBLD programmes, and its utility in significantly enriching the hit rate of secondary screens has been proven.

The use of FTSA has steadily increased during the past few years, not only judged by its increasing occurrence in the literature, but also by the fact that a large number of vendors are now providing dedicated kits, instrument protocols and T_m analysis tools. This increase in utility has been aided by a significant diversification of FTSA applications beyond initial small molecule screening into other areas such as optimising therapeutic protein production and formulation studies. In addition, the recent identification of an alternative reporter dye to enable FTSA deployment for samples containing detergents where SYPRO Orange has tended to fail has expanded its use.

With the expansion of the principles behind FTSA to include in-tissue and in-cell analysis (*e.g.*, CETSA), the important ability to monitor target specificity, drug transport and activation, as well as dose-dependent target engagement has been achieved for a wide range of protein targets. In addition, CETSA has the potential to monitor drug efficacy at the target level in patients and help to determine appropriate drug dosage or even highlight when resistance at the protein engagement level has been developed.

In summary, the expanding use of FTSA in small molecule drug discovery is expected to continue, with its use going beyond simple stabilisation of targets to using the technique to provide information on the mechanism of action in more complex binding events. Technological improvements should allow for both increased throughput and a further reduction in the amount of reagents required, especially in the case of the expensive protein targets. Related to these efforts, it is important to note that while FTSA data acquisition is gaining in throughput and complexity, data analysis has begun to fall behind, with human involvement often still required to inspect the large number of melting curves, making improvements to the analysis software essential. An automated analysis capability for FTSA, akin to SPR or DLS, where complex operations are performed in the background and accurate affinities are obtained would be invaluable.

Acknowledgements

I would like to acknowledge the help and support of Dr Martin Redhead (UCB), Dr Scott Pollack and Dr John Unitt (Sygnature Discovery) in the preparation and writing of this chapter.

References

1. J. A. Schellman, *Biopolymers*, 1975, **14**, 999–1018.
2. J. A. Schellman, *Biopolymers*, 1976, **15**, 999–1000.
3. M. W. Pantoliano, E. C. Petrella, J. D. Kwasnoski, V. S. Lobanov, J. Myslik, E. Graf, T. Carver, E. Asel, B. A. Springer, P. Lane and F. R. Salemme, *J. Biomol. Screening*, 2001, **6**, 429–440.
4. F. H. Niesen, H. Berglund and M. Vedadi, *Nat. Protoc.*, 2007, **2**, 2212–2221.
5. L. Shan, L. Pang, R. Zhang, N. J. Murgolo, H. Lan and J. A. Hedrick, *PCSK9 Binds to Multiple receptors and can be Functionally Inhibited by an EGF-A Peptide*, 2008, vol. 375.
6. S. Boivin, S. Kozak and R. Meijers, *Protein Expression Purif.*, 2013, **91**, 192–206.
7. G. J. Crowther, A. J. Napuli, A. P. Thomas, D. J. Chung, K. V. Kovzun, D. J. Leibly, L. J. Castaneda, J. Bhandari, C. J. Damman, R. Hui, W. G. J. Hol, F. S. Buckner, C. L. M. J. Verlinde, Z. Zhang, E. Fan and W. C. van Voorhis, *J. Biomol. Screening*, 2009, **14**, 700–707.
8. D. E. Epps, R. W. Sarver, J. M. Rogers, J. T. Herberg and P. K. Tomich, *Anal. Biochem.*, 2001, **292**, 40–50.
9. G. V. Semisotnov, N. A. Rodionova, O. I. Razgulyaev, V. N. Uversky, A. F. Gripas' and R. I. Gilmanshin, *Biopolymers*, 1991, **31**, 119–128.
10. M. Kohlstaedt, I. von der Hocht, F. Hilbers, Y. Thielmann and H. Michel, *Acta Crystallogr., Sect. D: Biol. Crystallogr.*, 2015, **71**, 1112–1122.
11. J. Kean, R. M. Cleverley, L. O'Ryan, R. C. Ford, S. M. Prince and J. P. Derrick, *Mol. Membr. Biol.*, 2008, **25**, 653–661.
12. A. I. Alexandrov, M. Mileni, E. Y. T. Chien, M. A. Hanson and R. C. Stevens, *Structure*, 2008, **16**, 351–359.
13. W. M. Rabeh, F. Bossard, H. Xu, T. Okiyoneda and M. Bagdany, 2012, **148**, 150–163.

14. T. Menzen and W. Friess, *J. Pharm. Sci.*, 2013, **102**, 415–428.
15. G. Senisterra, I. Chau and M. Vedadi, *Assay Drug Dev. Technol.*, 2012, **10**, 128–136.
16. A. Zubriene, J. Matuliene, L. Baranauskiene, J. Jachno, J. Torresan, V. Michailoviene, P. Cimmperman and D. Matulis, *Int. J. Mol. Sci.*, 2009, **10**, 2662–2680.
17. S. Breuer, M. W. Chang, J. Yuan and B. E. Torbett, *J. Med. Chem.*, 2012, **55**, 4968–4977.
18. M. C. Lo, A. Aulabaugh, G. Jin, R. Cowling, J. Bard, M. Malamas and G. Ellestad, *Anal. Biochem.*, 2004, **332**, 153–159.
19. S. Egger, A. Chaikuad, K. L. Kavanagh, U. Oppermann and B. Nidetzky, *J. Biol. Chem.*, 2011, **286**, 23877–23887.
20. G. Herzog, A. C. Joerger, M. D. Shmueli, A. R. Fersht, E. Gazit and D. Segal, *J. Biol. Chem.*, 2012, **287**, 44330–44337.
21. D. Matulis, J. K. Kranz, F. R. Salemme and M. J. Todd, *Biochemistry*, 2005, **44**, 5258–5266.
22. N. Rosa, M. Ristic, S. A. Seabrook, D. Lovell, D. Lucent and J. Newman, *J. Biomol. Screening*, 2015, **20**, 898–905.
23. M. N. Schulz, J. Landstrom and R. E. Hubbard, *Anal. Biochem.*, 2013, **433**, 43–47.
24. K. Phillips and A. H. de la Pena, *Curr. Protoc. Mol. Biol.*, 2011, **94**, 10.28.1–10.28.15.
25. T. T. Waldron and K. P. Murphy, *Biochemistry*, 2003, **42**, 5058–5064.
26. G. A. Holdgate, *Expert Opin. Drug Discovery*, 2007, **2**, 1103–1114.
27. J. F. Brandts and L. N. Lin, *Biochemistry*, 1990, **29**, 6927–6940.
28. M. Vedadi, F. H. Niesen, A. Allali-Hassani, O. Y. Fedorov, P. J. Finerty, G. A. Wasney, R. Yeung, C. Arrowsmith, L. J. Ball, H. Berglund, R. Hui, B. D. Marsden, P. Nordlund, M. Sundstrom, J. Weigelt and A. M. Edwards, *Proc. Natl. Acad. Sci. U. S. A.*, 2006, **103**, 15835–15840.
29. M. Redhead, R. Satchell, V. Morkunaite, D. Swift, V. Petrauskas, E. Golding, S. Onions, D. Matulis and J. Unitt, *Anal. Biochem.*, 2015, **479**, 63–73.
30. T. W. Mayhood and W. T. Windsor, *Anal. Biochem.*, 2005, **345**, 187–197.
31. P. Cimmperman, L. Baranauskiene, S. Jachimoviciute, J. Jachno, J. Torresan, V. Michailoviene, J. Matuliene, J. Sereikaite, V. Bumelis and D. Matulis, *Biophys. J.*, 2008, **95**, 3222–3231.
32. F. Dupeux, M. Rower, G. Seroul, D. Blot and J. A. Marquez, *Acta Crystallogr., Sect. D: Biol. Crystallogr.*, 2011, **67**, 915–919.
33. R. R. Kroe, J. Regan, A. Proto, G. W. Peet, T. Roy, L. D. Landro, N. G. Fuschetto, C. A. Pargellis and R. H. Ingraham, *J. Med. Chem.*, 2003, **46**, 4669–4675.
34. B. J. Davis and D. A. Erlanson, *Bioorg. Med. Chem. Lett.*, 2013, **23**, 2844–2852.
35. R. E. Hubbard and J. B. Murray, *Methods Enzymol.*, 2011, **493**, 509–531.
36. K. DeSantis, A. Reed, R. Rahhal and J. Reinking, *Nucl. Recept. Signaling*, 2012, **10**, e002.

37. L. Li, J. J. Dantzer, J. Nowacki, B. J. O'Callaghan and S. O. Meroueh, *Chem. Biol. Drug Des.*, 2008, **71**, 529–532.
38. J. J. Lavinder, S. B. Hari, B. J. Sullivan and T. J. Magliery, *J. Am. Chem. Soc.*, 2009, **131**, 3794–3795.
39. J. Kopec and G. Schneider, *J. Struct. Biol.*, 2011, **175**, 216–223.
40. P. A. McDonnell, J. Yanchunas, J. A. Newitt, L. Tao, S. E. Kiefer, M. Ortega, S. Kut, N. Burford, V. Goldfarb, G. J. Duke, H. Shen, W. Metzler, M. Doyle, Z. Chen, C. Tarby, R. Borzilleri, W. Vaccaro, M. Gottardis, S. Lu, D. Crews, K. Kim, L. Lombardo and D. L. Roussell, *Anal. Biochem.*, 2009, **392**, 59–69.
41. J. R. Horn and B. K. Shoichet, *J. Mol. Biol.*, 2004, **336**, 1283–1291.
42. R. Zhang and F. Monsma, *Curr. Opin. Drug Discovery Dev.*, 2010, **13**, 389–402.
43. P. C. Weber and F. R. Salemme, *Curr. Opin. Struct. Biol.*, 2003, **13**, 115–121.
44. J. F. Darby, J. Landström, C. Roth, Y. He, G. J. Davies and R. E. Hubbard, *Angew. Chem., Int. Ed.*, 2014, **53**, 13419–13423.
45. D. E. Koshland, *Angew. Chem., Int. Ed. Engl.*, 1995, **33**, 2375–2378.
46. S. E. Giuliani, A. M. Frank and F. R. Collart, *Biochemistry*, 2008, **47**, 13974–13984.
47. O. Fedorov, B. Marsden, V. Pogacic, P. Rellos, S. Müller, A. N. Bullock, J. Schwaller, M. Sundström and S. Knapp, *Proc. Natl. Acad. Sci. U. S. A.*, 2007, **104**, 20523–20528.
48. M. Koszelak-Rosenblum, A. C. Krol, D. M. Simmons, C. C. Goulah, L. Wroblewski and M. G. Malkowski, *J. Biol. Chem.*, 2008, **283**, 24962–24971.
49. G. Andreotti, M. Monticelli and M. V. Cubellis, *Drug Test. Anal.*, 2015, **7**, 831–834.
50. J.-Q. Fan and S. Ishii, *FEBS J.*, 2007, **274**, 4962–4971.
51. Z. Yu, A. R. Sawkar and J. W. Kelly, *FEBS J.*, 2007, **274**, 4944–4950.
52. T. Okumiya, J. L. M. Keulemans, M. A. Kroos, N. M. E. Van der Beek, M. A. Boer, H. Takeuchi, O. P. Van Diggelen and A. J. J. Reuser, *Mol. Genet. Metab.*, 2006, **88**, 22–28.
53. R. E. Boyd, G. Lee, P. Rybczynski, E. R. Benjamin, R. Khanna, B. A. Wustman and K. J. Valenzano, *J. Med. Chem.*, 2013, **56**, 2705–2725.
54. S. Patnaik, W. Zheng, J. H. Choi, O. Motabar, N. Southall, W. Westbroek, W. A. Lea, A. Velayati, E. Goldin, E. Sidransky, W. Leister and J. J. Marugan, *J. Med. Chem.*, 2012, **55**, 5734–5748.
55. A. F. Rudolf, T. Skovgaard, S. Knapp, L. J. Jensen and J. Berthelsen, *PLoS One*, 2014, **9**, 2–6.
56. T. Anastassiadis, S. W. Deacon, K. Devarajan, H. Ma and J. R. Peterson, *Nat. Biotechnol.*, 2011, **29**, 1039–1045.
57. A. Ciulli and C. Abell, *Curr. Opin. Biotechnol.*, 2007, **18**, 489–496.
58. D. C. Rees, M. Congreve Miles, C. W. Murray and R. Carr, *Nat. Rev. Drug Discovery*, 2004, **3**, 660–672.
59. C. W. Murray and D. C. Rees, *Nat. Chem.*, 2009, **1**, 187–192.

60. P. J. Hajduk and J. Greer, *Nat. Rev. Drug Discovery*, 2007, **6**, 211–219.
61. W. A. Lea and A. Simeonov, *PLoS One*, 2012, **7**, e36219.
62. F. H. Niesen, L. Schultz, A. Jadhav, C. Bhatia, K. Guo, D. J. Maloney, E. S. Pilka, M. Wang, U. Oppermann, T. D. Heightman and A. Simeonov, *PLoS One*, 2010, **5**, e13719.
63. D. S. Auld, S. Lovell, N. Thorne, W. A. Lea, D. J. Maloney, M. Shen, G. Rai, K. P. Battaile, C. J. Thomas, A. Simeonov, R. P. Hanzlik and J. Inglese, *Proc. Natl. Acad. Sci. U. S. A.*, 2010, **107**, 4878–4883.
64. T. E. Carver, B. Bordeau, M. D. Cummings, E. C. Petrella, M. J. Pucci, L. E. Zawadzke, B. A. Dougherty, J. A. Tredup, J. W. Bryson, J. Yanchunas, M. L. Doyle, M. R. Witmer, M. I. Nelen, R. L. DesJarlais, E. P. Jaeger, H. Devine, E. D. Asel, B. A. Springer, R. Bone, F. R. Salemme and M. J. Todd, *J. Biol. Chem.*, 2005, **280**, 11704–11712.
65. S. N. Krishna, C. H. Luan, R. K. Mishra, L. Xu, K. A. Scheidt, W. F. Anderson and R. C. Bergan, *PLoS One*, 2013, **8**, e81504.
66. W. Yu, E. J. Chory, A. K. Wernimont, W. Tempel, A. Scopton, A. Federation, J. J. Marineau, J. Qi, D. Barsyte-Lovejoy, J. Yi, R. Marcellus, R. E. Iacob, J. R. Engen, C. Griffin, A. Aman, E. Wienholds, F. Li, J. Pineda, G. Estiu, T. Shatseva, T. Hajian, R. Al-Awar, J. E. Dick, M. Vedadi, P. J. Brown, C. H. Arrowsmith, J. E. Bradner and M. Schapira, *Nat. Commun.*, 2012, **3**, 1288.
67. E. Wahlberg, T. Karlberg, E. Kouznetsova, N. Markova, A. Macchiarulo, A.-G. Thorsell, E. Pol, A. Frostell, T. Ekblad, D. Oncu, B. Kull, G. M. Robertson, R. Pellicciari, H. Schuler and J. Weigelt, *Nat. Biotechnol.*, 2012, **30**, 283–288.
68. B. L. Grasberger, T. Lu, C. Schubert, D. J. Parks, T. E. Carver, H. K. Koblish, M. D. Cummings, L. V. LaFrance, K. L. Milkiewicz, R. R. Calvo, D. Maguire, J. Lattanze, C. F. Franks, S. Zhao, K. Ramachandren, G. R. Bylebyl, M. Zhang, C. L. Manthey, E. C. Petrella, M. W. Pantoliano, I. C. Deckman, J. C. Spurlino, A. C. Maroney, B. E. Tomczuk, C. J. Molloy and R. F. Bone, *J. Med. Chem.*, 2005, **48**, 909–912.
69. M. A. Ashwell, J. M. Lapierre, C. Brassard, K. Bresciano, C. Bull, S. Cornell-Kennon, S. Eathiraj, D. S. France, T. Hall, J. Hill, E. Kelleher, S. Khanapurkar, D. Kizer, S. Koerner, J. Link, Y. Liu, S. Makhija, M. Moussa, N. Namdev, K. Nguyen, R. Nicewonger, R. Palma, J. Szwaya, M. Tandon, U. Uppalapati, D. Vensel, L. P. Volak, E. Volckova, N. Westlund, H. Wu, R. Y. Yang and T. C. K. Chan, *J. Med. Chem.*, 2012, **55**, 5291–5310.
70. H. L. Silvestre, T. L. Blundell, C. Abell and A. Ciulli, *Proc. Natl. Acad. Sci. U. S. A.*, 2013, **110**, 12984–12989.
71. D. S. Goldberg, S. M. Bishop, A. U. Shah and H. A. Sathish, *J. Pharm. Sci.*, 2011, **100**, 1306–1315.
72. J. Lavinder and S. Hari, *J. Am. Chem. Soc.*, 2010, **131**, 3–6.
73. U. B. Ericsson, B. M. Hallberg, G. T. DeTitta, N. Dekker and P. Nordlund, *Anal. Biochem.*, 2006, **357**, 289–298.

74. J. Romanuka, H. van den Bulke, R. Kaptein, R. Boelens and G. E. Folkers, *Protein Expression Purif.*, 2009, **67**, 104–112.
75. G. A. Senisterra, E. Markin, K. Yamazaki, R. Hui, M. Vedadi and D. E. Awrey, *J. Biomol. Screening*, 2006, **11**, 940–948.
76. C. G. Tate, *Trends Biochem. Sci.*, 2012, **37**, 343–352.
77. G. Lebon, T. Warne, P. C. Edwards, K. Bennett, C. J. Langmead, A. G. W. Leslie and C. G. Tate, *Nature*, 2011, **474**, 521–525.
78. F. Forneris, R. Orru, D. Bonivento, L. R. Chiarelli and A. Mattevi, *FEBS J.*, 2009, **276**, 2833–2840.
79. C. P. M. van Mierlo and E. Steensma, *J. Biotechnol.*, 2000, **79**, 281–298.
80. D. Martinez Molina, R. Jafari, M. Ignatushchenko, T. Seki, E. A. Larsson, C. Dan, L. Sreekumar, Y. Cao and P. Nordlund, *Science*, 2013, **341**, 84–87.
81. R. Jafari, H. Almqvist, H. Axelsson, M. Ignatushchenko, T. Lundbäck, P. Nordlund and D. M. Molina, *Nat. Protoc.*, 2014, **9**, 2100–2122.
82. T. Kreisig, A. A. Prasse, K. Zscharnack, D. Volke and T. Zuchner, *Sci. Rep.*, 2014, **4**, 5613.
83. H. Almqvist, H. Axelsson, R. Jafari, C. Dan, A. Mateus, M. Haraldsson, A. Larsson, D. M. Molina, P. Artursson, T. Lundbäck and P. Nordlund, *Nat. Commun.*, 2016, **7**, 11040.
84. D. P. Minde, M. M. Maurice and S. G. D. Rüdiger, *PLoS One*, 2012, **7**, 1–9.
85. M. M. Savitski, F. B. M. Reinhard, H. Franken, T. Werner, M. F. Savitski, D. Eberhard, D. Martinez Molina, R. Jafari, R. B. Dovega, S. Klaeger, B. Kuster, P. Nordlund, M. Bantscheff and G. Drewes, *Science*, 2014, **346**, 1255784.
86. H. Franken, T. Mathieson, D. Childs, G. M. A. Sweetman, T. Werner, I. Tögel, C. Doce, S. Gade, M. Bantscheff, G. Drewes, F. B. M. Reinhard, W. Huber and M. M. Savitski, *Nat. Protoc.*, 2015, **10**, 1567–1593.

CHAPTER 9

Fluorescent Probes in Medicinal Chemistry

SILVIA ORTEGA-GUTIÉRREZ

Department of Organic Chemistry I, Universidad Complutense de Madrid, Av. Complutense s/n, E-28040 Madrid, Spain
Email: siortega@quim.ucm.es

9.1 Introduction

The continuous growth of fluorescent methods in the past decade has offered an unprecedented opportunity for the study and manipulation of biological processes within living systems.

From a medicinal chemistry perspective, fluorescent methods can be used to obtain information at two different levels: (i) about macromolecules such as proteins and nucleic acids, *i.e.*, the usual therapeutic targets, or (ii) focused on small molecules or metabolites, which can be mediators of important (patho)physiological functions, so that the alteration of their levels is in a direct correlation with a disease. Accordingly, the different techniques based on the use of fluorescence can make key contributions in basically all stages of the drug discovery process, from the initial target validation and hit identification campaigns to the final human translation phases. Hence, in this chapter, after a general introduction about the different types of fluorophores, the most state-to-the-art techniques, classified on the basis of their object of study (*i.e.*, macromolecules or metabolites), will be described and contextualized within their main applications.

9.2 Types of Fluorophores

The usual way of making a (macro)molecule fluorescent is by attaching a fluorescent group, which is called a fluorophore. These are generally polyaromatic or heterocyclic hydrocarbons able to emit photons after excitation. They are characterized by four main parameters: the absorption maximum (λ_{max}), the emission maximum (λ_{em}), the extinction coefficient (ε) and the fluorescence quantum yield (Φ). Ideal fluorophores will have a large value of Stokes shift (difference between λ_{max} and λ_{em}) to minimize reabsorption of emitted photons, large ε, high Φ, stability, and biocompatibility.

Depending on the spectral features of the fluorophores, we can distinguish three main types of fluorescence (Figure 9.1), which have different specific applications: one- and two-photon fluorescence, and Förster resonance energy transfer (FRET).

One-photon fluorophores are the most common ones, and they are characterized by the use of a single photon of UV-visible light (generally 350–550 nm) to reach the excited state. This class includes, among others, the extensively used fluorescein, coumarin, boron dipyrromethene (Bodipy), and rhodamine derivatives (Figure 9.2).[1,2] To date, thousands of one-photon fluorophores have been developed, of which about 2000 have been commercialized. However, they have some limitations regarding deep-tissue imaging due to the shallow depth of tissue penetration (less than 100 μm) of the short excitation light, the cellular absorption of the photons, and the autofluorescence induced by exciting endogenous fluorophores, such as amino acids and nucleotides. To overcome this issue, during the past decade, two-photon fluorophores, which utilize two near-infrared (NIR) photons as the excitation source, have emerged as a new, indispensable imaging tool for biomedical research. Two-photon fluorophores have several advantages over one-photon fluorophores, including the capability of imaging deep inside a tissue, higher spatial resolution, and longer observation times. Although some existing one-photon fluorophores have two-photon

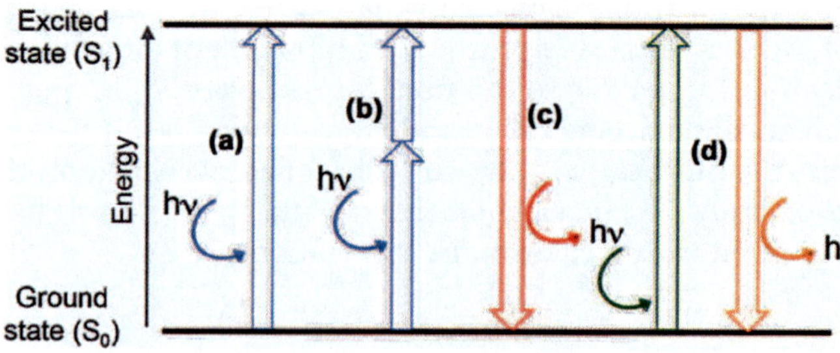

Figure 9.1 Energy diagram showing (a) one-photon absorption; (b) two-photon absorption; and (c) fluorescent emission for a molecule; (d) FRET occurs when the emission of a fluorophore (such as in (c)) is able to excite another nearby fluorophore, which, after relaxation, emits its characteristic fluorescence.

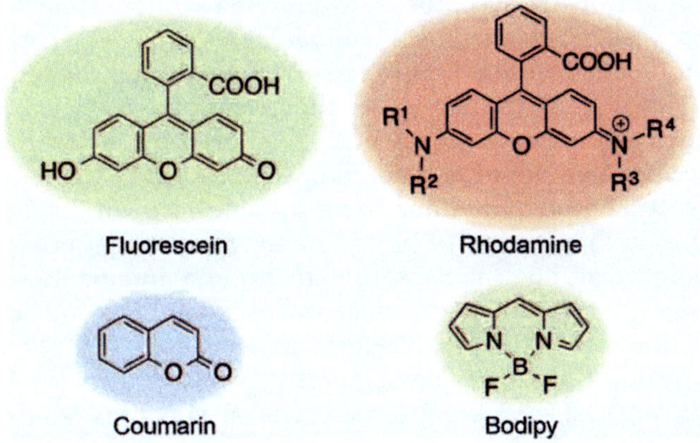

Figure 9.2 Representative one-photon fluorophores. The colour of the fluorophore's shadow indicates its wavelength of maximum emission (λ_{em}).

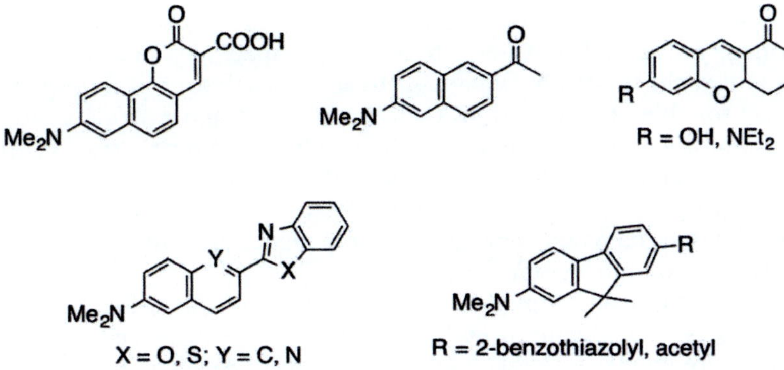

Figure 9.3 Representative two-photon fluorophores.

absorption capacity, this property is far from optimal because they generally show small two-photon absorption cross sections. This characteristic limits their use due to the high laser power required to obtain clear images.[3,4] Hence, in recent years, research efforts have focused on addressing this need, which has led to the development of novel efficient and application-specific two-photon fluorophores, the most representative of which are featured in Figure 9.3.

With respect to FRET, it is a physical phenomenon in which a donor chromophore in its excited state non-radiatively transfers its excitation energy to a neighbouring acceptor chromophore. If the two chromophores are fluorescent, FRET stands for fluorescence resonance energy transfer and, in this case, the RET is experimentally observed as an increase in the fluorescence intensity emission of the acceptor. On some occasions, the donor excitation initially required can lead to background noise due to undesirable direct excitation of the acceptor or to photobleaching. To avoid this limitation, it is possible to use a bioluminescent donor, which produces the initial photon emission that excites the acceptor without the requirement of external illumination. In this case, the technique is called BRET (bioluminescence resonance energy transfer). Since FRET is highly sensitive to the distance between donor and acceptor dipoles within the 1–10 nm range, FRET pairs, composed of a donor and a suitable acceptor, have been widely adopted to monitor a variety of biochemical activities that produce changes in molecular proximity, such as protein–protein interactions, conformational changes, intracellular ion concentrations, and enzyme activities.[5]

In addition to the usual organic small molecules used as fluorophores, a new class of novel materials has emerged in recent years for bioimaging. Quantum dots (QDs) are inorganic nanocrystals with sizes ranging from 2 to 10 nm that exhibit high fluorescence intensity and photostability. A QD usually consists of a semiconductor core and a shell, which protects the core from oxidation and enhances the quantum yield. QDs have a number of advantages over conventional fluorescent dyes: they are tens of times brighter and thousands of times more photostable, and they have broad luminescence excitation spectra and narrow symmetrical emission spectra with large Stokes shifts. Due to their unique spectral properties, QDs are ideal probes for multicolor imaging.[6] In addition, the spectral properties of QDs are tunable and depend on their composition and core size, thus providing a broad optical range of fluorescence from the near ultraviolet to the NIR region. However, although functionalized QDs have excellent photophysical properties, there are some challenges that need to be addressed before they can be readily applied in biomedicine, including issues such as toxicity and non-biodegradability.

In the following sections we will analyse how the different types of fluorophores have advanced and how they are key current contributors to important biomedical applications in the field of medicinal chemistry enabling the study of both macromolecules and metabolites.

9.3 Applications of Fluorescent Probes for the Study of Macromolecules

In the field of medicinal chemistry, macromolecules are usually synonymous with therapeutic targets. Accordingly, in this section we will focus on the use of fluorescent probes and on the information they can provide regarding proteins and nucleic acids, which are the targets of the immense majority of the current drugs on the market.

9.3.1 Fluorescent Probes for the Study of Proteins

Considering their prominence as drug targets, proteins are the most important macromolecules whose study can be addressed by the use of fluorescent probes. Proteins can be visualized with two main strategies: (i) the protein is recombinantly expressed and contains a fluorescent tag or (ii) the protein is labelled post-translationally using a ligand-targeted fluorescent probe. Clearly, the first strategy has the advantage of tight spatiotemporal control, but its main limitation is that it cannot be used in native systems, such as animal or human samples of cells or tissues. Thus, each methodology can be best suited for specific purposes, so we will analyse their features and applications for both medicinal chemistry and drug discovery.

9.3.1.1 Recombinant Expression of Fluorescent Proteins

The landmark discovery, in 1994, that the *Aequorea victoria* jellyfish green fluorescent protein (GFP) could be genetically encoded as a fluorescent tag in prokaryotic and eukaryotic proteins[7] paved the way for the subsequent development of numerous variants with distinct biochemical and fluorescent properties. Currently, the available fluorescent proteins have expanded the colour palette throughout the visible spectrum to encompass blue, cyan, green, yellow, orange, red, and far red variants.[8,9] In general, the strategy consists of incorporation of the gene that encodes for the desired fluorescent protein into the genome of the organism in the region of DNA that codes for the target protein, usually controlled by the same regulatory sequence. Hence, when the target protein is expressed, it will be fused to the fluorescent protein and can be visualized using fluorescence microscopy. The use of this technique has redefined the understanding of many biological processes including protein folding, protein transport, and RNA dynamics, which in the past could not be studied. In recognition of its significance to the biomedical sciences, Chalfie, Shimomura, and Tsien were awarded the 2008 Nobel Prize in Chemistry for the discovery and development of the GFP.

The latest applications of this technique include the development of infrared fluorescent proteins that can be expressed in live cells and even in whole animals, which are suitable for whole-body imaging.[8,10] In addition, recent improved variants of far red, NIR, and IR fluorescent proteins have enabled the real time observation of highly dynamic cellular processes. For example, the tight spatiotemporal coordination between cell apoptosis and embryonic morphogenesis and the dynamics of apoptosis during tumorigenesis have been observed in Drosophila.[11] Also, the imaging of cell cycle progression has been achieved in distinct cell types and even in tumour rodent xenografts.[8,12] Recent work has also been directed towards the real time imaging of neuronal activity. In this field, the development of genetically encoded calcium ion and voltage indicators (GECIs and GEVIs, respectively) is revolutionizing the way neuroscientists can now directly analyse the activity of a single neuron. However, this area still requires important efforts because, although remarkably efficient GECIs have been described,[13,14] the development of GEVIs has evolved at a slower pace,[15,16] and technical challenges, such as the temporally precise monitoring of voltage transients in individual neurons in the mammalian brain, still need to be successfully addressed.[8]

All of the above-mentioned applications have in common that the protein under study is expressed fused to a specific fluorescent protein that has been selected looking for the most appropriate features according to the particular application. This approach can be very powerful for visualization of even single molecules, but it does not provide information about protein association or the formation of oligomers. The existence of physiologically relevant oligomers of proteins, in particular of receptors such as the superfamily of G protein-coupled receptors (GPCRs), has been a hot topic of discussion in the last few years, as there was much speculation about not only their sheer existence but also their significance and function.[17] In 2009, the first study that confirmed that glycoprotein hormone receptors were able to di/oligomerize *in vivo* and that the oligomer formation was necessary for fully functional signalling was published.[18] Soon, other studies, using different biochemical and biophysical methodologies, were attempting to unravel the complex scenario of GPCR heteromers. Among them, advanced fluorescence microscopy techniques have been key contributors to our current knowledge of aspects that were not well known until now, such as GPCR complex stoichiometry and stability, and its physiological significance in native systems.[19] In particular, FRET and BRET have played a fundamental role in the advancement of this field, because the straightforward genetic fusion of fluorescent (for example, the cyan fluorescent protein, CFP) or bioluminescent (for instance, the enzyme luciferase) donors, and the appropriate acceptor proteins, to the C-terminal tail of GPCRs allows the easy detection of close proximity by measuring the fluorescence emitted by the RET effect. Recent applications of this methodology include the discovery of the relevance of the heteromers formed by the GPCRs GPR55 and CB_2 in cancer,[20] the orexin OX_1 and the corticotropin-releasing factor receptors in the stress associated to cocaine-seeking behaviour,[21] and the CB_1 and $5\text{-}HT_{2A}$ receptors in cognitive impairment induced by the psychoactive cannabinoid Δ^9-tetrahydrocannabinol.[22]

The use of fluorescent proteins has rapidly integrated into basic research and has allowed the advancement of cellular and deep-tissue *in vivo* imaging to an extent that was inconceivable only a few years ago. Although some limitations still exist, such as the artefacts that can arise from the oligomerization observed for some fluorescent proteins or the big size (in general between 25 and 30 KDa), which may affect target protein function and/or trafficking,[8] engineering efforts are on-going to solve these shortcomings. However, the main limitation, which is inherent to the methodology itself, is that it cannot be applied to the study of native systems (*i.e.*, biological samples coming from animal models or humans), because it requires artificial modification of the target protein. To overcome this weakness, the alternative is to use fluorescent ligands or small-molecule fluorescent probes that bind to the target protein. In this case, there is no need for modification of the protein and it can be directly detected and visualized in native systems.

9.3.1.2 Study of Proteins using Ligand-targeted Fluorescent Probes

Small-molecule fluorescent probes are attractive and versatile materials for cell-imaging reagents. They can be easily used in native biological systems allowing rapid detection of their corresponding target. In addition, the binding affinity, excitation and emission wavelength, and even localization to a specific organelle are features that can be optimized by using well-established design strategies. Thus, a diverse array of fluorescent probes can be designed to target various biomolecules and organelles. Among the most important proteins as drug targets, GPCRs and enzymes clearly stand out, as these two classes account for almost two thirds of the total number of drug targets currently engaged by marketed drugs. Hence, the development of probes for their study in native systems with biological significance has attracted the attention of many research groups. In the next section, the probes directed towards the different protein classes will be reviewed together with their most relevant applications in the context of drug discovery.

9.3.1.2.1 Development of Fluorescent Probes for the Study of Receptors. Among all the cellular receptors, the members of the GPCR superfamily are the targets of about 40% of currently prescribed drugs. However, it is believed that their therapeutic potential is still underexploited considering their increasingly complex biology, the number of orphan GPCRs (those for which the endogenous ligand is not known), and the GPCRs with unclear physiological functions. In this context, fluorescence-based methods are gaining increasing importance,[23,24] and, in particular, the development of fluorescent ligands that permit the real-time monitoring of ligand–receptor interactions, as well as the visualization and location of GPCRs. Hence, a number of fluorescent probes for the study of class A GPCRs have been described, including adenosine, adrenergic, angiotensin, histamine, muscarinic, prostanoid, serotonin, cannabinoid, and vasopressin receptors.[25–32] From a medicinal chemistry point of view, these probes can make decisive contributions at the different stages of the drug discovery process, including hit identification, target validation, and even patient diagnosis. For example, the development of fluorescent binding assays constitutes an alternative to the use of radioligands in the search for new ligands for adenosine A_3 receptors[33] or the μ-opioid receptor.[34] Also, small-molecule fluorescent probes can help to shed light on important questions about receptor localization, levels of expression, and target engagement, all of which are critical questions for target validation. However, to successfully realise this application, it is necessary that the probe performs efficiently in native systems and not only in transfected cells. This requirement is not always easy to accomplish, as on many occasions, probes that achieve good results in cells with receptor overexpression are not efficacious enough to detect protein expression at endogenous levels or do not have the sufficient sensitivity to uncover changes provoked by

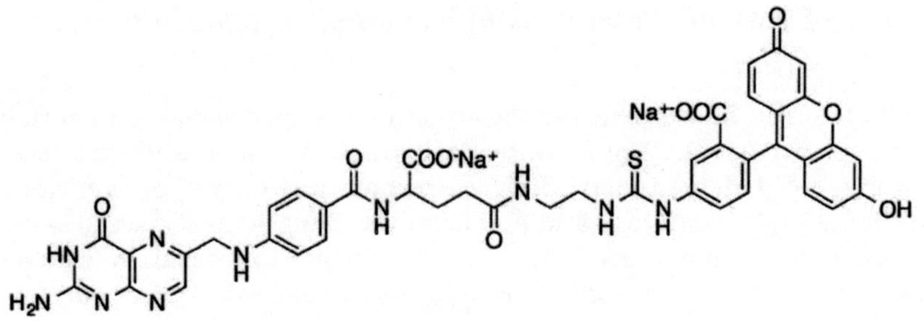

Figure 9.4 Structure of the folate–fluorescein conjugate EC17.

(physio)pathological stimuli. Thus, among the numerous fluorescent probes reported so far, only a few have been described that work in native systems. Among them, fluorescent probes that enable the visualization of estradiol,[35] cannabinoid,[29,30,36,37] and neuropeptide Y[38] receptor subtypes in native cells stand out. Moving forward, some fluorescent ligands have even evolved to become ligand-targeted imaging agents, demonstrating success in clinical trials by identifying and localizing tumour masses, and by selecting patients for corresponding ligand-targeted treatment.[39,40] There have been several preclinical studies and a few clinical trials of approved therapeutic antibodies conjugated to NIR fluorophores; these include conjugates based on cetuximab and panitumimab to target the epidermal growth factor receptor, trastuzumab to visualize the human epidermal growth factor receptor 2, and bevacizumab to detect the vascular endothelial growth factor.[40] Due to their large size, antibody conjugates may penetrate less effectively into tumour tissue compared with small molecules. This fact has prompted the development of this latter class of probes. For example, folate receptor α (FRα) is overexpressed in many epithelial cancers, including cancers of the ovary, breast, lung, kidney, and colon. In these patients, folate-targeted fluorescent dyes such as the folate–fluorescein conjugate EC17 (Figure 9.4) can be used as a tool to aid surgeons in the identification of malignant disease during surgery.

In a pilot study, this conjugate was found to be safe in an intravenously injected formulation, showing a pharmacodynamic profile that facilitated fluorescence imaging over the time course from 2 to 8 h after injection, and did not interfere with the standard surgical procedure. Moreover, the use of the fluorescent conjugate significantly helped the surgeons to detect tumour deposits compared to the use of visual observation alone, and all of the resected fluorescent lesions were subsequently confirmed by pathology to be malignant.[41]

9.3.1.2.2 Development of Fluorescent Probes for the Study of Enzymes. The study of the functional state of enzymes has been transformed by the development of fluorescent activity-based probes (ABPs). Although these probes are included in the wider field of activity-based protein profiling

(ABPP), which also includes ABPs suitable for proteomics, in this chapter we will focus on the fluorescent ABPs. For a broader coverage of ABPs, including applications for proteomics, the reader is referred to some excellent recent and comprehensive reviews about this exciting area of research.[42–44] From the pioneering work developed by Bogyo and Cravatt back in the early 2000s,[45,46] ABPP has become an ever-evolving field with more refined probes and methodologies being described every year. Essentially, an ABP probe contains three main elements: (i) a reactive group directed towards the conserved mechanistic feature of a set of enzymes; (ii) a reporter tag for visualization, identification, and/or purification of targeted proteins; and (iii) a linker region that can modulate the reactivity and specificity of the reactive group while avoiding potential steric interferences between the reporter and the reactive group.

The selection of the reactive group depends on the catalytic mechanism and the specific amino acids of the enzyme active site. In this regard, the use of fluorophosphonates for the study of serine hydrolases and the use of epoxide and acetoxymethyl ketone (AOMK) probes for targeting cysteine proteases constitute paradigmatic examples.[45,46] The number of directed ABPs, *i.e.*, probes designed to target a specific enzyme class, has dramatically increased over the last few years, and currently there are available probes for cathepsins, legumains, caspases, protein arginine methyltransferases, proteasomal proteases, kinases, phosphatases, and glycosidases, to name just a few[44,47] (some representative fluorescent probes are shown in Figure 9.5). Sometimes, in the search for broader coverage, probes that target enzymes that belong to mechanistically diverse families are desired. Towards this aim, electrophilic ABPs have been developed to target conserved catalytic aspartate, glutamate, cysteine, lysine, tyrosine, and serine residues across a mechanistically diverse set of enzymes.[44]

Conversely, specific probes for detection of individual enzymes can also be needed. In this regard, researchers have developed tailored ABPs that react with a limited number of enzymes, and, in some cases, even with a single enzyme. These probes can be used to localize enzyme activities by cellular imaging, as well as to detect low-abundance enzymes that are masked by more abundant enzymes. In this type of probe, a binding group that imparts selectivity (usually based on a known synthetic or natural product ligand) is introduced either as part of the reactive group or in the linker. Recent examples include probes for labelling histidine kinases,[48,49] flavine-dependent oxidases such as both monoamine oxidases (MAOs) A and B[50] or even MAO-B exclusively,[51] the integral membrane hydrolase KIAA1363,[52] and the Woodward's reagent K-based ABP for labelling the macrophage migration inhibitory factor (MIF)[53] (Figure 9.6).

A drawback encountered in first generation ABPs was the fact that the size, nature and/or charge of the fluorophore could diminish or even prevent the cellular uptake of the probes, making necessary the use of cell homogenates that disrupt intracellular interactions and specific subcellular localizations. To overcome this problem, currently used probes usually have small groups

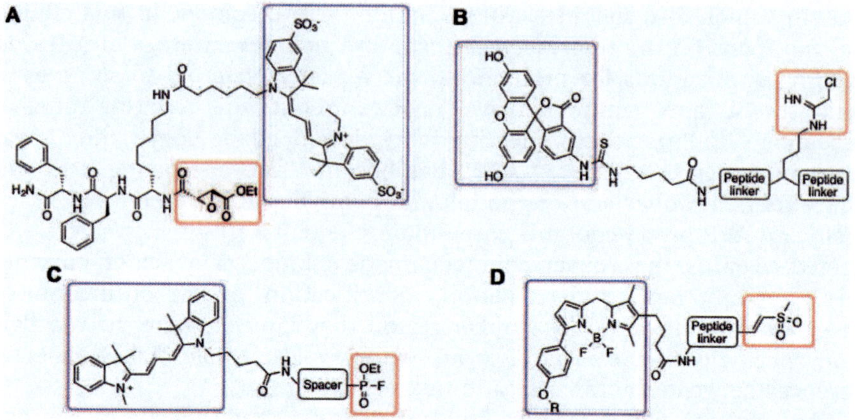

Figure 9.5 Structure of representative fluorescent ABPs for the study of (A) cysteine proteases, (B) protein arginine methyltransferases, (C) serine hydrolases, and (D) proteasomal proteases. Reactive groups (epoxide, cloroacetamidine, fluorophosphonate, and vinyl sulfone, respectively) are highlighted in red and fluorescent reporters in blue.

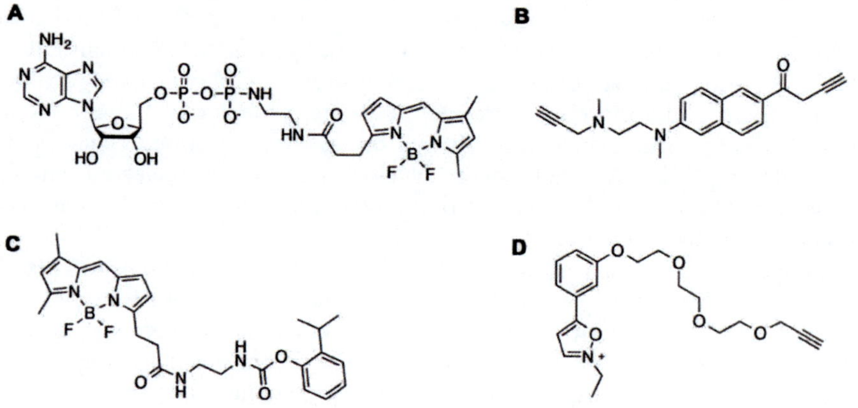

Figure 9.6 Structure of representative ABPs for labelling (A) histidine kinases, (B) MAO-B enzyme, (C) KIAA1363 hydrolase, and (D) MIF. The last probe is not fluorescent, so the terminal alkyne is used to attach a fluorophore tagged with an azide group by click chemistry.

suitable for bioorthogonal reactions. In this way, the probe, which is cell permeable, can be directly used in native cells and the fluorophore (containing the complementary group needed for the bioorthogonal reaction to proceed) is introduced in a subsequent step. The most popular pairs for these reactions are alkyne/azide, triphenylphospine/azide, and strain cyclooctyne/azide, although recent reactions have also included inverse-electron demand Diels Alder reactions between strained alkenes and tetrazines.[54–56]

At the present moment, the most important application of fluorescent ABPs in medicinal chemistry is their use in high-throughput assays based on fluorescence polarization that allow for the rapid primary screening of thousands of compounds in order to identify initial hits for the enzyme under study.[44] The advantage of this assay, in addition to speed, automation, and miniaturization, is that the assay does not require a deep biological or structural knowledge of the enzyme nor the existence of known inhibitors; only a source of enzyme and a fluorescent ABP are needed to set up this kind of assay. This fact makes this type of assay especially suited to the identification of hits for poorly known and unexplored enzymes. Recent applications of this methodology include the discovery of the first inhibitors of lysophospholipases 1 and 2,[57] protein arginine deiminase isoforms 1, 2, 3, and 4,[58] protein phosphatase methylesterase-1 (PME-1),[59] serine hydrolase α,β-hydrolase domain-containing protein 10 (ABHD10),[59] type II platelet-activating factor acetylhydrolase (PAFAH2),[60] or bacterial histidine kinases.[49] Considering that ABPs react covalently with their target(s), broad-spectrum fluorescent ABPs can also be used in-gel in the hit to lead and lead optimization phases to improve properties such as potency, selectivity and even *in vivo* effectiveness.[44] Further translation has recently been described for a protease-activated fluorescent imaging probe, LUM015, which allows the detection of tumour sites *in vivo* in a mouse model of soft tissue sarcoma (STS) and *ex vivo* in a first-in-human phase 1 clinical trial. LUM015 consists of a commercially available fluorescence quencher molecule (QSY21) attached through a GGRK peptide to a 20 KDa polyethylene glycol (PEG) and a Cy5 fluorophore. LUM015 is non-fluorescent, but upon proteolytic cleavage by cathepsins K, L, and S (proteases typically found at tumour sites), the quencher is released and fluorescent signal is turned on. In mice, intravenous injection of LUM015 labelled tumour cells, and residual fluorescence within the tumour bed predicted local recurrence. In 15 patients with STS or breast cancer, intravenous injection of LUM015 before surgery was safe and well tolerated at a dose that allows for detection of tumour-specific fluorescence. Imaging of resected human tissues showed that fluorescence from the tumour was significantly higher than fluorescence from normal tissues. These results support future clinical trials of LUM015 using intraoperative imaging of the tumour bed and comparing imaging results with histopathology. A feasibility study of intraoperative imaging with the LUM system has recently started for breast cancer patients (NCT02438358), and it may be possible to use the probe for patients with other cancers.[61]

9.3.2 Fluorescent Probes for the Study of Nucleic Acids

For decades, researchers have fluorescently labelled nucleic acids with a variety of labelling strategies to gain insight into the fate of these nucleic acids. Fluorescent probes can be sequence specific (*i.e.*, binding of the probes occurs only at a predefined primary sequence) or non-specific (the probe interacts all over the polynucleotide chain). These latter probes were the first that were developed, and their binding to the nucleic acid backbone can be due to non-covalent or covalent interactions. The usual non-covalent non-specific interactions with nucleic acids are either groove binding or insertion between base pairs, which is called intercalation. Groove binding dyes interact with the major or minor groove of the α-helix of double stranded DNA. The most commonly used fluorophores that bind in the minor groove are 4′,6-diamidino-2-phenylindole (DAPI) and Hoeshst stains (Figure 9.7). Intercalating dyes are often cationic molecules with planar aromatic rings that insert between the base pairs of the helix structure due to π-stacking interactions together with hydrogen bonding, van der Waals, and hydrophobic interactions. Most of these fluorophores exhibit a strong fluorescence enhancement upon binding. This class of agents can be monomers (intercalator) or dimers (bis-intercalator) for stronger binding and higher brightness. In addition, it is possible that monomers and dimers permit differentiation between different polynucleotides, such as single stranded and double stranded DNA or RNA, since intercalation is dependent on the secondary and tertiary structure of the nucleic acid. For example, cyanine dye TOPRO-3 (Figure 9.7) does not bind RNA, whereas its dimer TOTO-3 does. Other widely used fluorophores in this class are the anthracene DRAQ5, ethidium bromide, and propidium iodide (Figure 9.7).[62]

Targeting specific sequences is beneficial for obtaining site-specific sequence-directed labelling, without significant off-target binding. In enzymatic labelling, a functional group (or fluorophore) is added to nucleic acids by an enzyme-mediated reaction between the nucleic acid and a dedicated cofactor. There are enzymes, such as M.HhaI, the adenine-specific DNA methyl transferase (MTase) from *H. hemolyticus*, which recognize the double-strand DNA sequence 5′-GCGC-3′ and subsequently catalyze a covalent bond formation between the activated methyl group from a specifically designed cofactor (usually *S*-adenosyl-L-methionine, SAM) and the exocyclic amino group of cytosine. Different MTases will recognize different sequences, which are normally 4 to 6 base pairs long. The cofactor determines the functional group that is added in the enzymatic reaction to the base, so the synthesis of new cofactor analogues is important to expand the range of molecules that can be attached to the activated group (Figure 9.8). For example, the use of a SAM analogue bearing a terminal alkyne (and the appropriate MTase) enables this moiety to be transferred from the SAM analogue to the specific sequence of the DNA. Then, the triple bond can be used for the subsequent bioorthogonal introduction of an azide-bearing fluorophore, achieving an exquisite degree of selectivity. This strategy can be used to label not only specific sequences of DNA but also RNA (single or double stranded RNA, transfer RNA, and small RNAs).[62–65] Despite the potential of enzymatic labelling, its use for live cell imaging is still in its infancy and further optimization (and simplification) is on-going in many research laboratories.

Actually, a more conventional approach relies on the use of oligonucleotide probes that have been chemically modified with a fluorophore, so that

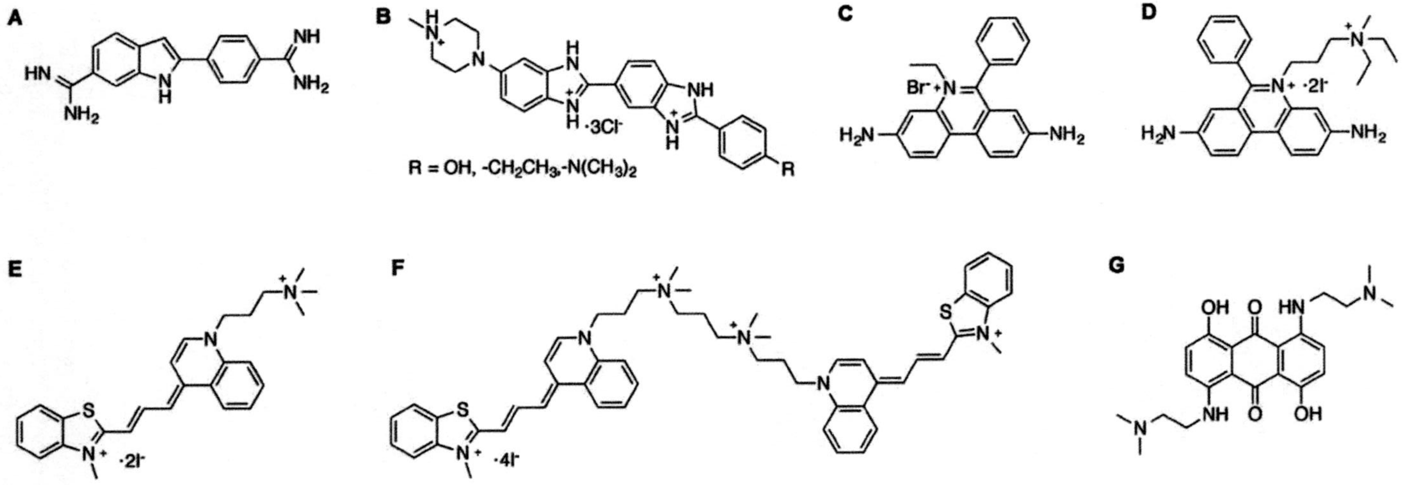

Figure 9.7 Structure of representative fluorescent probes for the study of nucleic acids. (A) DAPI, (B) Hoechst dyes, (C) ethidium bromide, (D) propidium iodide, (E) TOPRO-3, (F) TOTO-3, and (G) DRAQ5.

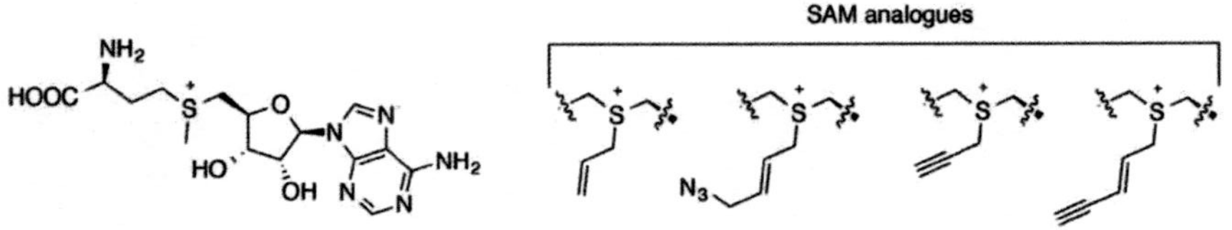

Figure 9.8 Structure of the cofactor SAM and its most widely used analogues.

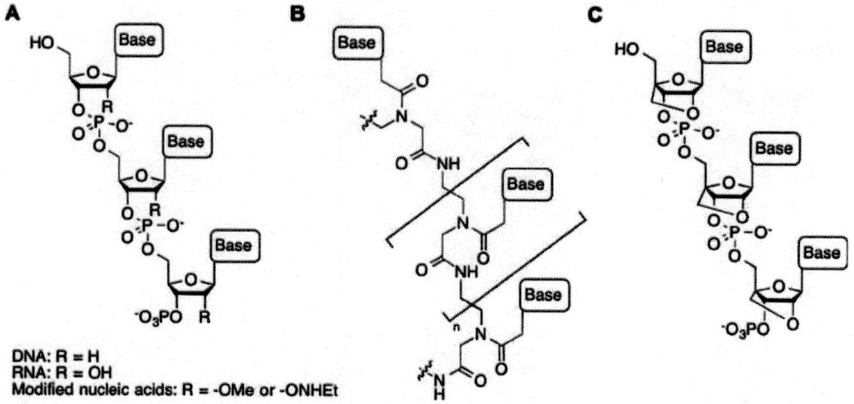

Figure 9.9 Structure of (A) DNA, RNA, 2′-*O*-methyl, and 2′-*O*-aminoethyl nucleic acids, (B) PNAs, and (C) LNAs.

they will bind specifically to their complementary sequences and this interaction can be revealed by fluorescence detection. This is the principle of fluorescence *in situ* hybridization (FISH), where the location of a target DNA or RNA sequence can be visualized by fluorescence microscopy in fixed and permeabilized cells. For application in live cells, chemically modified oligonucleotides (more stable and cell permeable) such as peptide (or polyamide) nucleic acids (PNAs), locked nucleic acids (LNAs), and 2′-*O*-methyl and 2′-*O*-aminoethyl modified nucleic acids are needed (Figure 9.9).[62] This technique has evolved to the use of two fluorophores attached at the opposite ends of an oligonucleotide. These two fluorophores are based on the FRET principle or the use of quencher/dye pairs so that in the unbound state, the oligonucleotide probe is non-fluorescent and the signal only appears after a conformational change due to binding to the sequence target that causes either both functionalities to get closer (FRET principle) or to be pushed away from each other (quencher/dye principle). This class of probes, referred to as molecular beacons, are characterized by low backgrounds and suitability for live-cell microscopy. Variations of this technique include the use of different fluorophores and quencher/dye pairs, oligonucleotides with a variety of conformations, and the number of oligonucleotides used (multiplexing).[66,67]

Finally, instead of using oligonucleotides tagged with fluorophores, it is also possible to use small molecules, such as hairpin pyrrole-imidazole polyamides, which recognize specific sequences in the DNA.[62,68,69]

These approaches offer new powerful tools for sequence-specific nucleic acid detection, and can be directly applied both in basic and translational research. Regarding the former, it is appreciated that one of the main bottlenecks associated with the clinical translation of DNA- and mRNA-based therapies is the low efficiencies in transfection. Hence, designing new delivery vehicles is crucial for the future success of these treatments and fluorescent-labelled nucleic acids are the best tools to test, compare, and optimize transfection reagents in live cell imaging. With respect to clinical translation, fluorescent probes directed towards nuclei acids are playing an essential role in diagnostic applications including DNA mapping, epigenome studies, and the detection of disease-relevant mutations.

9.4 Fluorescent Probes for the Study of Metabolites

The pressing need to monitor changes in the abundance as well as spatio-temporal patterns of small-molecule analytes in a wealth of applications has stimulated the development of numerous sophisticated biosensors. The use of fluorescence-based molecular sensors offers advantages such as high sensitivity, real time analysis, readily available instruments, multiple detection modes, and high throughput analyses. There are two main classes of fluorescent probes for the detection of metabolites. The first one encompasses the small molecule (usually organic) fluorophores, whereas the second relies on the design of tailored FRET biosensors. Although the development of the latter entails higher complexity, its clear advantage is the specificity towards the analyte under study. However, both approaches show broad applications in biomedicine.

9.4.1 Detection of Metabolites Using Small Molecule Fluorophores

The specific recognition of small molecules in complex media represents a current challenge in the life sciences. Bodipy fluorophore has several features that make it a good candidate for the development of fluorescent sensors in biological systems, such as its relatively high molar absorption coefficient and quantum yield, narrow and high-intensity emission peaks, relative inertness under physiologically relevant conditions, and resistance to photobleaching. In addition, its spectroscopic and photophysical properties can be finely tuned by substitution on the dipyrromethene core. These features have made bodipy one of the most popular scaffolds for the development of fluorescent probes for metabolites. In this regard, medium-sized bodipy-based libraries have been synthesized and screened for specific fluorescent responses after interaction with a particular metabolite. This strategy has enabled the identification of a bodipy-based probe for the detection of glucagon in live cells and tissues,[70] selective turn-on sensors for fructose (Fructose Orange)[71] or cysteine,[72] and a ratiometric sensor for glutathione[73] (Figure 9.10).

Besides bodipy, other fluorophores have also been incorporated into probes tailored to detect different metabolites. For example, functionalization of the fluorescent spinol core with aldehyde groups for amino acid binding provided a new class of fluorescent sensors (Figure 9.10) for the enantioselective recognition of free amino acids in living cells,[74] and a

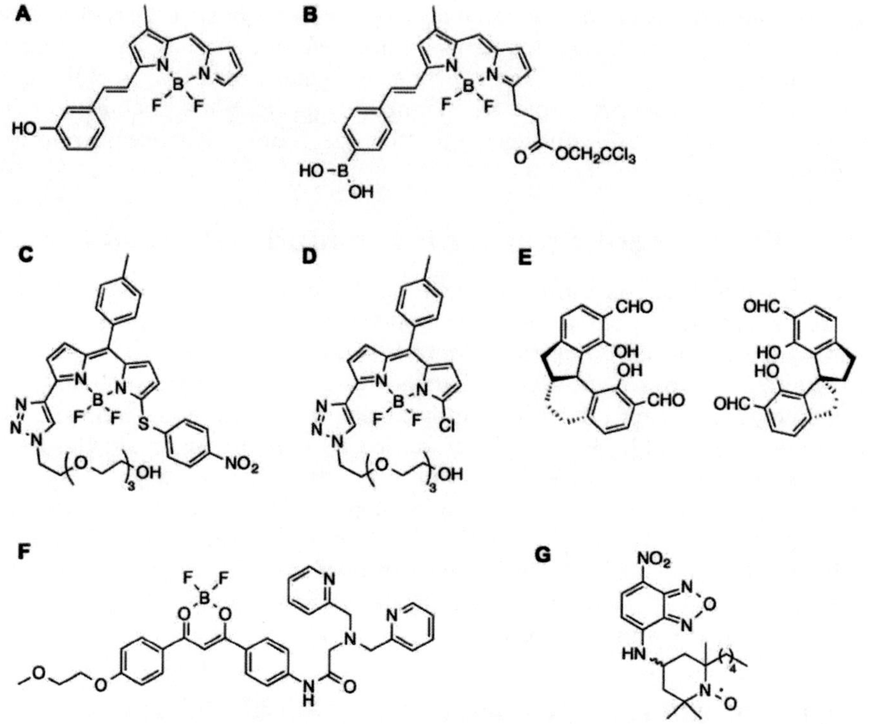

Figure 9.10 Structure of representative fluorescent probes for the detection of metabolites such as (A) glucagon, (B) fructose, (C) cysteine, (D) glutathione, (E) amino acids, (F) cadmium, and (G) lipid radicals.

variety of fluorophores have shown utility for the specific detection of biothiols.[75] Analogously, the difluoroboron dibenzoyl-based fluorescent probe for Cd^{2+} detection (Figure 9.10) displays high selectivity and sensitivity towards Cd^{2+} over other bivalent metals in aqueous solution under physiological conditions and also in living cells.[76] A more recent example describes the first high-sensitivity and specific fluorescent probe for lipid radicals based on nitrobenzoxadiazole (NBD). In this case, 2,2,6-trimethyl-4-(4-nitrobenzo[1,2,5]oxadiazol-7-ylamino)-6-pentylpiperidine-1-oxyl (NBD-Pen, Figure 9.10) enables the direct detection of lipid radicals in living cells and in animal models.[77] Finally, the use of more complex metal–organic frameworks (MOFs) based on lanthanides is starting to be incorporated into this field. A recent example of their application is a mixed crystal that contains europium and terbium that allows the selective detection of lysophosphatidic acid (LPA), a biomarker for ovarian cancer and other gynaecologic cancers, in physiologically relevant ranges.[78]

9.4.2 FRET Biosensors for the Detection of Small Molecules

Fluorescent biosensors based on FRET are well suited to the measurement of small-molecule concentrations in cells and *in vivo*. Such FRET biosensors are

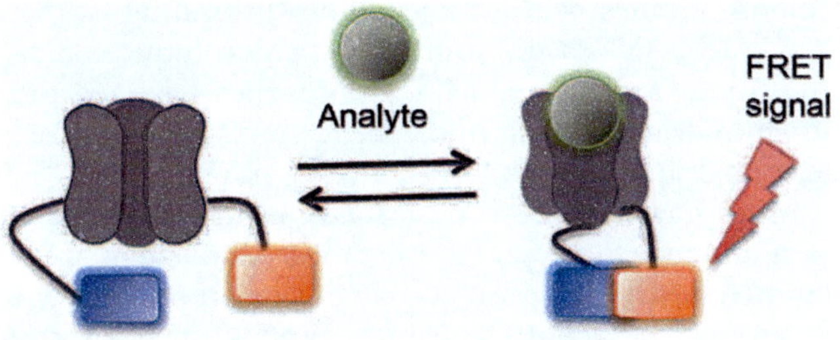

Figure 9.11 Schematic representation of a FRET biosensor for the detection of a small molecule analyte.

usually based on a receptor protein that undergoes a conformational change upon binding to the analyte of interest. Expressing such a receptor protein as a fusion protein with two fluorescent proteins that constitute a FRET pair can result in the generation of a FRET sensor. In the presence of the analyte, the FRET signal appears as emitted fluorescence and is proportional to the quantity of the analyte (see Figure 9.11 for a schematic representation). Successful examples of FRET sensors include those for the detection of neurotransmitters such as glutamate or γ-aminobutyric acid, ions such as Ca^{2+}, carbohydrates, cyclic nucleotides, ATP and others.[79,80]

9.4.3 Development of Cellular Organelle-targeting Fluorescent Probes

Irrespective of the methodology used for detecting an analyte of interest, the measurement of organelle-specific levels with subcellular resolution is fundamental. This challenge can be addressed in the case of biosensors with the incorporation of a genetically encoded localization sequence for expression in specific intracellular compartments, thus enabling organelle-targeted detection. In the case of small molecule fluorescent probes, they require the incorporation of a group that directs the probe to a specific cellular localization where the analyte of interest can be found. Ideally, before reaching the target organelle, the probe will be non-fluorescent in order to minimize background, and only after reaching the desired intracellular localization will the presence of the analyte turn on the fluorescent signal. This approach improves the subcellular mapping of metabolites enhancing both the sensitivity and selectivity of the fluorescent probe. The limiting factor is, of course, the availability of such organelle-targeting groups. For example, triphenylphosphonium (TPP) has long been used for mitochondria targeting. Besides the lipophilic cationic vectors, natural and synthetic peptides have also been reported to exhibit mitochondria-targeting features. Multiple mitochondria-targeting fluorescent probes for metal ions based on the same concept, the intrinsic lipophilic cationic fluorophore core, have been reported (see Figure 9.12 for some representative examples). In general, in the absence of the metal ion, the emission of the fluorophore is efficiently quenched by other parts of the molecule until binding of the analyte (for example, Hg^{2+}, K^{+}, Zn^{2+}, Cu^{+}, and Fe^{2+}), which induces a significant increase in the fluorescence. Analogous approaches have led to the development of mitochondria-targeting fluorescent probes to measure H_2O_2, HClO, NO, or reactive sulphur species levels, among others.[81]

Specific probes have also been described for digestive organelles such as endosomes, autophagosomes, and lysosomes. All of them are characterized by their low pH, which provides the acidic environment required by the hydrolytic enzymes they contain. Hence, probes for lysosomes and other acidic organelles are equipped with lipophilic weakly basic moieties such as lipophilic amines. Once the probes diffuse inside the acidic organelles, the amines are protonated and the probes become positively charged and cannot leave the organelle. Various fluorescent probes have been developed on the basis of this mechanism, including those that can detect metal ions such as Fe^{3+}, Zn^{2+}, and Cu^{2+} and reactive species like H_2O_2, NO, HClO or H_2S (see Figure 9.12 for some representative examples).[81]

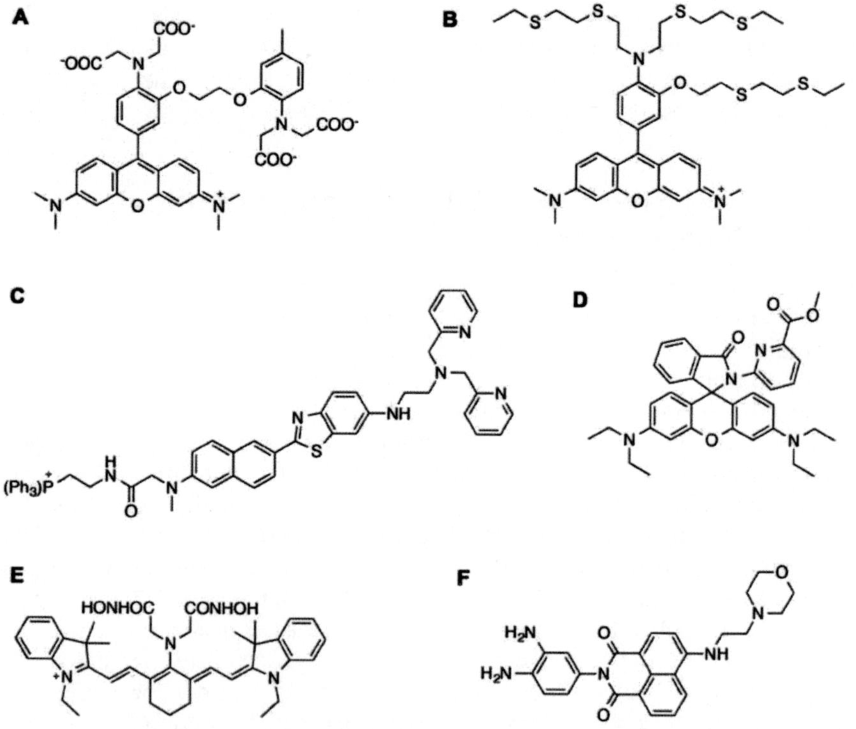

Figure 9.12 Rhodamine-based fluorescent probes for detection of mitochondrial (A) Ca^{2+} and (B) Hg^{2+}. (C) Two-photon probe for detection of mitochondrial Zn^{2+}. (D) Fluorescent turn-on probe for imaging lysosomal Fe^{3+}. (E) NIR fluorescent turn-on probe for the detection and imaging of Cu^{2+} in lysosomes. (F) Fluorescent probe for lysosomal NO imaging.

Given its crucial functions, numerous fluorescent probes have been reported that target the various elements of the plasma membrane. Since the lipid bilayer constitutes the major component of the plasma membrane, the most general method to design these probes is by integrating a membrane anchor, usually a lipophilic carbon chain, such as a string of fatty acids, into a known fluorescent probe through a linker. Many membrane-targeting probes have been designed and synthesized according to this method. Besides the lipophilic carbon chain, specific components on the plasma membrane, such as receptors, may also be used as targets for probe design. These probes include those for detecting metal ions such as Na^{+}, Ca^{2+}, Mg^{2+}, and Zn^{2+}, and other signalling molecules such as ATP and NO.

Selective probes for the endoplasmic reticulum (ER) are amphipathic, lipophilic cationic dyes with moderate-sized conjugated systems. Recent examples include reported ER-targeting fluorescent probes for Cu^{2+} and Zn^{2+}. With respect to the Golgi apparatus (GA), selective probes have been described to measure Fe^{2+}, Zn^{2+}, and Cu^{+}.

Although significant advances have been made in the field of organelle-targeting fluorescent probes, significant challenges still remain. It is desirable that future work in this area will translate into direct applications in the drug discovery process and in the medicinal chemistry field.

9.5 Future Perspectives

Fluorescence-based probes and fluorescence imaging have emerged as one of the most powerful techniques to monitor the localization, concentration, and activities of biomolecules within the context of living systems. In this chapter, we have summarized the most important classes of fluorescent probes and their application in the different stages of the drug discovery process, ranging from the initial phases such as hit identification to the final clinical validation of therapies.

Although the tremendous advances experienced by this field in the last few decades are undeniable, significant challenges still lie ahead. For example, further development of probes with better quantum yields in the red and far-red spectral regions, suitable for *in vivo* fluorescence imaging, are needed. Also, labelling strategies aimed at facilitating single-molecule studies would enable the study of receptors in native biological systems with minimal alterations of their native structure. Finally, the design and synthesis of quantitative and multicolour imaging probes is a nascent field, which will make fundamental contributions to the understanding of the complex nature of biology in living systems. All of this progress will require innovations in organic chemistry to enable the synthesis of new fluorophore derivatives with enhanced properties and also in molecular biology to develop improved genetically encoded fluorophores. Hopefully, in the next few years, we will witness how innovative research in this field develops novel tools with clinical applications that facilitate the monitoring of real time changes in the levels of metabolites and proteins of therapeutic interest.

Abbreviations

ABHD10	α,β-hydrolase domain containing protein 10
ABP	activity-based probe
ABPP	activity-based protein profiling
AOMK	acetoxymethyl ketone
Bodipy	boron dipyrromethene
ER	endoplasmic reticulum
FISH	fluorescence *in situ* hybridization
FRET	fluorescence (or Förster) resonance energy transfer
GECI	genetically encoded calcium ion indicator
GEVI	genetically encoded voltage indicator
GPCR	G protein-coupled receptor
IR	infrared
MAO	monoamine oxidase
MTase	methyl transferase
NIR	near-infrared
PAFAH2	type II platelet-activating factor acetylhydrolase
PME-1	protein phosphatase methylesterase-1
QD	quantum dot
SAM	*S*-adenosyl-L-methionine
STS	soft tissue sarcoma

Acknowledgements

Related work in our laboratories was supported by grants from the Spanish Ministerio de Economía y Competitividad (MINECO, SAF2013-48271-R-C02-01 and SAF2016-78792-R).

References

1. L. D. Lavis and R. T. Raines, *ACS Chem. Biol.*, 2014, **9**, 855–866.
2. M. S. Goncalves, *Chem. Rev.*, 2009, **109**, 190–212.
3. H. M. Kim and B. R. Cho, *Chem. Rev.*, 2015, **115**, 5014–5055.
4. T. Terai and T. Nagano, *Curr. Opin. Chem. Biol.*, 2008, **12**, 515–521.
5. B. T. Bajar, E. S. Wang, S. Zhang, M. Z. Lin and J. Chu, *Sensors*, 2016, **16**, DOI: 10.3390/s16091488.
6. R. Bilan, I. Nabiev and A. Sukhanova, *ChemBioChem*, 2016, **17**, 2103–2114.
7. M. Chalfie, Y. Tu, G. Euskirchen, W. W. Ward and D. C. Prasher, *Science*, 1994, **263**, 802–805.
8. E. A. Rodriguez, R. E. Campbell, J. Y. Lin, M. Z. Lin, A. Miyawaki, A. E. Palmer, X. Shu, J. Zhang and R. Y. Tsien, *Trends Biochem. Sci.*, 2016, DOI: 10.1016/j.tibs.2016.1009.1010.
9. P. J. Cranfill, B. R. Sell, M. A. Baird, J. R. Allen, Z. Lavagnino, H. M. de Gruiter, G. J. Kremers, M. W. Davidson, A. Ustione and D. W. Piston, *Nat. Methods*, 2016, **13**, 557–562.

10. D. Yu, M. A. Baird, J. R. Allen, E. S. Howe, M. P. Klassen, A. Reade, K. Makhijani, Y. Song, S. Liu, Z. Murthy, S. Q. Zhang, O. D. Weiner, T. B. Kornberg, Y. N. Jan, M. W. Davidson and X. Shu, *Nat. Methods*, 2015, **12**, 763–765.
11. T. L. To, B. J. Piggott, K. Makhijani, D. Yu, Y. N. Jan and X. Shu, *Proc. Natl. Acad. Sci. U. S. A.*, 2015, **112**, 3338–3343.
12. E. A. Rodriguez, G. N. Tran, L. A. Gross, J. L. Crisp, X. Shu, J. Y. Lin and R. Y. Tsien, *Nat. Methods*, 2016, **13**, 763–769.
13. M. Inoue, A. Takeuchi, S. Horigane, M. Ohkura, K. Gengyo-Ando, H. Fujii, S. Kamijo, S. Takemoto-Kimura, M. Kano, J. Nakai, K. Kitamura and H. Bito, *Nat. Methods*, 2015, **12**, 64–70.
14. H. Dana, B. Mohar, Y. Sun, S. Narayan, A. Gordus, J. P. Hasseman, G. Tsegaye, G. T. Holt, A. Hu, D. Walpita, R. Patel, J. J. Macklin, C. I. Bargmann, M. B. Ahrens, E. R. Schreiter, V. Jayaraman, L. L. Looger, K. Svoboda and D. S. Kim, *eLife*, 2016, **5**, DOI: 10.7554/eLife.12727.
15. Y. Gong, C. Huang, J. Z. Li, B. F. Grewe, Y. Zhang, S. Eismann and M. J. Schnitzer, *Science*, 2015, **350**, 1361–1366.
16. H. H. Yang, F. St-Pierre, X. Sun, X. Ding, M. Z. Lin and T. R. Clandinin, *Cell*, 2016, **166**, 245–257.
17. V. Casado, A. Cortes, J. Mallol, K. Perez-Capote, S. Ferre, C. Lluis, R. Franco and E. I. Canela, *Pharmacol. Ther.*, 2009, **124**, 248–257.
18. A. Rivero-Muller, Y. Y. Chou, I. Ji, S. Lajic, A. C. Hanyaloglu, K. Jonas, N. Rahman, T. H. Ji and I. Huhtaniemi, *Proc. Natl. Acad. Sci. U. S. A.*, 2010, **107**, 2319–2324.
19. H. F. Vischer, M. Castro and J. P. Pin, *Mol. Pharmacol.*, 2015, **88**, 561–571.
20. E. Moreno, C. Andradas, M. Medrano, M. M. Caffarel, E. Perez-Gomez, S. Blasco-Benito, M. Gomez-Canas, M. R. Pazos, A. J. Irving, C. Lluis, E. I. Canela, J. Fernandez-Ruiz, M. Guzman, P. J. McCormick and C. Sanchez, *J. Biol. Chem.*, 2014, **289**, 21960–21972.
21. G. Navarro, C. Quiroz, D. Moreno-Delgado, A. Sierakowiak, K. McDowell, E. Moreno, W. Rea, N. S. Cai, D. Aguinaga, L. A. Howell, F. Hausch, A. Cortes, J. Mallol, V. Casado, C. Lluis, E. I. Canela, S. Ferre and P. J. McCormick, *J. Neurosci.*, 2015, **35**, 6639–6653.
22. X. Vinals, E. Moreno, L. Lanfumey, A. Cordomi, A. Pastor, R. de La Torre, P. Gasperini, G. Navarro, L. A. Howell, L. Pardo, C. Lluis, E. I. Canela, P. J. McCormick, R. Maldonado and P. Robledo, *PLoS Biol.*, 2015, **13**, e1002194.
23. H. Tian, A. Furstenberg and T. Huber, *Chem. Rev.*, 2016, DOI: 10.1021/acs.chemrev.1026b00084.
24. Z. Ma, L. Du and M. Li, *J. Med. Chem.*, 2014, **57**, 8187–8203.
25. A. J. Vernall, S. J. Hill and B. Kellam, *Br. J. Pharmacol.*, 2014, **171**, 1073–1084.
26. D. Alonso, H. Vázquez-Villa, A. M. Gamo, M. F. Martínez-Esperón, M. Tortosa, A. Viso, R. Fernández de la Pradilla, E. Junquera, E. Aicart, M. Martín-Fontecha, B. Benhamú, M. L. López-Rodríguez and S. Ortega-Gutiérrez, *ACS Med. Chem. Lett.*, 2010, **1**, 249–253.

27. A. M. Gamo, J. A. Gonzalez-Vera, A. Rueda-Zubiaurre, D. Alonso, H. Vazquez-Villa, L. Martin-Couce, O. Palomares, J. A. Lopez, M. Martin-Fontecha, B. Benhamu, M. L. Lopez-Rodriguez and S. Ortega-Gutierrez, *Chemistry*, 2016, **22**, 1313–1321.
28. L. Martin-Couce, M. Martin-Fontecha, S. Capolicchio, M. L. Lopez-Rodriguez and S. Ortega-Gutierrez, *J. Med. Chem.*, 2011, **54**, 5265–5269.
29. L. Martin-Couce, M. Martin-Fontecha, O. Palomares, L. Mestre, A. Cordomi, M. Hernangomez, S. Palma, L. Pardo, C. Guaza, M. L. Lopez-Rodriguez and S. Ortega-Gutierrez, *Angew. Chem., Int. Ed.*, 2012, **51**, 6896–6899.
30. M. Martin-Fontecha, T. Eiwegger, T. Jartti, A. Rueda-Zubiaurre, K. Tiringer, J. Stepanow, T. Puhakka, B. Ruckert, S. Ortega-Gutierrez, M. L. Lopez-Rodriguez, M. Akdis, C. A. Akdis and O. Palomares, *J. Allergy Clin. Immunol.*, 2014, **133**, 926–929.
31. H. Vázquez-Villa, J. A. González-Vera, B. Benhamú, A. Viso, R. Fernández de la Pradilla, E. Junquera, E. Aicart, M. L. López-Rodríguez and S. Ortega-Gutiérrez, *J. Med. Chem.*, 2010, **53**, 7095–7106.
32. F. Ciruela, K. A. Jacobson and V. Fernandez-Duenas, *ACS Chem. Biol.*, 2014, **9**, 1918–1928.
33. L. A. Stoddart, A. J. Vernall, J. L. Denman, S. J. Briddon, B. Kellam and S. J. Hill, *Chem. Biol.*, 2012, **19**, 1105–1115.
34. L. S. Schembri, L. A. Stoddart, S. J. Briddon, B. Kellam, M. Canals, B. Graham and P. J. Scammells, *J. Med. Chem.*, 2015, **58**, 9754–9767.
35. F. J. Cespedes-Guirao, A. B. Ropero, E. Font-Sanchis, A. Nadal, F. Fernandez-Lazaro and A. Sastre-Santos, *Chem. Commun.*, 2011, **47**, 8307–8309.
36. R. R. Petrov, M. E. Ferrini, Z. Jaffar, C. M. Thompson, K. Roberts and P. Diaz, *Bioorg. Med. Chem. Lett.*, 2011, **21**, 5859–5862.
37. M. Bai, M. Sexton, N. Stella and D. J. Bornhop, *Bioconjugate Chem.*, 2008, **19**, 988–992.
38. E. Schneider, M. Keller, A. Brennauer, B. K. Hoefelschweiger, D. Gross, O. S. Wolfbeis, G. Bernhardt and A. Buschauer, *ChemBioChem*, 2007, **8**, 1981–1988.
39. M. Srinivasarao, C. V. Galliford and P. S. Low, *Nat. Rev. Drug Discovery*, 2015, **14**, 203–219.
40. J. Parrish-Novak, E. C. Holland and J. M. Olson, *Cancer J.*, 2015, **21**, 206–212.
41. G. M. van Dam, G. Themelis, L. M. Crane, N. J. Harlaar, R. G. Pleijhuis, W. Kelder, A. Sarantopoulos, J. S. de Jong, H. J. Arts, A. G. van der Zee, J. Bart, P. S. Low and V. Ntziachristos, *Nat. Med.*, 2011, **17**, 1315–1319.
42. X. Chen, Y. K. Wong, J. Wang, J. Zhang, Y. M. Lee, H. M. Shen, Q. Lin and Z. C. Hua, *Proteomics*, 2016, DOI: 10.1002/pmic.201600212.
43. L. E. Sanman and M. Bogyo, *Annu. Rev. Biochem.*, 2014, **83**, 249–273.
44. M. J. Niphakis and B. F. Cravatt, *Annu. Rev. Biochem.*, 2014, **83**, 341–377.
45. Y. Liu, M. P. Patricelli and B. F. Cravatt, *Proc. Natl. Acad. Sci. U. S. A.*, 1999, **96**, 14694–14699.

46. D. Greenbaum, K. F. Medzihradszky, A. Burlingame and M. Bogyo, *Chem. Biol.*, 2000, **7**, 569–581.
47. P. Yang and K. Liu, *ChemBioChem*, 2015, **16**, 712–724.
48. K. E. Wilke, S. Francis and E. E. Carlson, *J. Am. Chem. Soc.*, 2012, **134**, 9150–9153.
49. K. E. Wilke, S. Francis and E. E. Carlson, *ACS Chem. Biol.*, 2015, **10**, 328–335.
50. J. M. Krysiak, J. Kreuzer, P. Macheroux, A. Hermetter, S. A. Sieber and R. Breinbauer, *Angew. Chem., Int. Ed.*, 2012, **51**, 7035–7040.
51. L. Li, C. W. Zhang, J. Ge, L. Qian, B. H. Chai, Q. Zhu, J. S. Lee, K. L. Lim and S. Q. Yao, *Angew. Chem., Int. Ed.*, 2015, **54**, 10821–10825.
52. J. W. Chang, R. E. Moellering and B. F. Cravatt, *Angew. Chem., Int. Ed.*, 2012, **51**, 966–970.
53. Y. Qian, M. Schurmann, P. Janning, C. Hedberg and H. Waldmann, *Angew. Chem., Int. Ed.*, 2016, **55**, 7766–7771.
54. J. Li and P. R. Chen, *Nat. Chem. Biol.*, 2016, **12**, 129–137.
55. H. W. Shih, D. N. Kamber and J. A. Prescher, *Curr. Opin. Chem. Biol.*, 2014, **21**, 103–111.
56. D. M. Patterson, L. A. Nazarova and J. A. Prescher, *ACS Chem. Biol.*, 2014, **9**, 592–605.
57. J. J. Hulce, C. Joslyn, A. E. Speers, S. J. Brown, T. Spicer, V. Fernandez-Vega, J. Ferguson, B. F. Cravatt, P. Hodder and H. Rosen, *Probe Reports from the NIH Molecular Libraries Program*, National Center for Biotechnology Information (US), Bethesda (MD), 2010.
58. C. J. Dreyton, J. E. Jones, B. A. Knuckley, V. Subramanian, E. D. Anderson, S. J. Brown, V. Fernandez-Vega, C. Eberhart, T. Spicer, A. M. Zuhl, J. Ferguson, A. E. Speers, C. Wang, D. L. Boger, P. Thompson, B. F. Cravatt, P. Hodder and H. Rosen, *Probe Reports from the NIH Molecular Libraries Program*, National Center for Biotechnology Information (US), Bethesda (MD), 2010.
59. A. M. Zuhl, J. T. Mohr, A. E. Speers, D. A. Bachovchin, J. M. Berlin, T. Spicer, V. Fernandez-Vega, S. J. Brown, J. Ferguson, G. C. Fu, B. F. Cravatt, P. Hodder and H. Rosen, *Probe Reports from the NIH Molecular Libraries Program*, National Center for Biotechnology Information (US), Bethesda (MD), 2010.
60. A. Adibekian, K. L. Hsu, A. E. Speers, E. S. Monillas, S. J. Brown, T. Spicer, V. Fernandez-Vega, J. Ferguson, B. J. Bahnson, B. F. Cravatt, P. Hodder and H. Rosen, in *Probe Reports from the NIH Molecular Libraries Program*, National Center for Biotechnology Information (US), Bethesda (MD), 2010.
61. M. J. Whitley, D. M. Cardona, A. L. Lazarides, I. Spasojevic, J. M. Ferrer, J. Cahill, C. L. Lee, M. Snuderl, D. G. Blazer, 3rd, E. S. Hwang, R. A. Greenup, P. J. Mosca, J. K. Mito, K. C. Cuneo, N. A. Larrier, E. K. O'Reilly, R. F. Riedel, W. C. Eward, D. B. Strasfeld, D. Fukumura, R. K. Jain, W. D. Lee, L. G. Griffith, M. G. Bawendi, D. G. Kirsch and B. E. Brigman, *Sci. Transl. Med.*, 2016, **8**, 320ra324.

62. K. Rombouts, K. Braeckmans and K. Remaut, *Bioconjugate Chem.*, 2016, **27**, 280–297.
63. M. Tomkuviene, E. Kriukiene and S. Klimasauskas, *Adv. Exp. Med. Biol.*, 2016, **945**, 511–535.
64. A. Plotnikova, A. Osipenko, V. Masevicius, G. Vilkaitis and S. Klimasauskas, *J. Am. Chem. Soc.*, 2014, **136**, 13550–13553.
65. D. Gillingham and R. Shahid, *Curr. Opin. Chem. Biol.*, 2015, **25**, 110–114.
66. J. Zheng, R. Yang, M. Shi, C. Wu, X. Fang, Y. Li, J. Li and W. Tan, *Chem. Soc. Rev.*, 2015, **44**, 3036–3055.
67. T. Kuang, L. Chang, X. Peng, X. Hu and D. Gallego-Perez, *Trends Biotechnol.*, 2016, DOI: 10.1016/j.tibtech.2016.1009.1003.
68. T. Vaijayanthi, T. Bando, G. N. Pandian and H. Sugiyama, *ChemBioChem*, 2012, **13**, 2170–2185.
69. K. Nozeret, F. Loll, C. Escude and A. S. Boutorine, *ChemBioChem*, 2015, **16**, 549–554.
70. J. S. Lee, N. Y. Kang, Y. K. Kim, A. Samanta, S. Feng, H. K. Kim, M. Vendrell, J. H. Park and Y. T. Chang, *J. Am. Chem. Soc.*, 2009, **131**, 10077–10082.
71. D. Zhai, S. C. Lee, M. Vendrell, L. P. Leong and Y. T. Chang, *ACS Comb. Sci.*, 2012, **14**, 81–84.
72. L. Y. Niu, Y. S. Guan, Y. Z. Chen, L. Z. Wu, C. H. Tung and Q. Z. Yang, *Chem. Commun.*, 2013, **49**, 1294–1296.
73. L. Y. Niu, Y. S. Guan, Y. Z. Chen, L. Z. Wu, C. H. Tung and Q. Z. Yang, *J. Am. Chem. Soc.*, 2012, **134**, 18928–18931.
74. C. Zeng, X. Zhang and L. Pu, *Chemistry*, 2016, DOI: 10.1002/chem.201605028.
75. L. Y. Niu, Y. Z. Chen, H. R. Zheng, L. Z. Wu, C. H. Tung and Q. Z. Yang, *Chem. Soc. Rev.*, 2015, **44**, 6143–6160.
76. L. Xin, Y. Z. Chen, L. Y. Niu, L. Z. Wu, C. H. Tung, Q. X. Tong and Q. Z. Yang, *Org. Biomol. Chem.*, 2013, **11**, 3014–3019.
77. K. Yamada, F. Mito, Y. Matsuoka, S. Ide, K. Shikimachi, A. Fujiki, D. Kusakabe, Y. Ishida, M. Enoki, A. Tada, M. Ariyoshi, T. Yamasaki and M. Yamato, *Nat. Chem. Biol.*, 2016, **12**, 608–613.
78. S. Y. Zhang, W. Shi, P. Cheng and M. J. Zaworotko, *J. Am. Chem. Soc.*, 2015, **137**, 12203–12206.
79. S. Mukherjee, V. Jansen, J. F. Jikeli, H. Hamzeh, L. Alvarez, M. Dombrowski, M. Balbach, T. Strunker, R. Seifert, U. B. Kaupp and D. Wachten, *eLife*, 2016, **5**, DOI: 10.7554/eLife.14052.
80. A. Masharina, L. Reymond, D. Maurel, K. Umezawa and K. Johnsson, *J. Am. Chem. Soc.*, 2012, **134**, 19026–19034.
81. W. Xu, Z. Zeng, J. H. Jiang, Y. T. Chang and L. Yuan, *Angew. Chem., Int. Ed.*, 2016, **55**, 13658–13699.

CHAPTER 10

Transmission Cryo-electron Microscopy in Drug Discovery

CARLOS FERNÁNDEZ RODRÍGUEZ,[a] OSCAR LLORCA*[b] AND FABRIZIO MARTINO*[a]

[a] Centro de Investigaciones Biológicas – CIB, C/Ramiro de Maeztu 9 28040, Madrid, Spain; [b] Spanish National Cancer Research Centre (CNIO) Structural Biology Programme, Melchor Fernández Almagro 3 28029, Madrid, Spain
*Email: ollorca@cnio.es; fabriziom@cib.csic.es

10.1 Introduction

Transmission electron microscopy (TEM) has been used to study the three dimensional structure of biological assemblies for more than forty years. In the early seventies, Aaron Klug and colleagues used transmission electron microscopy to investigate the structure of viruses.[1] In TEM, the image of the sample of interest is formed in the microscope from the electrons scattered from the sample. Biological samples are very fragile and get easily damaged when exposed to the electron beam of the microscope. In order to reduce the damage induced by the electron beam on the sample, the sample is frozen and embedded into a thin layer of ice.[2] To further reduce damage to the sample, electron microscopy imaging is performed in low dose conditions where images are taken with less than 40 electrons per Å^2 per second.[2] This generates images with a very low signal-to-noise ratio. The signal comes from the biological molecules of interest and the noise comes from the inelastically scattered electrons produced when the beam passes through the frozen solvent that surrounds the molecules and the hardware of the detection apparatus. In order to increase the signal of the molecule of interest over the noise generated by the inelastic scattering of the electrons in the ice, people developed the so called "single particle" technique.[3] Assuming that the noise is randomly distributed in the image, in order to solve the structure of a 3D biological molecule, one has to collect images of millions of molecules of the same object and superimpose them in order to increase the signal coming from the object and subtract the noise. Until very recently, close-to-atomic-resolution cryo-EM structures were restricted to highly symmetrical biological assemblies such as viruses and filaments, and only the structure of very large asymmetric biological machineries, such as ribosomes, could be solved at resolutions at best between 5 Å and 6 Å. Protein complexes smaller than 1 MDa in molecular weight (MW) could only be analysed at resolutions worse than 10 Å. Recent advances in cryo-EM have now allowed scientists to investigate the structure of asymmetric biological protein complexes of almost any size at close-to-atomic resolution.

We will now briefly discuss resolution. First of all, for historical reasons, the EM community uses Å to measure resolution instead of metres. It is important to talk about resolution because in order to visualise protein alpha helixes, it is necessary to obtain electron density maps with resolutions better than 6 Å; to resolve beta-sheets, it is necessary to reach resolutions better than 4.5 Å; and to be able to recognise protein side chains, one needs to obtain electron density maps with a resolution close to 3.0 Å. The better the resolution, the more defined the details of the electron density maps and the more accurate the interpretation of such maps becomes. Electron density maps generated by cryo-EM usually present a gradient of resolution with parts that are better defined than others. The nominal resolution that is typically used to define the overall resolution is not the absolute resolution of the electron density map, but rather represents the cross-correlation between two independent 3D models of the same density map in Fourier space. 3D cryo-EM maps usually present resolution ranges better and worse than the nominal average resolution reported. So, structures that have a nominal resolution of 4 Å often present regions that have resolutions of around 3 Å, which can be modelled with a quite satisfactory degree of accuracy. This is the minimum target resolution that we would need to achieve to perform structure based drug design. In the rest of the chapter "high-resolution" structures will refer to structures with a resolution better than 4 Å.

10.2 Advances in Cryo-electron Microscopy

The development of Direct Electron Detectors (DEDs) has been one of the major advances in cryo-EM in recent years. DEDs are complementary metal-oxide-semiconductor (CMOS)-based cameras that are able to directly record electrons. In cameras that use older technology, electrons are converted into photons upon hitting a scintillating plate, and the photons are recorded by

CMOS or Charged Coupled Device (CCD) detectors. The signal recorded from these detectors had a poor signal-to-noise ratio due to the scattering of the electrons on the fluorescent plate and in the following conversion into signal. Direct electron detectors can directly acquire electrons without the need of converting them into photons. The images produced by DEDs present, therefore, a better signal-to-noise ratio compared to images produced with previous technology cameras.[4,5] The new generation DEDs record individual electron events at the time that they reach the detector.[6] This acquisition mode is referred to as the "counting mode" and increases the Detector Quantum Efficiency (DQE) of the cameras, producing images with very high signal-to-noise ratios.

Another advantage of modern DEDs, compared to detectors with older technology, is that DEDs can record movie frames. Each image acquired is fractionated into sub-frames. The K2 Summit® detector (GATAN) has a read out of 400 Frames Per Second (FPS), even though for most of its applications, only 40 FPS are recorded. Scientists are therefore able to collect images with long exposures (*i.e.*, up to 8 sec) and dose as low as one electron per pixel. The user can then align the movie frames *in silico* and correct the images for particle movement caused by the drift of the microscope grid holder and by the vibrations induced by the incident beam on the ice. Alignment of movie frames is very important in order to get high resolution images.[6,7] Another advantage of collecting frames of movies is that the total exposure dose is distributed through the movie frames and the user can sum frames with a desired total dose. Different steps of the processing of cryo-EM data require different doses. For example, picking and CTF correction might work better at high doses (*i.e.*, 40–50 e/A^2), 3D classification and removal of bad particles usually require medium to high doses (*i.e.*, 30–40 e/A^2) and high-resolution information can be obtained with as little as 2 e/A^2.[8] The decision of what total dose to use can be made post-data acquisition and during the processing procedure, and changed as necessary. This is a great advantage considering that data acquisition is limited by the availability of high-end microscopes. DEDs are an essential component of modern electron microscopes. Nevertheless, this technology is still quite costly and DEDs can comprise a considerable portion of the total cost of an electron microscope. A big effort has been made in reducing the thickness of the parts that are hit by electrons and, at the same time, maintaining the efficiency of detection. New and improved DEDs are likely to be released in the near future.

Automation of data acquisition is another important improvement in cryo-EM. It is now possible to collect data for several consecutive days almost without interruption. Modern microscopes can now collect up to 1000 images in 24 hours. An experienced user would need about three working days to achieve the same target. One important improvement in automation has been the development of automatic loading systems of cryo-EM specimens into the microscope. One of the most advanced automatic sample loading systems in electron microscopy is produced by the FEI Company and it is called the Autoloader™. This system allows users to load up to 12 cryo-EM grids into the microscope. The grids can then be exchanged in a fast and highly controlled manner that reduces ice contamination during the process. In microscopes equipped with automatic loading systems, the hardware that holds the grid in the electron beam is sealed in a closed and controlled environment inside the microscope. This minimises the temperature differences between the different components of the microscope and generates an improved stability of the stage that holds the sample in the electron beam. Microscopes that have an automatic loading system present less drift than older generation microscopes equipped with a side entry holder. Automatic loading systems also allow minimal user intervention reducing the training required to operate the microscope. Another advantage of the auto-loading system is that the user can screen grids and choose the best ones for collecting data with minimal time loss. The auto-loading systems of modern microscopes have not reached the efficiency levels of the ones used for X-ray crystallography beam lines, but they definitely represent a big step forward in cryo-EM data collection.

The cryo-EM field really misses a semi-automated system to assess the quality of the samples. At present, any quality assessment of the samples requires several hours, if not days, of analysis. The Stahlberg group developed a new and very promising software package that goes in this direction.[9] Even though this is a very encouraging step, a lot more still has to be done on this issue.

In order to recover phase information from the electrons scattered by the object, microscopists usually collect data at negative defocus. The more defocused the image, the higher the dumping of the high resolution information of the objects we image. Objects larger than 1 MDa MW, such as ribosomes or viruses, have acceptable contrast, even when imaged very close to focus. For objects smaller than 1 MDa MW, it is more difficult to combine contrast and resolution. To overcome this limitation, FEI, together with Radostin Danev at the Max Planck Institute of Biochemistry in Martinsried, developed the so called Volta-Phase plate.[10] The Volta-Phase plate creates phase contrast of the images at focus, allowing the high-resolution structures of very small objects, such as hemoglobin (44 kDa MW), to be solved.[11] The usage of such hardware is not routine and still requires very experienced users, but it will probably soon become more standard.

Poor beam coherence deteriorates high-resolution information even when images are taken close to focus. The use of a new generation of field emission guns (FEG) and condenser lenses allows more accurate control of the electron beam. Indeed, modern microscopes can produce an electron beam that is at the same time very focused, highly parallel, highly stable and highly coherent. This allows irradiation of an area as large as the detector size, reducing the radiation damage of the sample and allowing the acquisition of images in areas that are very close to each other. This speeds up the data acquisition process of a factor of up to 5×.

10.3 Examples of High-resolution Cryo-EM Structures Suitable for Drug Discovery

Cryo-EM provided medium to high resolution structures of macromolecular biological complexes even before the advent of DEDs. Cryo-EM applied to 2-D crystals provided the high resolution structure of Bacteriorhodopsin, first in 1990 and later in 2000.[12,13] In 2001, the Nogales lab described the binding site of taxol on tubulin from electron crystallography data.[14]

Highly symmetric biological assemblies such as viruses and actin or tubulin filaments have the advantage of being both large and symmetric. Bottcher and Chrowder solved the structure of hepatitis B virus at 7.4 Å resolution in 1997.[15] The advent of field emission guns as a better electron source than filaments allowed scientists to solve virus structures at quasi atomic resolution without the need for DEDs. Zhou and colleagues solved the structure of the polyhedrosis virus in 2008.[16] Other structures include the structure of human adenovirus,[17] the structure of the Sputnik virophage[18] and the structure of mature dengue virus.[19]

Early works on F-actin and tubulin reached resolutions that were better than 6 Å, even without DEDs. At this resolution the alpha helixes of the proteins become visible allowing a certain degree of structure modelling.[20–23] While DEDs were dispensable for solving structures of viruses at near atomic resolution, they proved to be necessary to achieve high-resolution structures of helical filaments such as tubulin.[24]

It is only since the advent of DEDs that complexes smaller than 1 MDa MW with little symmetry could be solved at resolutions better than 4 Å. Still, at present, only very few structures have been solved at resolutions better than 3 Å, the resolution that is required to unambiguously model the side chains of amino acids and the functional groups of small compounds. Among the structures that have been solved at a resolution better than 3 Å by cryo-EM, we should mention lactate dehydrogenase, a tetrameric protein of 145 kDa that was solved at 2.8 Å resolution by cryo-EM.[25] This structure revealed specific and well defined density for an inhibitor that was not modelled in a previous structure solved by X-ray crystallography. To the best of our knowledge, the first protein smaller than 100 kDa whose structure was solved at a resolution of around 2 Å is p97.[26] Subramaniam and colleagues solved the structure of full length p97 alone and bound to several small molecule inhibitors. The cryo-EM structure of p97 revealed specific interactions between the different domains of p97 and the drugs used in this study, especially at the interfaces between domains. This important information could not be retrieved from structures solved by X-ray crystallography because they were restricted to individual domains. This work is also a good example of how cryo-EM can study and resolve large movements between protein subunits. Indeed, the authors were able to solve three distinct conformations present during the ATPase reaction cycle. The cryo-EM structure of p97 is a very good example of how cryo-EM and X-ray crystallography can complement each other in structure-based drug design.

The structure of Beta galactosidase was solved at a slightly better resolution than the structure of p97.[27] The 2.2 Å structure of beta galactosidase showed a water molecule interacting with an inhibitor located in the active site of the enzyme, showing a level of detail comparable to what can be obtained by X-ray crystallography.

The highest resolution achieved so far by cryo-EM was achieved by the Subramaniam group with the structure of glutamate dehydrogenase solved at 1.8 Å resolution.[25] The authors were able to assign most of the protein side chains and most of the water molecules of the ordered inner core unambiguously. The quality of the map is so high that some of the rings of the aromatic amino acid present characteristic holes in electron density in the middle of the ring.

The smallest protein whose structure has been solved at about 3 Å resolution is the hemoglobin that is formed by 4 polypeptides of about 15 kDa each for a total of 64.5 kDa molecular weight.[11] This structure is a milestone in cryo-EM because it showed that cryo-EM can be used with biological molecules of any size and, therefore, that cryo-EM in principle could solve the structure of any protein. It is important to point out that the technology of the phase plate used in this work has not yet been implemented in an automated user-friendly workflow and therefore using such technology requires highly trained and skilled operators.

Interestingly, most of the structures that have been solved by cryo-EM at resolutions of better than 3 Å come from proteins that were already crystallized. These proteins are quite rigid, they can be easily produced and they present high stability in solution. These observations, and the limited number of structures published with resolutions of around or better than 3 Å published so far, strongly support the fact that the bottleneck of cryo-EM is still the preparation of the sample. Indeed, most biological samples are very flexible, poorly expressed and poorly stable.

The structure of the *P. falciparum* proteasome is another clear example of how cryo-EM can be employed in drug design (Figure 10.1).[28] In this work, the authors isolated a peptide motif with very strong inhibitory activity specifically for the 20S proteasome of the parasite and not for the human one. The authors solved the structure of the *P. falciparum* 20S proteasome in complex with the optimised inhibitor (Figure 10.1, right panels). The electron density map of the inhibitor allowed the authors to model the inhibitor unambiguously and the structure was used to clearly explain the specificity of the inhibitor for Plasmodium 20S and the low affinity for human 20S. The authors then tested the selected inhibitor in cells and infected mice; they observed almost complete reduction of the parasite burden following treatment and good tolerance to the drug. The authors showed that the selected inhibitor could be used to sensitise the parasite to traditional drugs towards which the parasite had been described as developing resistance. This work is one of the few examples of how cryo-EM can be used in structure-based drug design. It is important to point out that the structural investigation of proteasome has so far been restricted to human and yeast

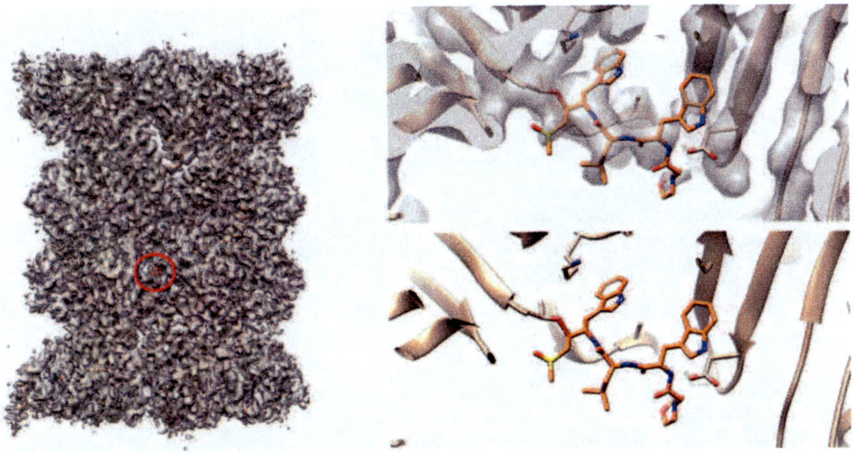

Figure 10.1 Electron density map and atomic model of the 20s proteasome from *P. falciparum* bound to the inhibitor Mor-WLW-sulphone highlighted by the red circle. Details of electron density map and atomic model of the inhibitor are presented in the right panels.

types, whereas cryo-EM opens up the opportunity of investigating the proteasome structure of, in principle, any pathogen.

Ribosomes have been extensively studied by cryo-EM to such an extent that they account for almost 20% of the structures deposited in the EM database. A few of the ribosome structures published so far might have a high impact in curing diseases caused by the parasites *P. falciparum* and *L. donovani.*[29–31] In 2014, Wong and colleagues showed, for the first time, the molecular mechanism of how the drug emetine inhibits the activity of the 80s ribosome of *P. falciparum.*[29] In 2017, Wong and colleagues published the structure of the 80s ribosome of *P. falciparum* bound to the drug mefloquine (Figure 10.2).[30] Shalev-Benami and colleagues solved the structure of the large ribosomal subunit of *L. donovani* parasite.[31] This structure will be very useful in the design and better understanding of future drugs that target the leismania ribosome because it revealed important differences between the parasite and the human ribosome. Cryo-EM is better suited than X-ray crystallography to study the structure of protein complexes from organisms such as parasites. This is because purification of proteins from organisms that present complex life cycles is very challenging and the amount of material purified from such organisms is often not sufficient for X-ray crystallography. Therefore, cryo-EM could be a very powerful tool to perform the structure-based drug design of neglected diseases.

Membrane proteins are a very attractive target for drug discovery. Even though X-ray crystallography is a well-established technique for investigating the structure of membrane proteins, obtaining enough material and sufficiently diffracting crystals of such proteins is very challenging and it is a bottle neck that has limited the number of structures solved so far. Cryo-EM

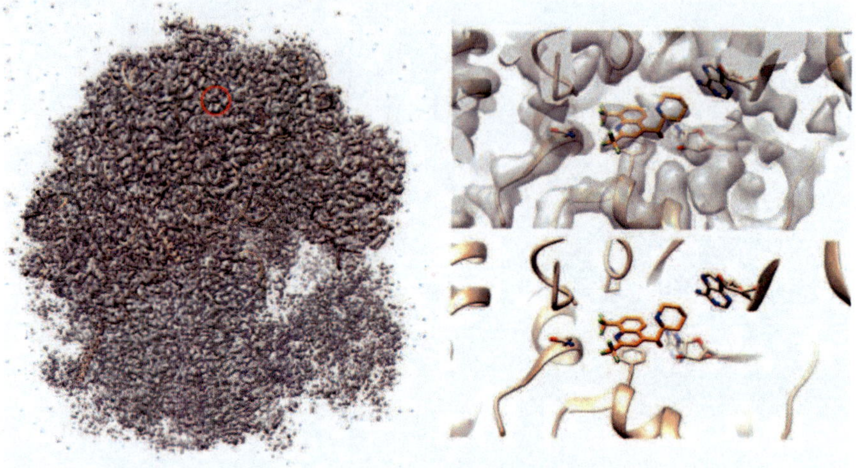

Figure 10.2 Electron density map and atomic model of the 80s ribosome from *P. falciparum* bound to mefloquine highlighted by the red circle. Details of the mefloquine binding pocket are shown in the right panels.

might substitute X-ray crystallography in the structure determination of membrane proteins. The amount of material needed for cryo-EM is lower than that needed for X-ray crystallography. Moreover, cryo-EM can accommodate and resolve a certain degree of flexibility that often describes physiological states of membrane proteins, such as the opening and closing of a channel. Flexibility often prevents proteins from crystallizing, however, it can itself be very limited by the packing of the protein into crystal lattices.

The *trans*-membrane acetylcholine receptor was one of the first structures to be solved at high-resolution by cryo-EM. Unwin and colleagues were able to trace secondary structure features of the acetylcholine receptor in 2003.[32] The images were collected on films and the processing took advantage of the helical arrangement of the receptor. This work was a breakthrough in the EM field and, only 10 years later, with the advent of DEDs, more cryo-EM high-resolution structures became available.

The structure of the gamma secretase was solved at 3.4 Å.[33] This structure is a milestone in the structural biology of membrane proteins for several reasons. First of all, gamma secretase is medically very interesting because dysfunction of this protein is linked with Alzheimer's disease. Secondly, the protein was expressed and purified from mammalian cells in amounts that would not be sufficient for crystallization. Moreover, gamma secretase was one of the first proteins smaller than 200 kDa whose structure was solved at near-atomic resolution by cryo-EM. In order to obtain a map suited for *de novo* modelling of the atomic structure of gamma secretase, the authors implemented a method for the *in silico* classification of heterogeneity that proved to be useful for other structures solved later.[33,34] The authors solved three different states that could be stabilised by different inhibitors.

The TRPV1 channel has been extensively studied by cryo-EM.[35–37] TRPV1 is the receptor for capsaicin, the pungent agent from chili peppers, and has a widely validated role in pain physiology. Indeed, TRPV1 and other somatosensory TRP channels are considered important targets for analgesic drugs. In 2013, the groups of Julius and Cheng obtained structures of TRPV1 in complex with capsaicin or a combination of the agonist resinferatoxin and the double knot toxin (Figure 10.3).[35,36] Although the nominal resolution was 3.4 Å, the electron density maps describing the compounds were quite poor and allowed only a rough positioning of the compounds (Figure 10.3, top right panel). In 2016, the same groups published the same structures with greatly improved densities for both the protein and the drugs (Figure 10.3, bottom right panel).[37] The 2016 structures showed the exact location of the small molecules bound to the protein. Interestingly, the presence of phospholipids in their preparation allowed the authors to propose a model of how the protein interacts with membranes. The TRPV1 cryo-EM structures discussed above are a clear example of how sample preparation can affect the final quality of electron density maps generated from cryo-EM. Indeed, when TRPV1 was embedded into nanodiscs the overall electron density maps improved so much that almost all of the side chains of the proteins were visible (Figure 10.3, top right panel). Similarly, the densities for the small molecules bound to the protein were clearly visible only when nanodiscs were used (Figure 10.3, bottom right panel).

The cryo-EM structure of the multidrug resistance protein MRP1 is one of the latest structures to be released in the EMDB (electron microscopy data bank).[38] MRP1 is an ATP-binding cassette transporter that confers cell resistance to a broad spectrum of drugs such as antidepressants, opiates and antibiotics. The structure of MRP1 presented in this work was solved with

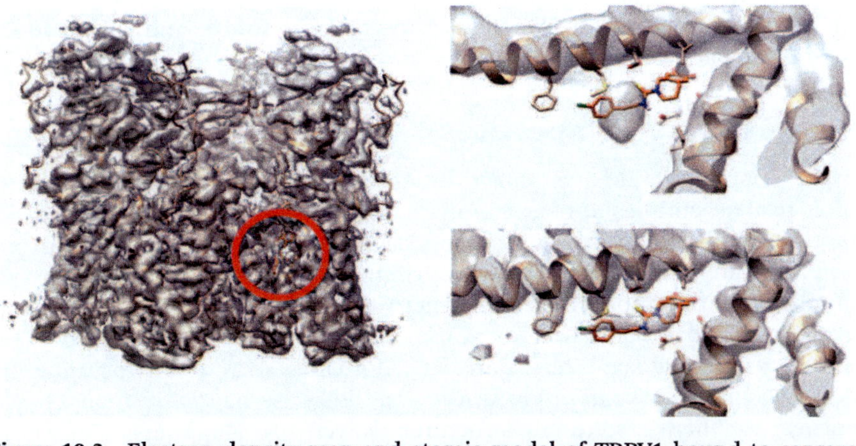

Figure 10.3 Electron density map and atomic model of TRPV1 bound to capsazepine highlighted by the red circle. Right top panel: detail of capsazepine from TRPV1 purified with amphipols. Right bottom panel: detail of capsazepine from TRPV1 purified with nanodiscs.

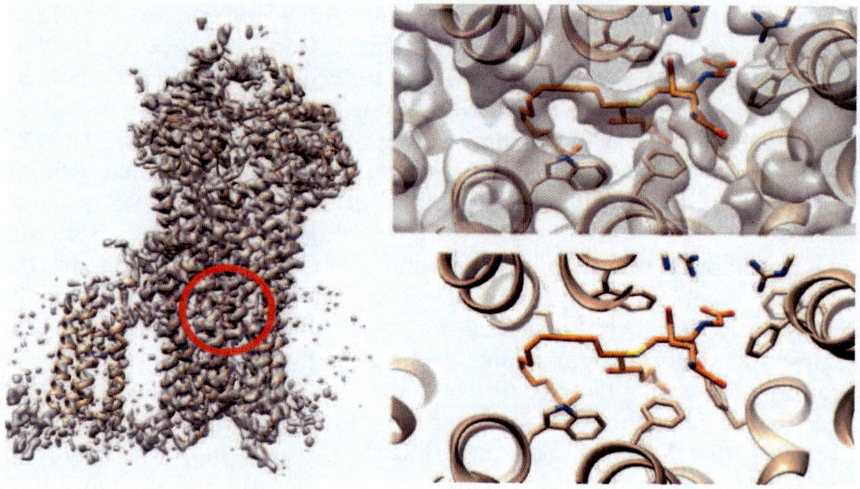

Figure 10.4 Electron density map and atomic model of MRPV1 bound to leucotriene C4 highlighted by the red circle. Details of the leucotriene C4 binding pocket are shown in the right panels.

some MRP1 substrates, such as leucotriene C4 (Figure 10.4, right panels), and it provides an explanation of why the same protein can recognise many different molecules. This work is not only important because it explains how MRP1 can recognise multiple substrates, but also because it confirms that cryo-EM can provide electron density maps that can be used to unambiguously model the atomic structures of not only proteins, but also small molecules. Because of the high medical relevance of membrane proteins and because of the advantages that cryo-EM presents over X-ray crystallography, it is feasible to believe that more and more cryo-EM structures of membrane proteins will be released in the near future and that cryo-EM might overtake X-ray crystallography in this field.

10.4 Future Perspectives

Even though cryo-EM has shown much potential in solving structures of biological assemblies at near atomic resolution, it is important to point out that obtaining structures at resolutions high enough to *de novo* build proteins and small molecules is not yet routine. The articles published so far demonstrate that modern electron microscopes have the right technology to achieve this goal, but if we look at the number of structures released in the EMDB, we clearly see that we are far from what X-ray crystallography has achieved in recent years. Indeed, the PDB database has more than 115 000 entries for high-resolution structures solved by X-ray crystallography, whereas there are fewer than 1500 solved by cryo-EM. So, we are facing an order of magnitude difference. This large difference is likely explained by the fact that high-resolution single particle cryo-EM has become possible only since the advent of DEDs in 2013. Only 5 structures with better than 4 Å resolution were published in 2013, and the number has increased slowly from 16 in 2014, to 79 in 2015 and 215 in 2016.

Even though people showed that it is possible, it is fair to admit that there are several issues that limit the use of cryo-EM on a routine basis. The high cost of purchasing and maintaining high-end microscopes and the microscopy hardware, such as DEDs, is the most limiting issue with the technique. The high cost and slow development of the microscopes are also linked to the fact that the production of such machines is in a quasi monopoly situation with only 2 companies controlling 100% of the market. The newest developments in the technique and the new interest from pharma-companies will hopefully introduce new players into the market, which will hopefully boost the research and development in cryo-EM and decrease the costs of purchasing and maintaining the microscopes.

The problem of high costs can be partially overcome by the creation of central facilities such as the eBIC at the Diamond Light Source and the Cryo-Electron Microscopy Platform for Structural Biology at the ESRF. More facilities are appearing and it is likely that more facilities will grow in the near future.

Automation is definitely needed to boost progress in the field. Almost any aspect of cryo-EM needs better automation. At present, there is a complete lack of high-throughput technology for the preparation of EM grids. The preparation of the grids still relies on a very old technique that is based on blotting an excess of sample placed on the grid by mechanical pressure with two pads of filter paper. A few labs are working on modern techniques that make use of nanocapillaries or nanosprays to deposit the sample on the grid.[39,40] Being able to control the thickness of the ice and the particle distribution in the ice is one of the pitfalls of cryo-EM.

There is also a lack of high-throughput or semi high-throughput technologies to screen for grid quality. The most modern microscope can load up to 12 grids at a time, but there is no software that can help the user in the evaluation of the quality of the grids. Screening different conditions is therefore still a laborious and slow process. While in crystallography, users are able to screen hundreds of crystallisation conditions in a few hours, for cryo-EM, we can test about ten conditions per day at best. One limitation is that there is no automated protocol to assess the quality of the sample once placed in the cryo-EM grids. This is also due to the fact that the quality of the data can be assessed only after extensive processing. People are implementing strategies that try to solve this problem.[9] More work has to be done, but we should take X-ray synchrotron beam lines as an example where a crystal can be loaded onto the beam by robots and the diffraction data can be collected automatically.

Data processing has become much more user friendly in the last few years and new software that requires very little user intervention is being developed.[41,42] New software is taking advantage of the power of graphic processing units (GPUs).[43] Medium- to top-end graphic cards can be purchased at very accessible prices, and one graphic card can accommodate the processing capability of a few hundred core computing cluster in a very affordable local machine that costs a few thousand euros. Because of this implementation, nowadays, practically every laboratory can afford extensive cryo-EM processing.

Because of the improvement in the quality of the resolution achieved, cryo-EM data can now use the tools for atomic model building and refinement developed for X-ray crystallography.[44,45] Even though processing of cryo-EM data has improved a lot in the past few years, we still lack the richness of software tools that have been developed for X-ray crystallography. It is feasible to predict that the number of software developers will increase with increasing interest in the technique.

Acknowledgements

Carlos Fernández Rodríguez's PhD is supported by the fellowship BES-2015-071348 awarded by the Spanish government; Oscar Llorca's research is supported by the grant SAF2014-52301-R awarded by the Spanish government; Fabrizio Martino's research is supported by the grant SAF2014-59993-JIN awarded by the Spanish government.

References

1. D. J. De Rosier and A. Klug, *Nature*, 1968, **217**, 130–134.
2. R. M. Glaeser, *Annu. Rev. Phys. Chem.*, 1985, **36**, 243–275.
3. J. Frank, *Annu. Rev. Biophys. Biomol. Struct.*, 2002, **31**, 303–319.
4. R. S. Ruskin, Z. Yu and N. Grigorieff, *J. Struct. Biol.*, 2013, **184**, 385–393.
5. G. McMullan, A. R. Faruqi, D. Clare and R. Henderson, *Ultramicroscopy*, 2014, **147**, 156–163.
6. X. Li, P. Mooney, S. Zheng, C. R. Booth, M. B. Braunfeld, S. Gubbens, D. A. Agard and Y. Cheng, *Nat. Methods*, 2013, **10**, 584–590.
7. S. H. Scheres, *Elife*, 2014, **3**, e03665.
8. T. Grant and N. Grigorieff, *Elife*, 2015, **4**, e06980.
9. N. Biyani, R. D. Righetto, R. McLeod, D. Caujolle-Bert, D. Castano-Diez, K. N. Goldie and H. Stahlberg, *J. Struct. Biol.*, 2017, **198**, 124–133.
10. R. Danev and W. Baumeister, *Elife*, 2016, **5**, e13046.
11. M. Khoshouei, M. Radjainia, W. Baumeister and R. Danev, *bioRxiv*.
12. R. Henderson, J. M. Baldwin, T. A. Ceska, F. Zemlin, E. Beckmann and K. H. Downing, *J. Mol. Biol.*, 1990, **213**, 899–929.
13. S. Subramaniam and R. Henderson, *Nature*, 2000, **406**, 653–657.
14. J. P. Snyder, J. H. Nettles, B. Cornett, K. H. Downing and E. Nogales, *Proc. Natl. Acad. Sci.*, 2001, **98**, 5312–5316.
15. B. Böttcher, S. A. Wynne and R. A. Crowther, *Nature*, 1997, **386**, 88–91.
16. X. Yu, L. Jin and Z. H. Zhou, *Nature*, 2008, **453**, 415–419.
17. H. Liu, L. Jin, S. B. S. Koh, I. Atanasov, S. Schein, L. Wu and Z. H. Zhou, *Science*, 2010, **329**, 1038–1043.

18. X. Zhang, S. Sun, Y. Xiang, J. Wong, T. Klose, D. Raoult and M. G. Rossmann, *Proc. Natl. Acad. Sci. U. S. A.*, 2012, **109**, 18431–18436.
19. X. Zhang, P. Ge, X. Yu, J. M. Brannan, G. Bi, Q. Zhang, S. Schein and Z. H. Zhou, *Nat. Struct. Mol. Biol.*, 2012, **20**, 105–110.
20. T. Fujii, A. H. Iwane, T. Yanagida and K. Namba, *Nature*, 2010, **467**, 724–728.
21. V. E. Galkin, A. Orlova, M. R. Vos, G. F. Schröder and E. H. Egelman, *Structure*, 2015, **23**, 173–182.
22. E. A. Montabana and D. A. Agard, *Proc. Natl. Acad. Sci. U. S. A.*, 2014, **111**, 3407–3412.
23. G. M. Alushin, G. C. Lander, E. H. Kellogg, R. Zhang, D. Baker and E. Nogales, *Cell*, 2014, **157**, 1117–1129.
24. R. Zhang, G. M. Alushin, A. Brown and E. Nogales, *Cell*, 2015, **162**, 849–859.
25. A. Merk, A. Bartesaghi, S. Banerjee, V. Falconieri, P. Rao, M. I. Davis, R. Pragani, M. B. Boxer, L. A. Earl, J. L. S. Milne and S. Subramaniam, *Cell*, 2016, **165**, 1698–1707.
26. S. Banerjee, A. Bartesaghi, A. Merk, P. Rao, S. L. Bulfer, Y. Yan, N. Green, B. Mroczkowski, R. J. Neitz, P. Wipf, V. Falconieri, R. J. Deshaies, J. L. S. Milne, D. Huryn, M. Arkin and S. Subramaniam, *Science*, 2016, **351**, 871–875.
27. A. Bartesaghi, A. Merk, S. Banerjee, D. Matthies, X. Wu, J. L. S. Milne and S. Subramaniam, *Science*, 2015, **348**, 1147–1151.
28. H. Li, A. J. O'Donoghue, W. A. van der Linden, S. C. Xie, E. Yoo, I. T. Foe, L. Tilley, C. S. Craik, P. C. A. da Fonseca and M. Bogyo, *Nature*, 2016, **530**, 233–236.
29. W. Wong, X. Bai, A. Brown, I. S. Fernandez, E. Hanssen, M. Condron, Y. H. Tan, J. Baum and S. H. W. Scheres, *Elife*, 2014, **3**, e03080.
30. W. Wong, X.-C. Bai, B. E. Sleebs, T. Triglia, A. Brown, J. K. Thompson, K. E. Jackson, E. Hanssen, D. S. Marapana, I. S. Fernandez, S. A. Ralph, A. F. Cowman, S. H. W. Scheres and J. Baum, *Nat. Microbiol.*, 2017, **2**, 17031.
31. M. Shalev-Benami, Y. Zhang, D. Matzov, Y. Halfon, A. Zackay, H. Rozenberg, E. Zimmerman, A. Bashan, C. L. Jaffe, A. Yonath and G. Skiniotis, *Cell Rep.*, 2016, **16**, 288–294.
32. A. Miyazawa, Y. Fujiyoshi and N. Unwin, *Nature*, 2003, **423**, 949–955.
33. X. Bai, E. Rajendra, G. Yang, Y. Shi and S. H. W. Scheres, *Elife*, 2015, **4**, e11182.
34. X. Bai, C. Yan, G. Yang, P. Lu, D. Ma, L. Sun, R. Zhou, S. H. W. Scheres and Y. Shi, *Nature*, 2015, **525**, 212–217.
35. M. Liao, E. Cao, D. Julius and Y. Cheng, *Nature*, 2013, **504**, 107–112.
36. E. Cao, M. Liao, Y. Cheng and D. Julius, *Nature*, 2013, **504**, 113–118.
37. Y. Gao, E. Cao, D. Julius and Y. Cheng, *Nature*, 2016, **534**, 347–351.
38. Z. L. Johnson and J. Chen, *Cell*, 2017, **168**, 1075–1085e9.
39. X. Feng, Z. Fu, S. Kaledhonkar, Y. Jia, B. Shah, A. Jin, Z. Liu, M. Sun, B. Chen, R. A. Grassucci, Y. Ren, H. Jiang, J. Frank and Q. Lin, *Structure*, 2017, **25**, 663–670e3.

40. S. A. Arnold, S. Albiez, A. Bieri, A. Syntychaki, R. Adaixo, R. A. McLeod, K. N. Goldie, H. Stahlberg and T. Braun, *J. Struct. Biol.*, 2017, **197**, 220–226.
41. S. H. W. Scheres, *J. Struct. Biol.*, 2012, **180**, 519–530.
42. A. Punjani, J. L. Rubinstein, D. J. Fleet and M. A. Brubaker, *Nat. Methods*, 2017, **14**, 290–296.
43. D. Kimanius, B. O. Forsberg, S. H. Scheres and E. Lindahl, *Elife*, 2016, **5**, e18722.
44. A. Brown, F. Long, R. A. Nicholls, J. Toots, P. Emsley and G. Murshudov, *Acta Crystallogr. D. Biol. Crystallogr.*, 2015, **71**, 136–153.
45. P. D. Adams, P. V. Afonine, G. Bunkóczi, V. B. Chen, I. W. Davis, N. Echols, J. J. Headd, L.-W. Hung, G. J. Kapral, R. W. Grosse-Kunstleve, A. J. McCoy, N. W. Moriarty, R. Oeffner, R. J. Read, D. C. Richardson, J. S. Richardson, T. C. Terwilliger and P. H. Zwart, *Acta Crystallogr. D. Biol. Crystallogr.*, 2010, **66**, 213–221.

CHAPTER 11

Molecular Imaging

EUGEN KUBALA,*[a,b,c] MARION I. MENZEL,[c] BENEDIKT FEUERECKER,[a] STEFFEN J. GLASER[b] AND MARKUS SCHWAIGER[a]

[a] Technische Universität München, Department of Nuclear Medicine, Klinikum rechts der Isar, Ismaninger Straße 22, Munich 81675, Germany; [b] Technische Universität München, Department of Chemistry, Lichtenbergstraße 4, Garching bei München 85748, Germany; [c] GE Global Research, Freisinger Landstraße 50, Garching bei München 85748, Germany
*Email: eugen.kubala@tum.de

Molecular imaging appeared in the early twenty-first century as a discipline at the intersection of molecular biology and *in vivo* imaging. It differs from conventional imaging by its usage of probes known as biomarkers or tracers that interact physically or chemically with their surroundings and target certain metabolic processes or abnormalities. Nowadays, there is a broad variety of molecular imaging modalities that are based on different principles. Some imaging technologies, such as magnetic resonance imaging, were at first developed for clinical use only, however, in the last decade, there has been a trend to apply these to preclinical disease models as well. Other modalities have taken a more traditional preclinical to clinical application path. Preclinical research provides an excellent opportunity to study and evaluate novel therapeutics. In this chapter, we divide the imaging modalities into three groups based on functional principles: (1) modalities based on magnetic resonance imaging (MRI) and novel hyperpolarized probes, (2) modalities based on radioactive probes, such as single photon emission

Drug Discovery Series No. 61
Biophysical Techniques in Drug Discovery
Edited by Angeles Canales

Published by the Royal Society of Chemistry, www.rsc.org

Table 11.1 Comparison of the fundamental properties of current molecular imaging modalities used in the experimental and clinical environment.

Imaging technique	Physical basis	Spatial resolution	Time resolution	Penetration depth	Sensitivity [mol]	Safety	Clinical translatability
MRI	Radio-frequency waves	> 50 μm	s–min	Unlimited	10^{-9}	+++	+++
Hyperpolarized MRSI	Radio-frequency waves	1–10 mm	s–min	Unlimited	10^{-3}	++	+++
PET	High-energy γ-rays	1–2 mm	s–min	Unlimited	10^{-15}	+	+++
SPECT	X-rays	1–2 mm	s–min	Unlimited	10^{-14}	+	+++
Bioluminescence	NIR-light	μm–mm	s	μm–cm	10^{-12}	+++	
Fluorescence imaging	NIR-light	μm–mm	s	μm–cm	10^{-12}	+++	+
MSOT	NIR-light/ultrasound	>50 μm	s	μm–cm	10^{-12}	+++	++

computed tomography (SPECT), positron emission tomography (PET), and (3) optical imaging modalities including bioluminescence imaging (BLI), fluorescence imaging (FI) and multi-spectral optoacoustic tomography (MSOT) (Table 11.1).

11.1 Magnetic Resonance Imaging

Magnetic resonance imaging (MRI) is an imaging modality based on the ability of certain atomic nuclei to absorb and emit energy in the form of radio-frequency signals when placed in an external high magnetic field. Nowadays, the most common nucleus in MRI is hydrogen (^1H), due to its high natural abundance in the human body as well as high signal-to-noise ratio. Its versatility has made MRI a widely-used tool in clinics as well as in research.

In the first decade of the new millennium, cardiac MRI, body MRI, fetal imaging, and functional MRI became a routine part of the clinical examination. MRI has also become the state of the art technology for imaging diseases of the central nervous system (CNS) and heart. The current research in CNS diseases focuses on studying stroke, demyelinating diseases, dementia, cerebrovascular disease, infectious diseases, epilepsy and multiple sclerosis.[1] In cardiac research, MRI plays an important role alongside other imaging modalities, such as cardiac CT, echocardiography, and nuclear medicine. Its application allows the precise localization of vascular occlusion, and quantification of perfusion, oxygenation deficits, membrane breakdown, and cytotoxic edema.[2] In the liver, pancreas, and bile duct, MRI is used to detect and characterize lesions, and focal and diffuse disorders.[3] MR-enterography provides information for inflammatory bowel disease assessment and small bowel tumor detection.[4] MR-colonography can detect large polyps in the intestine that are at increased risk of colorectal cancer.[5]

MRI has many advantages compared to other imaging modalities. It features a very high spatial resolution and allows morphological imaging as well as functional imaging. However, the major disadvantage of MRI is its low sensitivity of 10^{-3} to 10^{-5} mol L^{-1}. This is caused by the fact that the signal intensity is based on the difference between the nuclear spin state populations. At a magnetic field strength of a typical clinical MRI scanner (1.5 T), the difference between the high and low energy states is approximately 4.5 ppm. One strategy to increase the sensitivity is to raise the magnetic field strength. In the last few years, we have seen clinical MRI scanners with a magnetic field strength of up to 7 T. Another approach involves a process called hyperpolarization.

11.1.1 Hyperpolarization

There are three commonly used methods to achieve the hyperpolarized state: (1) the parahydrogen technique, (2) optical pumping and (3) dynamic nuclear polarization. Parahydrogen-induced polarization (PHIP) is based on the chemical reaction of a substrate containing a ^{13}C-nucleus with a parahydrogen, a state of hydrogen (H_2) that has its nuclei oriented in such a way that their magnetic moments cancel. During the chemical reaction, the spin order of the parahydrogen is transferred to the nuclear polarization of the ^{13}C nucleus *via* a diabatic field cycling scheme.[6] PHIP has been successfully applied for angiography measurements and measuring tissue perfusion.[7] Even though the instrumentation required for PHIP is relatively straightforward and cheap, clinical application has been limited, mainly due to the technically challenging development of the catalysts and hydrogenation chemistry.[8]

Optical pumping is used to hyperpolarize the noble gases ^{3}He or ^{129}Xe. Although the theory was described in the early 1960s, progress in the development of high-power lasers has allowed large-scale studies only recently.[9] The first images of mouse lung using hyperpolarized ^{129}Xe were published in 1994 demonstrating faster acquisition and better resolution than conventional MRI or emission tomography.[10] Three years later, the first images of the human lung were acquired.[11] In the majority of human studies, ^{3}He has been used due to its higher polarization levels and the lack of any significant safety concerns. However, due to the increasing price and low availability of ^{3}He, it has been replaced by the naturally abundant and relatively cheap ^{129}Xe.[12]

Optical pumping is based on the transfer of angular momentum by spin-exchange collisions between the electron spins of optically pumped rubidium atoms (Rb) and the nuclear spins of the noble gasses ^{3}He or ^{129}Xe.[13,14] In the high magnetic field, the circularly polarized laser light drives the electronic transition of the Rb atoms ($S_{\frac{1}{2}}$–$P_{\frac{1}{2}}$) to selectively pump ground-state Rb electrons to the $+\frac{1}{2}$ or $-\frac{1}{2}$ state entirely. In this state, the optically pumped Rb electrons can transfer their polarization to the nuclei of the ^{3}He or ^{129}Xe atoms *via* binary collisions or *via* van der Waals interactions.[10,15]

The most common application of hyperpolarized noble gasses is the functional visualization of lung airspaces to detect abnormalities, such as chronic obstructive pulmonary disease (COPD), asthma and cystic fibrosis.[16–18] Conventional clinical MRI of lungs is a very challenging procedure due to a very low proton density and inhomogeneous environment created by the many air–tissue interfaces.[19] These complications can be compensated for by using hyperpolarized gases because the optical pumping allows the nuclear polarization of the noble gas to be increased by five orders of magnitude.[20,21] Hyperpolarized gases are also used in diffusion MRI, allowing quantitative evaluation of the lung microstructure.[22] After inhalation, approximately 2% of ^{129}Xe dissolves in the lung's parenchyma. Measurement of the apparent diffusion coefficient (ADC) has proven to reveal sub-clinical smoking-related emphysematous changes, even before high-resolution CT.[23] It can also detect the subtle gravity-dependent gradient and age-related changes in alveolar size.[24,25] Clinical studies suggest that diffusion MRI provides a novel opportunity to quantify gas delivery, exchange, and transport in the lung.[26,27]

Dynamic nuclear polarization (DNP) was first described by Albert W. Overhauser in 1953.[28] Later that year, his hypothesis was confirmed experimentally by Carver and Slichter.[29] In 1958, Abragam and Proctor described this phenomenon for electrons in liquids and named it the "Solid effect".[30] At temperatures below 4 K, the electron-spin polarization reaches nearly 100% and is more than three orders of magnitude higher than the nuclear-spin polarization. This effect occurs due to the gyromagnetic ratio of the electron being three orders of magnitude greater than that of various nuclei. Weak interactions between nuclei and electrons, such as the Overhauser effect, solid effect, cross effect and thermal mixing, allow polarization transfer from the hyperpolarized electron spins to nuclear spins using microwave irradiation with a frequency close to the corresponding electron paramagnetic resonance (EPR) frequency.[31,32]

It took almost half a century since its discovery for DNP to be first introduced into biological systems. The team led by Jan H. Ardenkjaer-Larsen designed a hyperpolarizer capable of the rapid dissolution of a frozen hyperpolarized sample with minimal hyperpolarization loss.[32] This technique was named dissolution DNP and it opened up a new field of research called ^{13}C Magnetic Resonance Spectroscopy and Imaging (^{13}CMRSI) that allowed the investigation and characterization of various diseases *in vivo* in real time.[33,34] As a source of hyperpolarization, stable carriers of an unpaired electron, such as the trityl radical tris(8-carboxy-2,2,6,6-tetra-(hydroxyethyl)-benzo-[1,2–4,5]-bis-(1,3)-dithiole-4-yl)-methyl sodium salt (OX063) and (2,2,6,6-tetramethylpiperidin-1-yl)oxyl (TEMPO), were designed. Using microwave irradiation, hyperpolarization is transferred from these stable radicals to the desired ^{13}C-labeled molecule resulting in increased polarization of up to five orders of magnitude.[35,36] However, as soon as the hyperpolarized ^{13}C-labeled molecule is in the liquid state, the gained polarization decays according to the longitudinal relaxation time (T_1) of the hyperpolarized ^{13}C-nuclei. Therefore, invention of a fast dissolution technique shortening the time between dissolution and administration of the hyperpolarized molecule to the biological models was crucial.[37]

For successful hyperpolarization, the position of the ^{13}C isotope within the molecule is essential. The molecule needs to fulfill several requirements: (1) the T_1 of the ^{13}C nucleus needs to be in the order of tens of seconds. A ^{13}C atom that has no covalent bonds with ^{1}H-nuclei usually fulfills this criterion the best. (2) The candidate molecule needs to be rapidly metabolized within a maximum of two to three times T_1. (3) The resulting metabolic product must have a significantly different chemical shift from the original molecule (in the order of ppm). (4) The sample mixture must form an amorphous glass in its solid state. This decreases the distance between the hyperpolarized unpaired electrons and ^{13}C nuclei, allowing the polarization transfer. If the candidate molecule does not form an amorphous glass naturally, it must be highly soluble in solvents that do, such as glycerol or dimethyl sulfoxide. In the end, these requirements limit the application of DNP to a relatively small number of molecules. For more information,

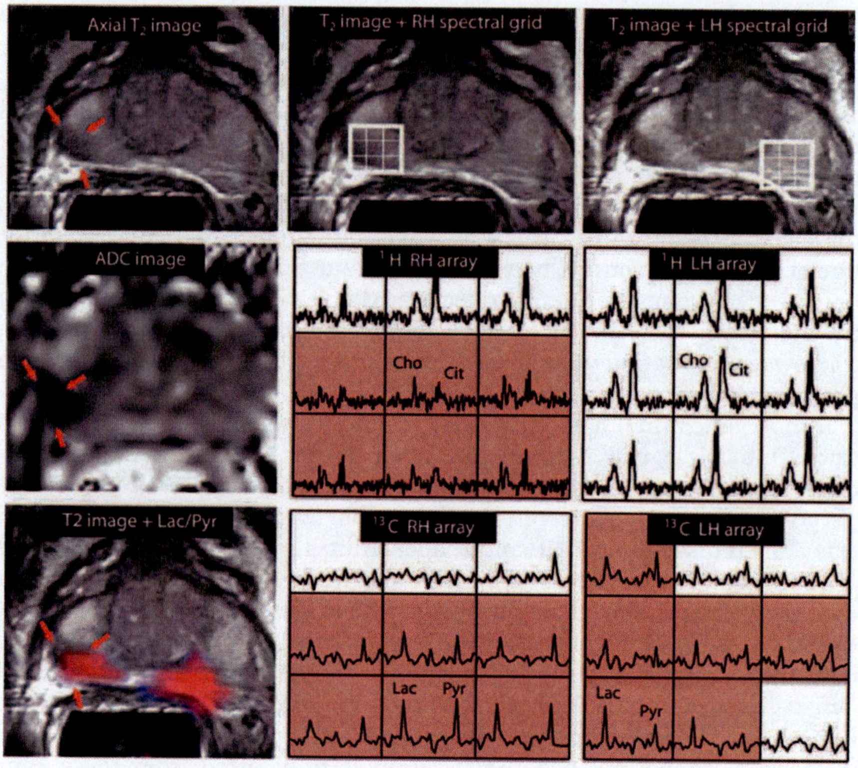

Figure 11.1 First published results of hyperpolarized [1-^{13}C]pyruvate application in patients. The axial T_2-weighted image shows a region with reduced apparent diffusion coefficient (ADC) connected to reduced signal intensity (red arrows). The ^{1}H spectral arrays from this area show reduced citrate (cit) and elevated choline (cho)/citrate. The T_2 image combined with [1-^{13}C]lactate/[1-^{13}C]pyruvate ratio shows the spatial resolution of the signal (highlighted in pink). The ^{13}C spectral arrays support these findings and show the RH and LH areas with temporal resolution. From S. J. Nelson *et al.*, Metabolic Imaging of Patients with Prostate Cancer Using Hyperpolarized [1-^{13}C]Pyruvate, *Sci. Transl. Med.*, 2013, 5, 198ra108. Reprinted with permission from AAAS.[53]

see Chaumeil *et al.*, 2015.[38] Even after choosing a suitable molecule, developing a successful hyperpolarization protocol can be technically very challenging.[33,39]

In the past fifteen years, many substrates have been successfully hyperpolarized. However, the most interesting and studied molecule is [1-^{13}C]pyruvate. It has been applied to image the Warburg effect in many preclinical studies of cancer,[40,41] as well as to measure the apparent diffusion coefficient, which indicated the distribution of intra- and extracellular localization of the hyperpolarized metabolite.[42] Hyperpolarized [2-^{13}C]pyruvate demonstrated the ability to follow the Krebs cycle.[43] Administration of hyperpolarized [1-^{13}C]fumarate, on the other hand, precisely identified cellular necrosis.[44] ^{13}CMRSI also proved its ability to visualize the pH of a tissue in real time,[45] and to detect hypoxia using hyperpolarized ^{13}C-bicarbonate and ^{13}C-labeled zymonic acid.[45,46] Hyperpolarized [1-^{13}C]sodium acetate allowed researchers to follow fatty acid metabolism, which was found to be altered in diabetes.[47,48] [5-^{13}C]glutamine and [1-^{13}C]glutamate demonstrated their application in monitoring glutaminolysis, which serves as an alternative carbon and energy source for cancer proliferation of, for example, hepatocellular carcinoma.[49,50] This method also showed the ability to distinguish between primary and secondary glioblastoma based on the presence of the IDH1 mutation. In this case, hyperpolarized [1-^{13}C]2-oxoglutarate was used as a tracer.[51] Many other molecules have been hyperpolarized including [1-^{13}C]ethyl pyruvate, [1-^{13}C]lactate, and ^{13}C-urea. For more information, see the reviews by Brindle *et al.* (2011), Keshari *et al.* (2014) and Chaumeil *et al.* (2015).[38,39,52]

After various tumor models showed high [1-^{13}C]pyruvate to [1-^{13}C]lactate turnover in preclinical studies, several cancer types were chosen and are currently being investigated by clinical studies. The first application of [1-^{13}C]pyruvate was achieved by Nelson *et al.* in 2013 (Figure 11.1). They were able to localize prostate cancer lesions in 31 patients and the results confirmed the safety of the agent as well as demonstrating elevation of [1-^{13}C]pyruvate metabolism in regions that were later identified and confirmed by biopsy to be tumorous.[53] In future years, we will see the results of currently ongoing clinical studies facilitating [1-^{13}C]pyruvate in studies of solid malignant tumors and prostate cancer, as well as cardiomyopathy and hypertension.[54]

11.2 Modalities Based on Radioactive Isotopes

Imaging modalities used in nuclear medicine, such as Positron Emission Tomography (PET) and Single Photon Emission Computed Tomography (SPECT), are based on the detection of gamma radiation emitted by a radioactive tracer and its biodistribution *in vivo*. Most tracers used in both techniques are selectively labeled with radioactive isotopes allowing the study of radiopharmaceutical characteristics, radiopharmaceutical distribution, receptor distribution, and application in a clinical context for the determination and differentiation of disease extent with high sensitivity. In principle, any labeled molecule can be used for imaging as well as for therapeutic purposes. However, characteristics such as radioactive decay, decay type and isotope half-life should be considered when choosing an isotope of a radiopharmaceutical.

In both PET and SPECT, commonly used detectors comprise a scintillation crystal that converts the gamma radiation into a visible scintillation light and a photodetector that collects the scintillation light and translates it into an electric pulse. These pulses are converted into images using various reconstruction algorithms, such as maximum likelihood expectation maximization (MLEM) or ordered subsets expectation maximum (OSEM).[55] It is noteworthy that gamma radiation passing through matter can lose energy *via* either (1) photoelectric absorption, (2) Compton scatter or (3) pair production, with (1) and (2) being the most abundant in nuclear medicine (for detailed information, please refer to Spanoudaki *et al.*, 2004).[55] As these interactions take place both in the patient and in the scintillation crystal itself, they can contribute to a reduction in image quality. For optimum image quality, a scintillator should take these processes into account, but also cover a range of energies and time resolutions. The materials used for scintillators can be divided into organic and inorganic materials, which show complementary properties. Organic scintillators are very fast but have poor light output, while inorganic materials show high light output, but have a slower decay constant. The most common scintillator crystals include thallium-doped sodium iodide (NaI(TI)), bismuth-germanate (BGO), polycrystalline cerium-activated lutetium-oxy-orthosilicate (LSO(Ce)), gadolinium orthosilicate (GSO(Ce)), CsI(TI), BaF_2 and $LaBr_3$. For optimal performance, the isotope's maximum emission wavelength should match the wavelength for which the photodetector has the highest sensitivity. Upon emission of the scintillation light, the traditional way to convert these signals into electrical pulses has been the use of a photomultiplier tube (PMT).[56] Inside a magnetic field, the performance of a PMT is severely degraded, limiting its application in simultaneous PET/MRI.[57] By the introduction of semiconductor photodetectors such as avalanche photodiodes (APDs), this issue can be resolved.[58] Silicon photomultipliers (SiPM) represent a further improvement. These detectors are solid state photodetectors, which, upon absorption of a photon, can produce very short pulses within the range of tens of nanoseconds, comparable to photomultiplier tubes.[59] They are built from APD arrays on a common silicon substrate and allow detection of single-photon events in sequentially connected silicon APDs. In recent years, the use of cadmium zinc telluride (CZT) semiconductors that directly convert electromagnetic radiation, such as gamma-rays, into electrical signals has yielded a new type of detector system that, notably, has found application in cardiac SPECT (*e.g.*, D-SPECT).[60,61]

11.2.1 Positron Emission Tomography: Principles and Applications

Positron emission tomography (PET) is one of the key imaging techniques in nuclear medicine. The modality is based on the detection of gamma radiation that is generated by β^+ decay. In this process, a proton inside a radionuclide nucleus is converted into a neutron while a positron and an electron neutrino are emitted at the same time. Upon interaction of the positron with an electron, two high energy photons, each with an energy of 511 keV, get emitted at an angle of 180° relative to each other (annihilation). These gamma rays can be recognized by detectors positioned around the object of interest.

The principle of PET is based on the fact that coincidence events of these gamma rays can be detected. Thereby, the time and spatial distributions of these decay events serve as essential information upon which the spatial distribution of the radiopharmaceuticals in the body can be reconstructed, and upon which a series of images can be generated.

A major step forward improving the performance of PET was made by the introduction of time of flight (TOF) PET scanners allowed by the development of lutetium oxyorthosilicate (LSO) crystals. LSO crystals provided a combination of higher light output and higher stopping efficiency for 511 keV photons compared to BGO crystals, which had a lower counting rate capability, higher random coincidence rates and lower spatial resolution. With LSO crystals, a tight coincidence timing window ($\leq$3 ns) was achievable, and they showed short decay rates. Further research led to the development of lutetium–yttrium oxyorthosilicate (LYSO), which had similar properties to those of LSO crystals, however, it allowed a system timing resolution in the range of 450–500 ps.[62] By combining PET with CT or MRI, the advantage of the high sensitivity of the functional imaging of PET can be symbiotically combined with the high morphological precision of CT and MRI, allowing the correct description of functional processes confined to a correct organ or structure.

There are numerous isotopes that qualify as PET compatible in nuclear medicine. All must fulfill narrow requirements for half-life and decay properties to give good image contrast (β^+ decay energy). The ideal half-life ranges between a few minutes and a few hours, and the isotope must decay in a way that no long-living radioactive isotopes remain in the body. Common isotopes include ^{11}C, ^{15}O, ^{18}F, ^{64}Cu, ^{68}Ga and ^{124}I, where ^{18}F (half-life *ca.* 109 min) and ^{68}Ga (half-life *ca.* 68 min) are currently the most used ones. The overall signal-to-background contrast depends, amongst other things, upon the tracer's uptake by the targeted tissue and its clearance kinetics. The sensitivity of PET allows detection of the tracer in nano- to picomolar concentrations irrespective of tissue depth. Usually, only low amounts of the tracer are necessary so that biological processes can be imaged without disturbance to the normal physiological processes and kinetics.[63–65] For some of the isotopes mentioned above, a local cyclotron and radiochemical synthesis unit are required as some of these traces have a very short half-life (*e.g.*, ^{15}O: 2 min).

Currently, the most commonly used radiopharmaceutical for PET is 2-deoxy-2-^{18}F-fluoro-β-D-glucose (^{18}F-FDG). This molecule behaves *in vivo* like regular glucose and is taken up into solid tumors of the Warburg type as well as any tissue with elevated glycolysis (*e.g.*, inflammation). In contrast to glucose, ^{18}F-FDG gets trapped inside a cell after phosphorylation by Hexokinase II. Therefore, ^{18}F-FDG accumulation provides information on glycolytic metabolic activity and allows the early detection of tumors as well as the monitoring of a cancer treatment response.[66] Nowadays, it is used for staging and primary diagnosis in order to detect breast, colorectal, esophageal, head and neck, lung, thyroid, and pancreatic cancer, melanoma, sarcoma, and lymphoma, and can provide valuable information in cases of an unknown primary tumor (CUP). However, its biggest strength is in tumor staging and recurrence screening. The clinical use of ^{18}F-FDG PET is limited in slowly growing or highly differentiated and less metabolically active tumors such as prostate cancer and hepatocellular carcinoma. Most studies show a clear correlation of metabolic response measured by ^{18}F-FDG-PET and successful treatment. ^{18}F-FDG-PET shows the best results in the follow-up of a therapy response for a broad variety of cancers such as gastrointestinal stromal tumors, lymphoma, breast cancer, ovarian cancer, sarcomas, Hodgkin disease, non-Hodgkin lymphoma, colorectal cancer, melanoma, prostate cancer, pancreatic cancer, thyroid cancer, testicular cancer and lung cancer. It also serves as an excellent modality to image recurrence in breast cancer, colorectal cancer, head and neck cancer, Hodgkin lymphoma, non-Hodgkin lymphoma and thyroid cancer. In non-oncologic questions, ^{18}F-FDG-PET has also shown its benefits, and, therefore, it has been approved for use in the diagnosis of Alzheimer's disease.[67] Even though ^{18}F-FDG-PET provides very high sensitivity, false positive findings can occur and must be considered in clinical reasoning. For some of these findings, physiological processes, such as activation of brown fat, colonic and gynecologic activity, inflammatory processes, and rebound thymic hyperplasia can be identified as the source.[68]

Because ^{18}F-FDG is not tumor-specific and has limitations in tumors with no or low glycolytic activity, more tumor specific radiopharmaceuticals have been developed. One method is to target amino acid uptake, which is enhanced in majority proliferating cancer types. Several radiolabeled natural and non-natural amino acid analogs have been developed, such as ^{11}C-methionine (^{11}C-Met), ^{18}F-fluoromethyl tyrosine (^{18}F-FMT), ^{18}F-fluoroethyl tyrosine (^{18}F-FET) and ^{18}F-1-amino-3-fluorocyclobu-tane-1-carboxylic acid (^{18}F-FACBC). In general, radiolabeled amino acids show a higher specific uptake in solid cancers, such as prostate cancer and epidermoid lung carcinoma, and are less inclined to show false positive signals due to, for example, inflammation compared to ^{18}F-FDG. Artificial amino acids that are designed not to be metabolically active show greater tumor specificity than natural amino acids. Radiolabeled amino acid derivatives, such as nucleoside analogs including ^{11}C-thymidine, ^{18}F-30-deoxy-30-fluorothymidine (^{18}F-FLT) and ^{18}F-5-fluoro-20-deoxyuridine (^{18}F-FUdR), were designed to be taken up and partially incorporated into DNA, serving as a marker of proliferation. Cancer proliferation is also associated with increased cell membrane phospholipid synthesis. Radiolabeled lipid precursors like ^{11}C-acetate, ^{11}C-choline, ^{18}F-fluorocho-line and ^{18}F-fluoroethylcholine were found to accumulate in tumors with low muscle background activity. Of these tracers, ^{11}C-choline showed promising results in the staging and diagnosis of prostate cancer. Nevertheless, ^{11}C-choline lost its prior significance with the advent of a more specific tracer that binds to the prostate membrane specific antigen (PSMA), thereby showing superior imaging properties. This membrane antigen is ubiquitously expressed on the membrane of most prostate cancers and metastases. ^{68}Ga-PSMA improved the diagnosis and staging, and prevented relapse of PSMA positive prostate cancers allowing for precise determination of tumor localization and disease extent in lymphatic tissue and also in bone metastasis (Figure 11.2).[69] Compared with ^{11}C-choline, studies showed that imaging with ^{68}Ga-PSMA identifies more lesions such as lymph nodes, especially at low PSA levels.[70]

Other frequently used tracers for somatostatin receptor imaging include ^{68}Ga-DOTATATE and ^{68}Ga-DOTANOC PET/CT. The somatostatin receptors on the membrane of malignant cells serve as a target for a construct consisting of an antibody coupled to a chelator (*e.g.*, DOTA), which tightly holds the radioactive substance (in the above cases, the beta plus emitter ^{68}Ga). These tracers are mainly used in the diagnosis and staging of neuroendocrine tumors including pituitary adenomas, pheochromocytomas, paragangliomas, neuroblastomas, and medullary thyroid cancers, as it has been shown that many of these malignancies show an overexpression of several somatostatin receptors (SSTR), notably SSTR-2. Several compounds exist, such as ^{68}Ga-DOTATATE or ^{68}Ga-DOTANOC PET/CT, that have different affinity to the SSTR receptors (Figure 11.2).[71]

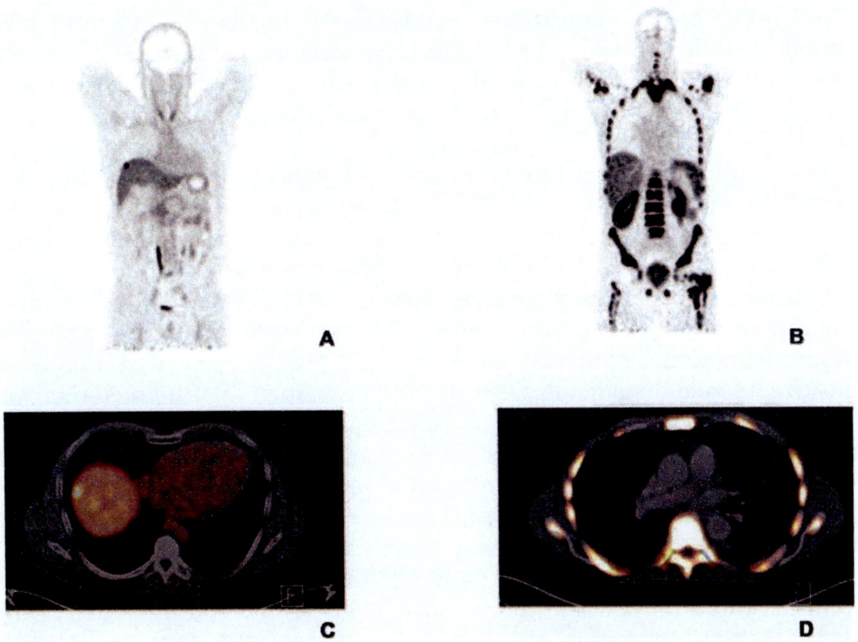

Figure 11.2 Examples of images obtained using novel positron emission imaging (PET) tracers. (A) Coronal slice using ^{68}Ga-DOTATATE-PET imaging and a corresponding fused PET/CT image showing a positive liver lesion (C). (B) PET image using ^{68}Ga-PSMA showing intense accumulation in multiple bone metastases as also demonstrated in (D), a fused PET/CT slice showing PSMA positive bone lesions in the ribs.

As hypoxia is a characteristic pathophysiological property of many types of cancer, imaging with PET tracers based on ^{18}F-labeled nitroimidazoles and ^{62}Cu-ATSM (copper complex with diacetyl-bis(N^4-methylthiosemicarbazone) can locate hypoxic regions. ^{18}F-fluoromisonidazole (^{18}F-FMISO) is probably the most prominent hypoxia imaging agent.[72] Other promising applications include imaging of angiogenesis using integrins, such as $\alpha_v\beta_3$, that are strongly connected to the facilitation of invasion, metastasis, and therapy resistance of cancer cells. Small peptides with the sequence arginine–glycine–aspartic (RGD) are ligands of integrins and can be used with radiolabeled tracers, such as ^{18}F-FRGD2 and ^{64}Cu-DOTA-labeled quantum dots.[73,74]

Beyond the application of PET in oncology, various PET tracers are also employed in other disciplines, such as cardiologic and neurologic clinical settings. As cardiac diseases represent one of the leading causes of death in the western hemisphere, the detection of acute and chronic heart diseases is of utmost importance. With regard to perfusion, PET is advantageous compared to traditional SPECT tracers as it allows quantification of myocardial blood flow, attenuation correction and superior image quality in coronary arterial disease. With regard to tracers used in perfusion imaging, ^{13}N–NH_3, ^{15}O–water, ^{82}Rubidium and ^{18}F-flurpiridaz have been investigated. ^{18}F-FDG-PET/MRI demonstrated its advantages in the estimation of myocardial viability but also in myocarditis, sarcoidosis and atherosclerotic plaque imaging. Current developments include imaging the cytokine receptor CXCR4 in atherosclerotic plaques with ^{68}Ga-Pentixafor as a possible alternative for ^{18}F-FDG in inflammation.[75]

In neurology, a broad line-up of tracers has been investigated in recent years. The employment of ^{18}F-FDG in the diagnostic workup of hypometabolism in epileptogenic foci, dementia, and Alzheimer's disease has proven to be a valuable tool.[76–78] ^{18}F-FDG is also highly sensitive for the early diagnosis of Alzheimer's disease, distinguishing conversion from mild cognitive impairment to Alzheimer's disease as well as differentiating it from other dementias.[79] Further, more specific tracers that detect amyloid positive plaques, such as ^{11}C-PIB, ^{18}F-florbetapir, ^{18}F-flutemetamol and ^{18}F-florbetaben, exist and are associated with amyloid positivity, as demonstrated in an ameta-analysis.[80] In patients with Alzheimer's disease, PET imaging is positive in the majority of cases. Specific tracers, such as the ^{11}C-labeled Pittsburgh compound B (^{11}C-PIB), selectively identify beta-amyloid with a high affinity and specificity for fibrillar beta-amyloid.[79] ^{18}F-florbetapir, ^{18}F-flutemetamol and ^{18}F-florbetaben, in contrast, offer the advantage of a longer half-life due to labeling with ^{18}F, and further, while there is a high correlation with ^{11}C-PIB, amino acid PET using, for example, ^{18}F-fluoro-ethyl-l-tyrosine (FET) or L-*methyl*-^{11}C-methionine, can add substantial clinical information in the delineation and discrimination of tumor tissue from inflammatory or reactive changes and tissue edema and contributes to improving diagnosis and staging, for example, of glioma.[81] In Parkinson's disease, ^{18}F-3,4-dihydroxyphenylalanine (^{18}F-DOPA) PET is used to evaluate dopamine synthesis and storage in the presynaptic terminals, and serves as marker of dopa decarboxylase activity and dopamine turnover.[79] PET can diagnose Parkinson's disease at an early stage before clinical symptoms occur. More tracers exist that also investigate the postsynaptic dopamine receptors, such as ^{11}C-SCH 23390 and ^{11}C-NNC 756.[79]

11.2.2 Single-Photon Emission Computed Tomography: Principles and Applications

Single-photon emission computed tomography is also among the most frequently used nuclear medicine techniques both in hospitals as well as in ambulatory practice. A broad variety of tracers are available allowing investigation of multiple clinical questions concerning, for instance, kidney function, myocardial perfusion, bone lesions, or distribution within lymphatic tissue. In comparison to PET, SPECT has about 2–3 orders magnitude lower sensitivity and lower spatial resolution. It benefits from a much broader lineup of radionuclides including ^{99m}Tc, ^{111}In, ^{67}Cu and ^{123}I, and therapeutically used ones such as ^{131}I, ^{177}Lu, ^{90}Y and ^{213}Bi. The ideal SPECT radiotracer fulfills the following criteria: (1) excellent tissue penetration, (2) high stability *in vivo*, (3) high affinity to the target structure, (4) safe human application, (5) specific uptake and retention in target cells and (6) rapid clearance from non-targeted tissues and organs.[82] Due to its properties, the most commonly used radioisotope in SPECT is ^{99m}Tc. This is because (1) its half-life of 6.02 hours is long enough for radiotracer preparation and patient imaging, but sufficiently short to keep radiation exposure of the patient low; (2) the γ-energy of 141 keV is close to the optimal value for commercially available gamma camera; (3) it is easily produced using affordable ^{99}Mo–^{99m}Tc generator systems and (4) its diverse coordination chemistry allows a large variety of different tracers to be generated.[83]

In daily clinical use, the following conjugates are frequently used: ^{99m}Tc-HDP for bone scintigraphy, ^{99m}Tc-MAG3 for functional assessment of both kidneys in renal scintigraphy, ^{99m}Tc-Sestamibi for parathyroid scintigraphy, ^{99m}Tc-MIBI for studies on perfusion of coronary vessels, ^{99m}Tc-MAA for perfusion scintigraphy of the lungs and evaluation of the liver–lung shunt pre-therapeutically before selective intravascular radiopeptide therapy of the liver and ^{99m}Tc-Pertechnetat for evaluation of the thyroid in thyroid scintigraphy. Numerous other tracers exist that are used to answer special clinical questions, such as detecting active or occult types of bleeding by using radioactively labeled erythrocytes that are collected from the patient. ^{99m}Tc-HIDA and PIPIDA are used in investigating liver diseases when questions regarding the function of the liver, or infections of the bile paths or liver transplants need to be investigated. Further, ^{99m}Tc labeled porridge can be used to investigate the function of the esophagus, as a time activity curve allows precise determination of the time that the porridge needs to pass through different parts of the esophagus. As many examinations in routine clinical diagnostics benefit from the use of the well-established and widely available isotope ^{99m}Tc, imaging using SPECT is used by a large community.

Generally, SPECT radiopharmaceuticals can be divided into two groups based on their biodistribution: (1) radionuclides biodistributed based on their chemical and physical properties and (2) targeted radionuclides with a specific interaction only at the site of interest.[84] The targeted radiopharmaceuticals usually consist of two main parts: a γ-radiation emitting radionuclide and a target biomolecule. If a metallic radionuclide such as ^{67}Ga-gallium or ^{67}Cu-copper is used, a bifunctional chelator is necessary as a linker between the metallic radioisotope and targeting biomolecule.[85] The target biomolecule navigates the radioisotope to the area of interest. The biomolecules are based on analogs of peptides, vitamins, nucleosides and specific antibodies or their fragments. One of the most interesting applications of SPECT is imaging human cancer based on targeting the variety of regulatory peptide receptors overexpressed on the surface of most cancer cells. Peptide-based target biomolecules are currently by far the largest group of radiopharmaceuticals in use. Short peptides consisting of less than 40 amino acid residues can be easily prepared using solid phase peptide synthesis. They show excellent tissue penetration and fast clearance, and they cause minimal immune response.[86]

There is ongoing continuous research focused on designing other peptide-based radiopharmaceuticals. Bombesin and neurotensin analogs target the gastrin-releasing peptide receptor (GRPR) and neurotensin receptor (NTR), respectively. GRPR is overexpressed in a high variety of cancers including prostate, breast, pancreatic, gastric, small cell lung, and colorectal cancer, as well as in epidermoid carcinoma of the anal canal.[87] Bombesin analogs labeled with radioisotopes such as ^{111}In or ^{177}Lu have been intensively investigated. The bombesin derivative DO3A-CH_2CO-G-4-aminobenzoyl-Q-W-A-V-G-H-L-M-NH_2 (AMBA) was found to be successful in both diagnostic and therapeutic applications.[88] Also, the "panbombesin" analog showed high affinity to all three bombesin receptor subtypes, possibly allowing a broader field of implementation.[89]

NTR is overexpressed in several neuroendocrine tumor types and pancreatic carcinomas.[90] Numerous radiopharmaceuticals based on neurotensin analogs and labeled with ^{99m}Tc, ^{111}In, and ^{188}Re have been synthesized.[91] As with other small neuropeptides, neurotensin analogs are susceptible to rapid degradation in plasma by peptidases. Modifications needed to be introduced to protect the three major cleavage sites. The resulting NT-XIX showed high tumor uptake and low uptake in organs, particularly low in kidneys.[92] Currently, there are many other peptide-based radiotracers based on cholecystokinin (CCK), gastrin, Glucagon-like peptide-1(GLP)/exendin, arginine–glycine–aspartic acid (RGD)–peptide and others available, promising various new opportunities for future precision diagnostics and medicines.[93]

11.3 Optical Molecular Imaging

Currently, the most promising optical technologies in molecular imaging are bioluminescence imaging (BLI) and fluorescent imaging (FI). There are many advantages of optical imaging compared to conventional imaging modalities. These include (1) high sensitivity, (2) relatively low cost and (3) a high degree of safety. Thus, optical molecular imaging has begun to play a crucial role in the imaging of tumor occurrence, progression and treatment development. However, because optical molecular imaging is based on visible and infrared light, its main disadvantage is the depth of light penetration. The photon's low energy gets quickly absorbed and scattered in the tissue, resulting in severe degradation of spatial resolution. This limits optical molecular imaging to relatively superficial areas of the human body or preclinical research in small animals.

11.3.1 Bioluminescence Imaging (BLI)

Bioluminescence imaging (BLI) is based on the detection of visible light emitted by a biochemical reaction. For the application of BLI *in vivo*, biological entities, such as bacteria, tumor cells, immune cells or genes, are tagged with a reporter gene encoded for luciferase, an enzyme that generates visible light through the oxidation of an enzyme-specific substrate.[94] Up to now, approximately thirty luciferases have been identified in different organisms, however, only nine have been deeply studied, and their structure and reaction mechanism have been described.[95] The most commonly used luciferase in BMI comes from the North American firefly, *Photinus Pyralis*. Since it was first reported in 1985, the reporter gene has been several times genetically modified to enhance its expression in mammalian cells.[96,97]

The most significant advantage of BLI, compared to other fluorescence-based optical imaging modalities is its non-existent background noise (the so-called auto-fluorescence), since there is no endogenous visible light emitted in mammalian cells and tissues. It therefore possesses an unprecedented sensitivity with respect to other conventional imaging modalities allowing identification of up to ten individual cancer cells embedded deep in optically scattering tissue.[95] BLI is also one of the cheapest molecular imaging techniques available at present. As with other optical imaging methods, BLI can be conducted in living organisms without anesthesia. However, immobilization of the studied subject is necessary to produce high-quality data. In comparison to the non-optical modalities, the data collection times for BLI are relatively short, in the order of minutes.[98] There are two conditions that BLI needs to fulfill for successful *in vivo* application: (1) non-toxicity of the substrate, and (2) easy cell-permeability and distribution through the tissues when applied exogenously. Luciferin showed only modest cell-permeability, and the study of radiolabeled ^{14}C-D-luciferin demonstrated a non-homogeneous biodistribution with low uptake into some organs, such as the brain.[99] Several studies have identified regions of the luciferase protein responsible for the color of emitted light and described amino acid substitutions at these sites that result in shifts of the emission spectra.[100] Having a collection of several luciferases with different emission wavelengths able to penetrate mammalian tissues is extremely useful for *in vivo* application.

The biggest limiting factor of BLI is its low levels of generated light. Furthermore, the emitted light is largely absorbed and scattered by the surrounding environment, leading to very poor spatial resolution and depth resolving capacity. The primary absorbing molecules in mammalian tissues are hemoglobin, melanin and other pigmented macromolecules. Hemoglobin strongly absorbs light in a wavelength range of 400–630 nm, limiting the effective imaging depth to a few millimeters in this spectral region.[101] On the other hand, light at wavelengths of 650–1200 nm has been shown to be transmitted and detected through several centimeters of opaque living tissues.[102] Another limiting factor of using BMI *in vivo* is the requirement of oxygen. In anaerobic environments, such as the gut's lumen or necrotic regions of tumors, luciferases were found to be non-luminescent.[103] The level of oxygen concentration required for bioluminescence is well below the level that mammalian cells need to survive. Thus, the loss of emitting light can be used to localize hypoxic areas with necrotic cells.[103]

The primary application of BLI has been to monitor tumor growth, cell trafficking, protein–protein interactions and response to treatment, including gene therapy and immunotherapy.[104] BLI has been successful in assessing treatment efficacy in a disease's early stages in drug screening studies. There are two methods for applying BLI to the study of protein–protein interactions: (1) bioluminescence resonance energy transfer (BRET) and (2) luciferase complementary imaging assay (LCI), also referred to as "Split LCI". BRET is based on the Förster resonance energy transfer (FRET) between two chromophores attached to the two proteins under study: (1) luciferase as the donor and (2) fluorescent protein as the acceptor of light. Observation of a fluorescent protein signal confirms the interaction between the proteins.[105] On the other hand, LCI is based on the fragmentation of luciferase into two inactive domains that are attached to each of the studied partner proteins. When the studied proteins interact, the luciferase fragments converge and luminescence can be observed.[106] These two methods have been applied to study immunoassay antigen–antibody interactions[107] (Figure 11.3).

The majority of preclinical studies have focused on the examination of the most important carcinogenic processes, such as angiogenesis, metastasis and treatment response.[95] BLI was used to investigate viral and nonviral DNA-based therapies allowing assessment of the efficacy, safety and long-term expression of foreign genes delivered to a mouse fetus in utero.[108] Other applications include target validation allowing the efficiency of tumor therapies to be tested, especially against micrometastases that require highly sensitive detection.[109] Both BRET and LCI have been used to study

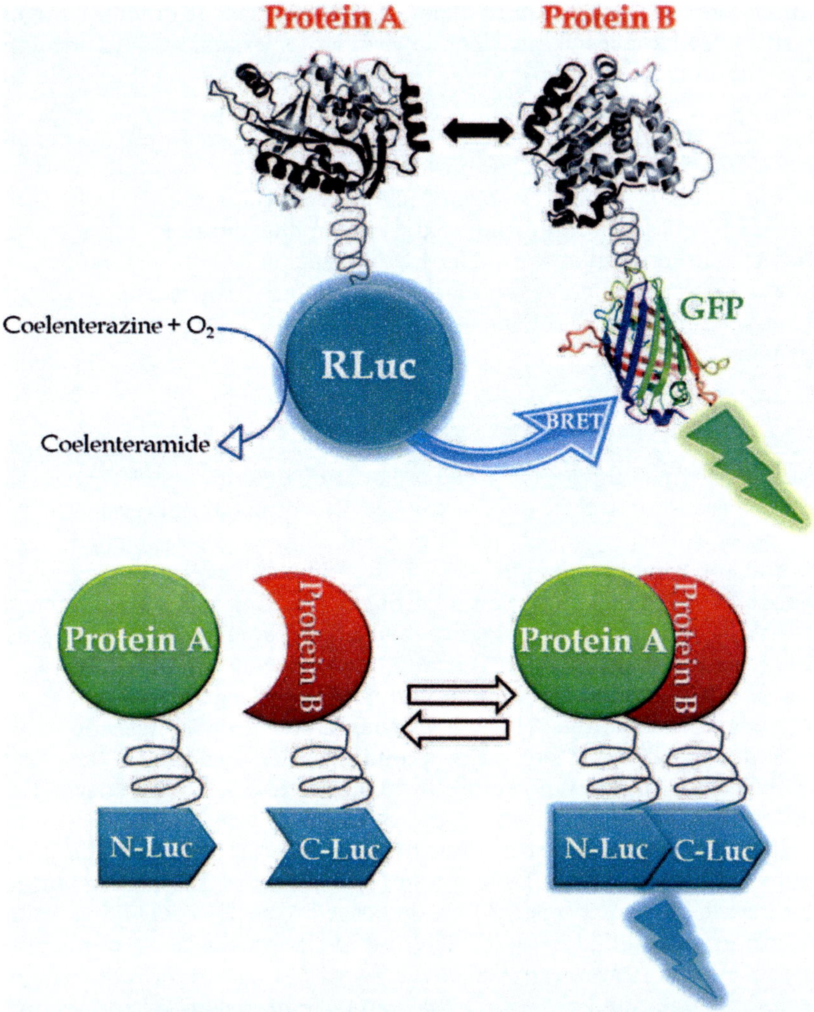

Figure 11.3 Methods used in bioluminescence imaging to study protein–protein interactions. (A) Bioluminescence resonance energy transfer (BRET). (B) Luciferase complementary imaging assay (LCI), also referred to as "Split LCI".
Reproduced from ref. 95 with permission from the Royal Society of Chemistry.

neurodegenerative diseases, such as Alzheimer's disease, Parkinson's disease and amyotrophic lateral sclerosis. The BRET pair of RLuc8 and CdSe or CdTe QDs was used to develop a drug against Alzheimer's disease for low-intensity laser therapy to suppress the oxidative stress caused by beta-amyloid.[110] BLI has shown a significant advantage compared to conventional methods in the study of infectious diseases. The conventional methods require euthanasia of experimental animals at different stages of the study. BLI makes it possible to avoid this expensive and inhumane treatment and allows re-imaging of the same animal during the whole study period.[103]

In summary, BLI has proven to be one of the most effective optical imaging methods to visualize biological processes. Nowadays, only nine bioluminescent systems have been fully understood, and each possesses its own set of limitations. In future years, new applications, as well as identification of currently unknown luciferins and luciferases from a range of unstudied organisms, such as the marine polychaetes *Chaetopterus*, *Harmothoe*, *Odontosyllis*, terrestrial oligochaete *Henlea*, sea salps and others, are anticipated.

11.3.2 Fluorescence Imaging

Fluorescence imaging (FI) is based on detecting light emitted by extrinsically administered fluorescence dyes or genetically-expressed fluorescent proteins after being excited by an external light source having a wavelength shorter than the emission wavelength. The emitted light is detected using a highly sensitive CCD camera equipped with filters rejecting light at the excitation (intrinsic) wavelength to avoid cross-talk. The most effective wavelengths lie in the near-infrared range (NIR) from 700 nm to 900 nm, due to their relatively low absorption by hemoglobin and water, and low tissue autofluorescence background. This allows the detection of fluorescence dyes at depths of several centimeters. Compared to other methods, FI has several advantages: (1) it does not require a substrate that has to be enzymatically modified to emit photons, (2) it allows the usage of several fluorophores simultaneously, (3) the same fluorophores can be used during *in vivo* monitoring and *ex vivo* validation and (4) there is a large variety of fluorophores available, which results in a decreased need for developing reporter gene expressing animal models, resulting in faster translation of preclinical data into clinics. However, the disadvantages of FI are: (1) the requirement of an external excitation light source that can be scattered and absorbed during propagation, resulting in lower fluorophore excitation, and (2) significant tissue auto-fluorescence causing high background noise levels.[111]

There are three major strategies to illuminate a tissue. Typically, illumination with a so-called continuous wave (CW) is used. This approach measures the intensity of the photons collected from the object's surface after propagation through the bulk tissue volume.[112] Another strategy applies modulated illumination intensity at frequencies of 100 MHz–1 GHz, which allows fluctuations in light attenuation and phase to be measured at the different frequencies. The third strategy uses ultrafast photon pulses in the range of 100 fs–100 ps, offering detection of photon kinetics with resolutions below 100 fs.[113]

Early 2D planar fluorescence imaging modalities such as fluorescence reflectance imaging (FRI) and transillumination fluorescence imaging (TFI) were based on simply detecting emitted light within a chosen range of wavelength. These modalities mainly differ by the relative location of the light source and detector. In FRI, the light source and the detector are placed on the same side of the investigated object. In TFI, the light source and detector are located on opposite sides of the object. Both approaches did not consider the effects of light attenuation and scattering. This resulted in low resolution and a lack of quantitative (depth-resolved) imaging performance.

Modern fluorescence molecular tomography (FMT) adds a third dimension by detecting tomographic trans-illumination information with a 360° viewing angle. Even though the light undergoes high scattering after just a few millimeters of propagation through the tissue, the correct fluorescence intensities can be later reconstructed using mathematical models.[114] Today's highly sensitive CCD cameras allow sub-millimeter resolution and sensitivity below one picomole of fluorescent dye.[115] Additional information, allowing accurate co-registration of the FMT data with anatomy, can be obtained by combining FMT with another imaging modality such as computed tomography. A hybrid system, fluorescence-mediated tomography–X-ray computer tomography (FMT-CT) allows high spatial sampling for both of the imaging modalities resulting in highly accurate image co-registration.[116,117]

FI agents can be classified into three categories: (1) unspecific, (2) targeted and (3) activatable fluorophores. Unspecific fluorescent probes, such as FITC-labeled albumin and indocyanine green (ICG), enable non-invasive visualization of permeability and perfusion characteristics. ICG has long been used in clinics. Targeted and activatable probes can visualize biological processes through binding to specific cellular or molecular targets. Freely circulating targeted contrast agents emit unspecific background noise. On the other hand, activatable probes are detectable only when bound to their specific target.[118]

The primary FI application is to image live biological processes in tumors, such as metabolism, proliferation, hypoxia, angiogenesis, invasiveness, inflammation, and apoptosis in real time. Similarly to SPECT, the design of FI tracers focused on imaging glycolysis by producing a fluorescent analog of D-glucose: 2-deoxy-D-glucose (2DG). Preclinical studies reported that 2DG efficiently visualized tumors based on GLUT-1 presence.[119] However, its performance showed no ability in monitoring treatment.[120] Tracers based on fluorescent analogs of bombesin and folic acid were designed to image tumor proliferation. Folic acid labeled with indocyanine green (ICG-FA) showed steady accumulation in folate receptor positive tumors for up to 48 hours.[121] Hypoxia can be detected by a carbonic anhydrase IX targeted fluorescence tracer called HypoxiSense680. The affected region can be detected as early as 3 h after injection with an optimal signal to noise ratio at 12–24 h. Carbonic anhydrase IX was shown to be a successful molecule for follow-up treatment with Temozolomide in glioma.[122]

Two methods have been successfully developed to image angiogenesis. Labeled vascular endothelial growth factor (Cy5.5-VEGF) and labeled monoclonal antibody Bevacizumab (Bevacizumab-IRD800) both target the vascular endothelial growth factor (VEGF) receptor commonly upregulated in cancer.[123,124] The other strategy involves labeled integrins, molecules strongly expressed in new-born vessels and tumors. Dozens of tracers, especially those based on RGD peptide, have been developed following this strategy showing good tumor accumulation and contrast to the background.[111]

The invasiveness and metastatic potential of a tumor can be targeted by imaging specifically up-regulated proteases, such as cathepsin and matrix metalloprotease (MMP) analogs such as ProSense 680 and MMPSense, respectively. Cathepsins are involved in (1) the degradation of vascular and epithelial basement membrane during angiogenesis, (2) dissolution of cell-cell junctions and (3) transport of cancer cells into the blood or lymph stream, and to distant organs.[125]

Several fluorescent probes were developed to study inflammation in animal models: Xenolight Rediject COX-2, NANQ-IMC6, SLX-Lipo-Cy5.5 and ProSense 750. Xenolight Rediject COX-2 and NANQ-IMC6 target Cyclooxygenase-2 (COX-2) activity in tumors. COX-2 is commonly over-expressed in tumors and inflammatory tissues, but not in healthy cells. It is associated with all cancer stages, and its expression increases with tumor malignancy. Imaging of COX2 activity can identify patients that might benefit from preventative treatment with aspirin decreasing colorectal cancer development.[126] Dysregulation of apoptosis is a common feature during primary tumor formation. Response to some anti-neoplastic treatments can cause an increase in the apoptotic cell rate. FI probes are mainly based on labeled Annexin V targeting phosphatidylserine (PS) translocation in the outer membrane and on probes, such as TCAPQ647, targeting caspase-3 activity.[127,128]

Translation of the preclinical results to clinical trials faces two main limitations: (1) the physical limitations of fluorescence imaging in humans, especially related to light penetration, that can be overcome by endoscopic or intra-operative imaging procedures and (2) the clinical approval of FI probes for human studies, especially in the absence of toxicology studies for the the novel tracers. However, in the next few years, we will keep seeing preclinical studies involving human cancer types for the validation of new targets and new targeted imaging strategies.

11.4 Multi-Spectral Optoacoustic Tomography

Optoacoustic imaging is based on detecting acoustic waves generated by absorption of very short light pulses. The basic underlying physical phenomemon was first described by Alexander Graham Bell in 1880.[129] Almost 100 years later in the early 1970s, the application of this method was proposed in biological research and medicine.[130] However, its real application in preclinical and clinical research has only been seen in the last decade. In multi-spectral optoacoustic tomography (MSOT), the tissue of interest is illuminated by nanosecond-duration near-infrared (NIR) light pulses generated by a tunable wavelength laser. Photoactive molecules, such as hemoglobin or melanin, and extrinsically administered probes absorb the light. In response, the local temperature transiently increases, causing a thermoelastic expansion, which is detected in the form of acoustic waves covering a broad range of frequencies from tens of KHz up to tens of MHz.[131]

In comparison with other imaging modalities, MSOT is a very fast imaging technology, enabling real-time imaging of fast events and dynamic processes, such as pharmacokinetics, also in 3D.[132] Additionally, it can simultaneously deliver anatomical, functional and molecular information within the depth of approximately 3 cm and at high spatial resolution (150 μm), which is usually possible only by combining several different technologies.[133] MSOT does not rely on detection of the emitted light, but rather on the direct absorption of light in all of the bio-chromes present in tissues, including intrinsic molecules and any extrinsically-administered contrast agents. It uses laser pulses of different wavelengths to select absorption characteristics of the various photoactive molecules to gain spatially-resolved maps of tissue spectroscopic profiles. Additionally, it allows widely available optical fluorescent agents but also non-fluorescent (purely absorbing) agents to be distinguished by probing their absorption spectra over other background chromophores. The photoactive molecules with a steep drop in their absorption spectrum have been shown to be the optimal choice for MSOT. The absorption spectrum of conventional organic fluorochromes provides a spectrally narrow absorption window in a wavelength range of 650–850 nm. However, *in vivo* application creates the issue of tissue heterogeneity, wherein the data sets obtained usually represent a combined contribution of all photoactive molecules present in the volume imaged, including strong background optoacoustic signal contributions from the blood haemoglobin. Therefore, mathematical modeling of light and ultrasound propagation at different illumination wavelengths is used to extract accurate information from the obtained data sets[134] (Figure 11.4).

The early *in vivo* application of MSOT was based on imaging vascularization. Since MSOT can distinguish between oxidized and deoxidized forms of hemoglobin, it was used to image oxygenation in myocardial infarction,[135] the brain,[136] orthotopic glioblastomas,[137] breast carcinoma[138] and melanoma lymph node metastases in rodents, showing more sensitive detection of micrometastases and in-transit metastases that were undetectable with FDG PET/CT.[139] Thanks to the design of a handheld MSOT imaging device, studies were performed on healthy volunteers to image angiography and microvascularity of healthy tissue[140] and hair bulb surroundings,[141] as well as on patients with melanoma[142] and sentinel lymph node metastasis.[143] Inflammatory processes within the intestinal wall were imaged in patients with Crohn's disease[144] and thyroid anatomy and morphology were studied.[145]

Since MSOT has the ability to simultaneously distinguish between different photoactive molecules within a tissue; much research has been done to develop more efficient probe molecules to enhance their generated optoacoustic conversion efficiency. Agents based on nanoparticles were successfully used to improve the visualization of tumor angiogenesis in gastrointestinal cancer cells[146] and inflammation in rheumatoid arthritis,[147]

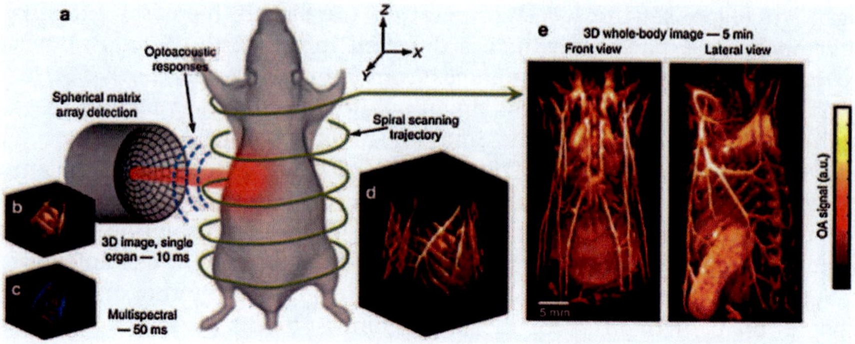

Figure 11.4 Spiral volumetric optoacoustic tomography (SVOT) as a new imaging approach using multi-spectral optoacoustic tomography (MSOT). (a) Whole-body tomographic data acquisition is performed along a spiral (helical) scanning trajectory by means of a spherical matrix ultrasound detection array, further capable of real-time three-dimensional (3D) imaging. (b) Representative single 3D image acquired from the mouse's abdomen. The system can render single (~1 cm^3) volumes at a frame rate of 100 Hz, only limited by the pulse repetition rate of the excitation laser. (c) Multi-spectral (or spectrally unmixed) 3D images can be generated for each position of the matrix array after acquiring volumetric image data at multiple excitation wavelengths of the laser. By fast sweeping of the laser wavelength, it only takes 50 ms to generate a volumetric multispectral dataset at five wavelengths. (d) Volumetric optoacoustic image rendered from a larger area by performing a partial scan for 5 s. (e) It takes ~5 min to acquire whole-body image data by combining all images acquired along the entire spiral trajectory. Reprinted from X. L. Dean-Ben *et al.*, Advanced optoacoustic methods for multiscale imaging of *in vivo* dynamics, *Light Sci. Appl.*, 2017, **6**, e16247, with permission from Nature Publishing Group.[158]

and to determine pH in pancreatic cancer.[148] Also, agents based on fluorescent dyes were successfully applied to study tumor vascularization,[149] renal functions,[150] pH in pancreatic cancer[151] and gliomas.[152] The clinically allowed fluorescent dye Indocyanine Green (ICG) was applied to study gastrointestinal motility in mice. Lutzweiler *et al.* showed the first application of MSOT to image vascularization and contrast agent biodistribution in humans by imaging finger joints.[153]

MSOT is indeed a very promising novel optical molecular imaging technique. Its broad ability to image the optical absorption properties of both intrinsic tissue chromophores and exogenous contrast agents, as well as detecting subtle tumors and early metastasis, and differentiating viable tumor areas from dead tumor tissue, provide it with vast applicability in preclinical research and clinical practice.[154] In recent years, we have also seen agents as well as preclinical scanners for the hybrid use of MSOT with CT,[155] PET,[156] MRI[137] and ultrasound.[136] The possibility of creating contract agents for advanced image-guided cancer therapeutics has also been proposed.[157]

11.5 Conclusions

Molecular imaging has been one of the most prominent modalities of the twenty-first century. Currently, there is already a broad variety of imaging modalities that are based on different principles; however, they serve the same purpose of visualizing molecular interactions in diseases in clinical everyday life as well as in preclinical and clinical research. Modalities that were at first developed for human use only are currently being translated into preclinical research. Moreover, novel modalities, such as MSOT and ^{13}C hyperpolarized MRI, which proved their functionality in preclinical research, are currently being translated to clinical usage. Understanding and non-invasive imaging of molecular interactions *in vivo* and in real-time provides an excellent opportunity to study and evaluate novel therapeutics as well as provide insight into real-time metabolism.

Acknowledgements

E. K. gratefully acknowledges the support of the Graduate School of Bioengineering (GSB) at Technische Universität München. This work was supported by the German Research Foundation (DFG) within the SFB collaborative research center 824 "Imaging for Selection, Monitoring, and Individualization of Cancer Therapies". E. K. would like to thank Dr Timo Schirmer, Prof. Daniel Razansky, and Dr Laura Jacobs.

References

1. S. Laughlin and W. Montanera, *Postgrad. Med.*, 1998, **104**, 73–76.
2. G. Captur, C. Manisty and J. C. Moon, *Heart*, 2016, **102**, 1429–1435.
3. A. Frydrychowicz, M. G. Lubner, J. J. Brown, E. M. Merkle, S. K. Nagle, N. M. Rofsky and S. B. Reeder, *J. Magn. Reson. Imaging*, 2012, **35**, 492–511.
4. G. Masselli and G. Gualdi, *Radiology*, 2012, **264**, 333–348.
5. F. M. Zijta, S. Bipat and J. Stoker, *Eur. J. Radiol.*, 2010, **20**, 1031–1046.
6. C. R. Bowers and D. P. Weitekamp, *Phys. Rev. Lett.*, 1986, **57**, 2645–2648.
7. K. Golman, O. Axelsson, H. Jóhannesson, S. Månsson, C. Olofsson and J. S. Petersson, *Magn. Reson. Med.*, 2001, **46**, 1–5.
8. P. Bhattacharya, E. Y. Chekmenev, W. H. Perman, K. C. Harris, A. P. Lin, V. A. Norton, C. T. Tan, B. D. Ross and D. P. Weitekamp, *J. Magn. Reson.*, 2007, **186**, 150–155.
9. B. M. Goodson, *J. Magn. Reson.*, 2002, **155**, 157–216.
10. M. S. Albert, G. D. Cates, B. Driehuys, W. Happer, B. Saam, C. S. J. Springer and A. Wishnia, *Nature*, 1994, **370**, 199–201.
11. J. P. Mugler, 3rd, B. Driehuys, J. R. Brookeman, G. D. Cates, S. S. Berr, R. G. Bryant, T. M. Daniel, E. E. de Lange, J. H. Downs, 3rd, C. J. Erickson, W. Happer, D. P. Hinton, N. F. Kassel, T. Maier, C. D. Phillips, B. T. Saam, K. L. Sauer and M. E. Wagshul, *Magn. Reson. Med.*, 1997, **37**, 809–815.

12. J. P. Mugler and T. A. Altes, *J. Magn. Reson. Imaging*, 2013, **37**, 313–331.
13. M. A. Bouchiat, T. R. Carver and C. M. Varnum, *Phys. Rev. Lett.*, 1960, **5**, 373–375.
14. B. C. Grover, *Phys. Rev. Lett.*, 1978, **40**, 391–392.
15. K. Golman, J. S. Petersson, J. H. Ardenkjaer-Larsen, I. Leunbach, L. G. Wistrand, G. Ehnholm and K. Liu, *J. Magn. Reson. Imaging*, 2000, **12**, 929–938.
16. T. A. Altes, P. L. Powers, J. Knight-Scott, G. Rakes, T. A. Platts-Mills, E. E. de Lange, B. A. Alford, J. P. Mugler, 3rd and J. R. Brookeman, *J. Magn. Reson. Imaging*, 2001, **13**, 378–384.
17. L. F. Donnelly, J. R. MacFall, H. P. McAdams, J. M. Majure, J. Smith, D. P. Frush, P. Bogonad, H. C. Charles and C. E. Ravin, *Radiology*, 1999, **212**, 885–889.
18. H. U. Kauczor, M. Ebert, K. F. Kreitner, H. Nilgens, R. Surkau, W. Heil, D. Hofmann, E. W. Otten and M. Thelen, *J. Magn. Reson. Imaging*, 1997, **7**, 538–543.
19. H. Hatabu, D. C. Alsop, J. Listerud, M. Bonnet and W. B. Gefter, *Eur. J. Radiol.*, 1999, **29**, 245–252.
20. W. Happer, E. Miron, S. Schaefer, D. Schreiber, W. A. van Wijngaarden and X. Zeng, *Phys. Rev. A: At., Mol., Opt. Phys.*, 1984, **29**, 3092–3110.
21. B. Driehuys, G. D. Cates, E. Miron, K. Sauer, D. K. Walter and W. Happer, *Appl. Phys. Lett.*, 1996, **69**, 1668–1670.
22. D. A. Yablonskiy, A. L. Sukstanskii, J. C. Leawoods, D. S. Gierada, G. L. Bretthorst, S. S. Lefrak, J. D. Cooper and M. S. Conradi, *Proc. Natl. Acad. Sci. U. S. A.*, 2002, **99**, 3111–3116.
23. S. B. Fain, S. R. Panth, M. D. Evans, A. L. Wentland, J. H. Holmes, F. R. Korosec, M. J. O'Brien, H. Fountaine and T. M. Grist, *Radiology*, 2006, **239**, 875–883.
24. S. Fichele, N. Woodhouse, A. J. Swift, Z. Said, M. N. J. Paley, L. Kasuboski, G. H. Mills, E. J. R. van Beek and J. M. Wild, *J. Magn. Reson. Imaging*, 2004, **20**, 331–335.
25. S. B. Fain, T. A. Altes, S. R. Panth, M. D. Evans, B. Waters, J. P. Mugler, 3rd, F. R. Korosec, T. M. Grist, M. Silverman, M. Salerno and J. Owers-Bradley, *Acad. Radiol.*, 2005, **12**, 1385–1393.
26. J. P. Mugler, 3rd, T. A. Altes, I. C. Ruset, I. M. Dregely, J. F. Mata, G. W. Miller, S. Ketel, J. Ketel, F. W. Hersman and K. Ruppert, *Proc. Natl. Acad. Sci. U. S. A.*, 2010, **107**, 21707–21712.
27. Z. I. Cleveland, G. P. Cofer, G. Metz, D. Beaver, J. Nouls, S. S. Kaushik, M. Kraft, J. Wolber, K. T. Kelly, H. P. McAdams and B. Driehuys, *PLoS One*, 2010, **5**, e12192.
28. A. W. Overhauser, *Phys. Rev.*, 1953, **92**, 411–415.
29. T. R. Carver and C. P. Slichter, *Phys. Rev.*, 1953, **92**, 212–213.
30. A. Abragam and W. G. Proctor, *Phys. Rev.*, 1958, **109**, 1441–1458.
31. A. Abragam and M. Goldman, *Rep. Prog. Phys.*, 2001, **41**, 395–467.
32. J. H. Ardenkjaer-Larsen, B. Fridlund, A. Gram, G. Hansson, L. Hansson, M. H. Lerche, R. Servin, M. Thaning and K. Golman, *Proc. Natl. Acad. Sci. U. S. A.*, 2003, **100**, 10158–10163.

33. F. A. Gallagher, M. I. Kettunen and K. M. Brindle, *Prog. Nucl. Magn. Reson. Spectrosc.*, 2009, **55**, 285–295.
34. R. E. Hurd, Y.-F. Yen, A. Chen and J. H. Ardenkjaer-Larsen, *J. Magn. Reson. Imaging*, 2012, **36**, 1314–1328.
35. J. H. Ardenkjaer-Larsen, B. Fridlund, A. Gram, G. Hansson, L. Hansson, M. H. Lerche, R. Servin, M. Thaning and K. Golman, *Proc. Natl. Acad. Sci. U. S. A.*, 2003, **100**, 10158–10163.
36. K. Golman, R. in 't Zandt and M. Thaning, *Proc. Natl. Acad. Sci. U. S. A.*, 2006, **103**, 11270–11275.
37. A. Comment, J. Rentsch, F. Kurdzesau, S. Jannin, K. Uffmann, R. B. van Heeswijk, P. Hautle, J. A. Konter, B. van den Brandt and J. J. van der Klink, *J. Magn. Reson.*, 2008, **194**, 152–155.
38. M. M. Chaumeil, C. Najac and S. M. Ronen, *Methods Enzymol.*, 2015, **561**, 1–71.
39. K. M. Brindle, S. E. Bohndiek, F. A. Gallagher and M. I. Kettunen, *Magn. Reson. Med.*, 2011, **66**, 505–519.
40. O. Khegai, R. F. Schulte, M. A. Janich, M. I. Menzel, E. Farrell, A. M. Otto, J. H. Ardenkjaer-Larsen, S. J. Glaser, A. Haase, M. Schwaiger and F. Wiesinger, *NMR Biomed.*, 2014, **27**, 1256–1265.
41. M. I. Menzel, E. V. Farrell, M. A. Janich, O. Khegai, F. Wiesinger, S. Nekolla, A. M. Otto, A. Haase, R. F. Schulte and M. Schwaiger, *J. Nucl. Med.*, 2013, **54**, 1113–1119.
42. F. Schilling, S. Duwel, U. Kollisch, M. Durst, R. F. Schulte, S. J. Glaser, A. Haase, A. M. Otto and M. I. Menzel, *NMR Biomed.*, 2013, **26**, 557–568.
43. M. A. Schroeder, H. J. Atherton, D. R. Ball, M. A. Cole, L. C. Heather, J. L. Griffin, K. Clarke, G. K. Radda and D. J. Tyler, *FASEB J.*, 2009, **23**, 2529–2538.
44. T. H. Witney, M. I. Kettunen, D. Hu, F. A. Gallagher, S. E. Bohndiek, R. Napolitano and K. M. Brindle, *Br. J. Cancer*, 2010, **103**, 1400–1406.
45. S. Düwel, C. Hundshammer, M. Gersch, B. Feuerecker, K. Steiger, A. Buck, A. Walch, A. Haase, S. J. Glaser, M. Schwaiger and F. Schilling, *Nat. Commun.*, 2017, **8**, 15126.
46. D. J. Scholz, M. A. Janich, U. Köllisch, R. F. Schulte, J. H. Ardenkjaer-Larsen, A. Frank, A. Haase, M. Schwaiger and M. I. Menzel, *Magn. Reson. Med.*, 2015, **73**, 2274–2282.
47. U. Koellisch, C. V. Gringeri, G. Rancan, E. V. Farell, M. I. Menzel, A. Haase, M. Schwaiger and R. F. Schulte, *Magn. Reson. Med.*, 2015, **74**, 1011–1018.
48. U. Koellisch, C. Laustsen, T. S. Norlinger, J. A. Ostergaard, A. Flyvbjerg, C. V. Gringeri, M. I. Menzel, R. F. Schulte, A. Haase and H. Stodkilde-Jorgensen, *Physiol. Rep.*, 2015, **3**, e12474.
49. E. Chiavazza, E. Kubala, C. V. Gringeri, S. Duwel, M. Durst, R. F. Schulte and M. I. Menzel, *J. Magn. Reson.*, 2013, **227**, 35–38.
50. F. A. Gallagher, M. I. Kettunen, S. E. Day, D. Hu, M. Karlsson, A. Gisselsson, M. H. Lerche and K. M. Brindle, *Magn. Reson. Med.*, 2011, **66**, 18–23.

51. M. M. Chaumeil, P. E. Z. Larson, S. M. Woods, L. Cai, P. Eriksson, A. E. Robinson, J. M. Lupo, D. B. Vigneron, S. J. Nelson, R. O. Pieper, J. J. Phillips and S. M. Ronen, *Cancer Res.*, 2014, **74**, 4247–4257.
52. K. R. Keshari and D. M. Wilson, *Chem. Soc. Rev.*, 2014, **43**, 1627–1659.
53. S. J. Nelson, J. Kurhanewicz, D. B. Vigneron, P. E. Z. Larson, A. L. Harzstark, M. Ferrone, M. van Criekinge, J. W. Chang, R. Bok, I. Park, G. Reed, L. Carvajal, E. J. Small, P. Munster, V. K. Weinberg, J. H. Ardenkjaer-Larsen, A. P. Chen, R. E. Hurd, L.-I. Odegardstuen, F. J. Robb, J. Tropp and J. A. Murray, *Sci. Transl. Med.*, 2013, **5**, 198ra108.
54. U.S. National Institutes of Health, *https://clinicaltrials.gov/ct2/results?term=hyperpolarized+pyruvate&Search=Search*, accessed February 2017.
55. V. C. Spanoudaki and S. I. Ziegler, *Handb. Exp. Pharmacol.*, 2008, 53–74.
56. G. F. Knoll, *Radiation detection and measurement*, John Wiley & Sons, 2010.
57. N. L. Christensen, B. E. Hammer, B. G. Heil and K. Fetterly, *Phys. Med. Biol.*, 1995, **40**, 691.
58. A. Kolb, C. Parl, F. Mantlik, C. C. Liu, E. Lorenz, D. Renker and B. J. Pichler, *Med. Phys.*, 2014, **41**, 081916.
59. L. Herrnsdorf, M. Caccia and S. Mattsson, *Radiat. Prot. Dosim.*, 2016, **169**, 430–435.
60. A. B. Alenazy, G. R. Wells and T. D. Ruddy, *Expert Rev. Med. Devices*, 2017, **14**, 213–222.
61. T. E. Schlesinger, J. E. Toney, H. Yoon, E. Y. Lee, B. A. Brunett, L. Franks and R. B. James, *Mater. Sci. Eng., R*, 2001, **32**, 103–189.
62. S. Surti, *J. Nucl. Med.*, 2015, **56**, 98–105.
63. L. H. Portnow, D. E. Vaillancourt and M. S. Okun, *Neurology*, 2013, **80**, 952–956.
64. M. E. Raichle, *Trends Neurosci.*, 2009, **32**, 118–126.
65. G. Muehllehner and J. S. Karp, *Phys. Med. Biol.*, 2006, **51**, R117–R37.
66. J. R. Spermon, L. F. De Geus-Oei, L. A. L. M. Kiemeney, J. A. Witjes and W. J. G. Oyen, *BJU Int.*, 2002, **89**, 549–556.
67. S. Tzimopoulou, V. J. Cunningham, T. E. Nichols, G. Searle, N. P. Bird, P. Mistry, I. J. Dixon, W. A. Hallett, B. Whitcher, A. P. Brown, M. Zvartau-Hind, N. Lotay, R. Y. K. Lai, M. Castiglia, B. Jeter, J. C. Matthews, K. Chen, D. Bandy, E. M. Reiman, M. Gold, E. A. Rabiner and P. M. Matthews, *J. Alzheimer's Dis.*, 2010, **22**, 1241–1256.
68. J. W. Fletcher, B. Djulbegovic, H. P. Soares, B. A. Siegel, V. J. Lowe, G. H. Lyman, R. E. Coleman, R. Wahl, J. C. Paschold, N. Avril, L. H. Einhorn, W. W. Suh, D. Samson, D. Delbeke, M. Gorman and A. F. Shields, *J. Nucl. Med.*, 2008, **49**, 480–508.
69. T. Maurer, M. Eiber, M. Schwaiger and J. E. Gschwend, *Nat. Rev. Urol.*, 2016, **13**, 226–235.
70. J. Schwenck, H. Rempp, G. Reischl, S. Kruck, A. Stenzl, K. Nikolaou, C. Pfannenberg and C. la Fougere, *Eur. J. Nucl. Med. Mol. Imaging*, 2017, **44**, 92–101.

71. B. Feuerecker, K. Scheidhauer, M. Schwaiger and M. Mustafa, *Clin. Transl. Imaging*, 2017, 1–11.
72. G. Mees, R. Dierckx, C. Vangestel and C. Van de Wiele, *Eur. J. Nucl. Med. Mol. Imaging*, 2009, **36**, 1674–1686.
73. M. Ginj, H. Zhang, B. Waser, R. Cescato, D. Wild, X. Wang, J. Erchegyi, J. Rivier, H. R. Macke and J. C. Reubi, *Proc. Natl. Acad. Sci. U. S. A.*, 2006, **103**, 16436–16441.
74. W. Cai, K. Chen, Z.-B. Li, S. S. Gambhir and X. Chen, *J. Nucl. Med.*, 2007, **48**, 1862–1870.
75. F. Hyafil, J. Pelisek, I. Laitinen, M. Schottelius, M. Mohring, Y. Döring, E. Van der Vorst, M. Kallmayer, K. Steiger and A. Poschenrieder, *J. Nucl. Med.*, 2016, jnumed-116.
76. K. Herholz, *Ann. Nucl. Med.*, 2003, **17**, 79–89.
77. M. J. Koepp and F. G. Woermann, *Lancet Neurol.*, 2005, **4**, 42–53.
78. E. Salmon, C. B. Ir and R. Hustinx, in *Seminars in nuclear medicine*, Elsevier, 2015, vol. 45, pp. 541–551.
79. I. Sarikaya, *Nucl. Med. Commun.*, 2015, **36**, 775–781.
80. R. Ossenkoppele, W. J. Jansen, G. D. Rabinovici, D. L. Knol, W. M. van Der Flier, B. N. M. Van Berckel, P. Scheltens, P. J. Visser, S. C. J. Verfaillie and M. D. Zwan, *J. Am. Med. Assoc.*, 2015, **313**, 1939–1950.
81. F. C. Gaertner, S. Fürst and M. Schwaiger, *Cancer Imaging*, 2013, **13**, 36.
82. R. Alford, M. Ogawa, P. L. Choyke and H. Kobayashi, *Mol. Biosyst.*, 2009, **5**, 1279–1291.
83. C. Müller and R. Schibli, in *Molecular Imaging in Oncology*, ed. O. Schober and B. Riemann, Springer Berlin Heidelberg, Berlin, Heidelberg, 2013, pp. 65–105.
84. S. Liu, *Adv. Drug Delivery Rev.*, 2008, **60**, 1347–1370.
85. S. Liu, *Adv. Drug Delivery Rev.*, 2008, **60**, 1347–1370.
86. M. Schottelius and H.-J. Wester, *Methods*, 2009, **48**, 161–177.
87. J. C. Reubi, *Endocr. Rev.*, 2003, **24**, 389–427.
88. M. E. Maddalena, J. Fox, J. Chen, W. Feng, A. Cagnolini, K. E. Linder, M. F. Tweedle, A. D. Nunn and L. E. Lantry, *J. Nucl. Med.*, 2009, **50**, 2017–2024.
89. H. Zhang, J. Chen, C. Waldherr, K. Hinni, B. Waser, J. C. Reubi and H. R. Maecke, *Cancer Res.*, 2004, **64**, 6707–6715.
90. J. C. Reubi, B. Waser, H. Friess, M. Buchler and J. Laissue, *Gut*, 1998, **42**, 546–550.
91. F. Alshoukr, C. Rosant, V. Maes, J. Abdelhak, O. Raguin, S. Burg, L. Sarda, J. Barbet, D. Tourwe, D. Pelaprat and A. Gruaz-Guyon, *Bioconjugate Chem.*, 2009, **20**, 1602–1610.
92. V. Maes, E. Garcia-Garayoa, P. Blauenstein and D. Tourwe, *J. Med. Chem.*, 2006, **49**, 1833–1836.
93. M. Fani and H. R. Maecke, *Eur. J. Nucl. Med. Mol. Imaging*, 2012, **39**, 11–30.

94. T. Wilson and J. W. Hastings, *Annu. Rev. Cell Dev. Biol.*, 1998, **14**, 197–230.
95. Z. M. Kaskova, A. S. Tsarkova and I. V. Yampolsky, *Chem. Soc. Rev.*, 2016, **45**, 6048–6077.
96. J. R. de Wet, K. V. Wood, D. R. Helinski and M. DeLuca, *Proc. Natl. Acad. Sci. U. S. A.*, 1985, **82**, 7870–7873.
97. B. A. Sherf and K. V. Wood, *Promega Notes*, 1994, **49**, 14–21.
98. C. H. Contag and M. H. Bachmann, *Annu. Rev. Biomed. Eng.*, 2002, **4**, 235–260.
99. F. Berger, R. Paulmurugan, S. Bhaumik and S. S. Gambhir, *Eur. J. Nucl. Med. Mol. Imaging*, 2008, **35**, 2275–2285.
100. J. W. Hastings, *Gene*, 1996, **173**, 5–11.
101. B. J. Tromberg, N. Shah, R. Lanning, A. Cerussi, J. Espinoza, T. Pham, L. Svaasand and J. Butler, *Neoplasia*, 2000, **2**, 26–40.
102. B. W. Rice, M. D. Cable and M. B. Nelson, *J. Biomed. Opt.*, 2001, **6**, 432–440.
103. C. H. Contag, P. R. Contag, J. I. Mullins, S. D. Spilman, D. K. Stevenson and D. A. Benaron, *Mol. Microbiol.*, 1995, **18**, 593–603.
104. A. A. Khalil, M. J. Jameson, W. C. Broaddus, T. D. Chung, S. E. Golding, S. M. Dever, E. Rosenberg and K. Valerie, *ISRN Mol. Imaging*, 2013, 689279.
105. Y. Xu, D. W. Piston and C. H. Johnson, *Proc. Natl. Acad. Sci. U. S. A.*, 1999, **96**, 151–156.
106. T. Azad, A. Tashakor and S. Hosseinkhani, *Anal. Bioanal. Chem.*, 2014, **406**, 5541–5560.
107. L. A. Frank and V. V. Krasitskaya, in *Bioluminescence: Fundamentals and Applications in Biotechnology*, ed. G. Thouand and R. Marks, Springer Berlin Heidelberg, Berlin, Heidelberg, 2014, vol. 1, pp. 175–197.
108. G. S. Lipshutz, C. A. Gruber, Y. Cao, J. Hardy, C. H. Contag and K. M. Gaensler, *Mol. Ther.*, 2001, **3**, 284–292.
109. M. Edinger, T. J. Sweeney, A. A. Tucker, A. B. Olomu, R. S. Negrin and C. H. Contag, *Neoplasia*, 1999, **1**, 303–310.
110. B. L. Bungart, L. Dong, D. Sobek, G. Y. Sun, G. Yao and J. C.-M. Lee, *Nanomedicine*, 2014, **10**, 15–17.
111. C. Martelli, A. Lo Dico, C. Diceglie, G. Lucignani and L. Ottobrini, *Oncotarget*, 2016, **7**, 48753–48787.
112. V. Ntziachristos, J. Ripoll, L. V. Wang and R. Weissleder, *Nat. Biotechnol.*, 2005, **23**, 313–320.
113. B. Chance, *Annu. Rev. Biophys. Biophys. Chem.*, 1991, **20**, 1–28.
114. N. Deliolanis, T. Lasser, D. Hyde, A. Soubret, J. Ripoll and V. Ntziachristos, *Opt. Lett.*, 2007, **32**, 382–384.
115. E. E. Graves, J. Ripoll, R. Weissleder and V. Ntziachristos, *Med. Phys.*, 2003, **30**, 901–911.
116. D. Hyde, R. de Kleine, S. A. MacLaurin, E. Miller, D. H. Brooks, T. Krucker and V. Ntziachristos, *Neuroimage*, 2009, **44**, 1304–1311.
117. R. B. Schulz, A. Ale, A. Sarantopoulos, M. Freyer, E. Soehngen, M. Zientkowska and V. Ntziachristos, *IEEE Trans. Med. Imaging*, 2010, **29**, 465–473.

118. J. Klohs, A. Wunder and K. Licha, *Basic Res. Cardiol.*, 2008, **103**, 144–151.
119. J. L. Kovar, W. Volcheck, E. Sevick-Muraca, M. A. Simpson and D. M. Olive, *Anal. Biochem.*, 2009, **384**, 254–262.
120. J.-C. Tseng, Y. Wang, P. Banerjee and A. L. Kung, *Mol. Imaging Biol.*, 2012, **14**, 553–560.
121. W. K. Moon, Y. Lin, T. O'Loughlin, Y. Tang, D.-E. Kim, R. Weissleder and C.-H. Tung, *Bioconjugate Chem.*, 2003, **14**, 539–545.
122. S. Kaluz, M. Kaluzova, S.-Y. Liao, M. Lerman and E. J. Stanbridge, *Biochim. Biophys. Acta*, 2009, **1795**, 162–172.
123. M. V. Backer, V. Patel, B. T. Jehning and J. M. Backer, *Bioconjugate Chem.*, 2006, **17**, 912–919.
124. A. G. T. Terwisscha van Scheltinga, G. M. van Dam, W. B. Nagengast, V. Ntziachristos, H. Hollema, J. L. Herek, C. P. Schroder, J. G. W. Kosterink, M. N. Lub-de Hoog and E. G. E. de Vries, *J. Nucl. Med.*, 2011, **52**, 1778–1785.
125. G.-J. Tan, Z.-K. Peng, J.-P. Lu and F.-Q. Tang, *World J. Biol. Chem.*, 2013, **4**, 91–101.
126. S. Foersch, C. Neufert, M. F. Neurath and M. J. Waldner, *Diagn. Ther. Endosc.*, 2013, 250641.
127. A. Agarwal, M. A. Mackey, M. A. El-Sayed and R. V. Bellamkonda, *ACS Nano*, 2011, **5**, 4919–4926.
128. K. Bullok and D. Piwnica-Worms, *J. Med. Chem.*, 2005, **48**, 5404–5407.
129. A. G. Bell, *Am. J. Sci.*, 1880, **Series 3 V**, 305–324.
130. A. Rosencwaig, *Science*, 1973, **181**, 657–658.
131. D. Razansky, N. C. Deliolanis, C. Vinegoni and V. Ntziachristos, *Curr. Pharm. Biotechnol.*, 2012, **13**, 504–522.
132. V. Ermolayev, X. L. Dean-Ben, S. Mandal, V. Ntziachristos and D. Razansky, *Eur. J. Radiol.*, 2016, **26**, 1843–1851.
133. J. Sharkey, L. Scarfe, I. Santeramo, M. Garcia-Finana, B. K. Park, H. Poptani, B. Wilm, A. Taylor and P. Murray, *Eur. J. Pharmacol.*, 2016, **790**, 74–82.
134. V. Ntziachristos and D. Razansky, *Chem. Rev.*, 2010, **110**, 2783–2794.
135. A. Taruttis, M. Wildgruber, K. Kosanke, N. Beziere, K. Licha, R. Haag, M. Aichler, A. Walch, E. Rummeny and V. Ntziachristos, *Photoacoustics*, 2013, **1**, 3–8.
136. I. Olefir, E. Mercep, N. C. Burton, S. V. Ovsepian and V. Ntziachristos, *J. Biomed. Opt.*, 2016, **21**, 86005.
137. A. B. E. Attia, C. J. H. Ho, P. Chandrasekharan, G. Balasundaram, H. C. Tay, N. C. Burton, K.-H. Chuang, V. Ntziachristos and M. Olivo, *J. Biophotonics*, 2016, **9**, 701–708.
138. A. Chekkoury, A. Nunes, J. Gateau, P. Symvoulidis, A. Feuchtinger, N. Beziere, S. V. Ovsepian, A. Walch and V. Ntziachristos, *Neoplasia*, 2016, **18**, 459–467.
139. V. Neuschmelting, H. Lockau, V. Ntziachristos, J. Grimm and M. F. Kircher, *Radiology*, 2016, **280**, 137–150.
140. X. L. Dean-Ben and D. Razansky, *Photoacoustics*, 2013, **1**, 68–73.

141. S. J. Ford, P. L. Bigliardi, T. C. P. Sardella, A. Urich, N. C. Burton, M. Kacprowicz, M. Bigliardi, M. Olivo and D. Razansky, *J. Invest. Dermatol.*, 2016, **136**, 753–761.
142. S. Y. Chuah, A. B. E. Attia, V. Long, C. J. H. Ho, P. Malempati, C. Y. Fu, S. J. Ford, J. S. S. Lee, W. P. Tan, D. Razansky, M. Olivo and S. Thng, *Skin Res. Technol.*, 2016, **23**, 221–226.
143. I. Stoffels, S. Morscher, I. Helfrich, U. Hillen, J. Leyh, N. C. Burton, T. C. P. Sardella, J. Claussen, T. D. Poeppel, H. S. Bachmann, A. Roesch, K. Griewank, D. Schadendorf, M. Gunzer and J. Klode, *Sci. Transl. Med.*, 2015, **7**, 317ra199.
144. M. J. Waldner, F. Knieling, C. Egger, S. Morscher, J. Claussen, M. Vetter, C. Kielisch, S. Fischer, L. Pfeifer, A. Hagel, R. S. Goertz, D. Wildner, R. Atreya, D. Strobel and M. F. Neurath, *Gastroenterology*, 2016, **151**, 238–240.
145. A. Dima and V. Ntziachristos, *Photoacoustics*, 2016, **4**, 65–69.
146. C. Bao, N. Beziere, P. del Pino, B. Pelaz, G. Estrada, F. Tian, V. Ntziachristos, J. M. de la Fuente and D. Cui, *Small*, 2013, **9**, 68–74.
147. J. Vonnemann, N. Beziere, C. Bottcher, S. B. Riese, C. Kuehne, J. Dernedde, K. Licha, C. von Schacky, Y. Kosanke, M. Kimm, R. Meier, V. Ntziachristos and R. Haag, *Theranostics*, 2014, **4**, 629–641.
148. M. K. Gurka, D. Pender, P. Chuong, B. L. Fouts, A. Sobelov, M. W. McNally, M. Mezera, S. Y. Woo and L. R. McNally, *J. Controlled Release*, 2016, **231**, 60–67.
149. N. Beziere, N. Lozano, A. Nunes, J. Salichs, D. Queiros, K. Kostarelos and V. Ntziachristos, *Biomaterials*, 2015, **37**, 415–424.
150. L. Scarfe, A. Rak-Raszewska, S. Geraci, D. Darssan, J. Sharkey, J. Huang, N. C. Burton, D. Mason, P. Ranjzad, S. Kenny, N. Gretz, R. Levy, B. Kevin Park, M. Garcia-Finana, A. S. Woolf, P. Murray and B. Wilm, *Sci. Rep.*, 2015, **5**, 13601.
151. C. W. Kimbrough, A. Khanal, M. Zeiderman, B. R. Khanal, N. C. Burton, K. M. McMasters, S. M. Vickers, W. E. Grizzle and L. R. McNally, *Clin. Cancer Res.*, 2015, **21**, 4576–4585.
152. N. C. Deliolanis, A. Ale, S. Morscher, N. C. Burton, K. Schaefer, K. Radrich, D. Razansky and V. Ntziachristos, *Mol. Imaging Biol.*, 2014, **16**, 652–660.
153. C. Lutzweiler, R. Meier, E. Rummeny, V. Ntziachristos and D. Razansky, *Opt. Lett.*, 2014, **39**, 4061–4064.
154. L. R. McNally, M. Mezera, D. E. Morgan, P. J. Frederick, E. S. Yang, I.-E. Eltoum and W. E. Grizzle, *Clin. Cancer Res.*, 2016, **22**, 3432–3439.
155. J. Cao, J. Campbell, L. Liu, R. P. Mason and A. R. Lippert, *Anal. Chem.*, 2016, **88**, 4995–5002.
156. W. Wu, W. Driessen and X. Jiang, *J. Am. Chem. Soc.*, 2014, **136**, 3145–3155.
157. J. Liu, X. Zheng, L. Yan, L. Zhou, G. Tian, W. Yin, L. Wang, Y. Liu, Z. Hu, Z. Gu, C. Chen and Y. Zhao, *ACS Nano*, 2015, **9**, 696–707.
158. X. L. Dean-Ben, T. F. Fehm, S. J. Ford, S. Gottschalk and D. Razansky, *Light: Sci. Appl.*, 2017, **6**, e16247.

Subject Index